VOLUME SEVENTY FIVE

CURRENT TOPICS IN MEMBRANES

Lipid Domains

CURRENT TOPICS IN MEMBRANES, VOLUME 75

Series Editors

ROBERT BALABAN
National Heart, Lung and Blood Institute
National Institutes of Health
Bethesda, Maryland, USA

SIDNEY A. SIMON
Department of Neurobiology
Duke University Medical Centre
Durham, North Carolina, USA

VOLUME SEVENTY FIVE

CURRENT TOPICS IN MEMBRANES

Lipid Domains

Edited by

ANNE K. KENWORTHY
Department of Molecular Physiology and Biophysics
Vanderbilt School of Medicine
Nashville, TN, USA

ELSEVIER

Amsterdam • Boston • Heidelberg • London
New York • Oxford • Paris • San Diego
San Francisco • Singapore • Sydney • Tokyo

Academic Press is an imprint of Elsevier

Academic Press is an imprint of Elsevier
225 Wyman Street, Waltham, MA 02451, USA
525 B Street, Suite 1800, San Diego, CA 92101-4495, USA
125 London Wall, London, EC2Y 5AS, UK
The Boulevard, Langford Lane, Kidlington, Oxford OX5 1GB, UK

First edition 2015

ISBN: 978-0-12-803295-4
ISSN: 1063-5823

For information on all Academic Press publications
visit our website at http://store.elsevier.com/

CONTENTS

CONTRIBUTORS

David A. Brown
East Carolina Diabetes and Obesity Institute and Department of Physiology, East Carolina University, Greenville, NC, USA

Garth Burn
Department of Physics and Randall Division of Cell and Molecular Biophysics, King's College London, London, UK

Michael B. Butler
Department of Physics and Astronomy, University of Maine, Orono, ME, USA

Nikki M. Curthoys
Department of Physics and Astronomy, University of Maine, Orono, ME, USA

Maurizio Del Poeta
Department of Molecular Genetics and Microbiology, Stony Brook University, Stony Brook, NY, USA

Amir M. Farnoud
Department of Molecular Genetics and Microbiology, Stony Brook University, Stony Brook, NY, USA

Kerney Jebrell Glover
Department of Chemistry, Lehigh University, Bethlehem, PA, USA

Juliette Griffié
Department of Physics and Randall Division of Cell and Molecular Biophysics, King's College London, London, UK

Samuel T. Hess
Department of Physics and Astronomy, University of Maine, Orono, ME, USA

Volker Kiessling
Department of Molecular Physiology and Biological Physics, Center for Membrane Biology, University of Virginia, Charlottesville, VA, USA

James B. Konopka
Department of Molecular Genetics and Microbiology, Stony Brook University, Stony Brook, NY, USA

Rasagna Kosaraju
Department of Biochemistry & Molecular Biology and East Carolina Diabetes and Obesity Institute, East Carolina University, Greenville, NC, USA

Diego Krapf
Department of Electrical and Computer Engineering and School of Biomedical Engineering, Colorado State University, Fort Collins, CO, USA

Stéphanie Lebreton
Unité de Trafic Membranaire et Pathogénèse, Institut Pasteur, Paris, France

Ilya Levental
Department of Integrative Biology and Pharmacology, University of Texas Health Science Center at Houston – Medical School, Houston, TX, USA

Kandice R. Levental
Department of Integrative Biology and Pharmacology, University of Texas Health Science Center at Houston – Medical School, Houston, TX, USA

Jennifer Lilieholm
Department of Physics and Astronomy, University of Maine, Orono, ME, USA

Erwin London
Department of Biochemistry and Cell Biology, Stony Brook University, Stony Brook, NY, USA

Michael Mlodzianoski
Department of Physics and Astronomy, University of Maine, Orono, ME, USA

Andrew J. Nelson
Department of Physics and Astronomy, University of Maine, Orono, ME, USA

Dylan M. Owen
Department of Physics and Randall Division of Cell and Molecular Biophysics, King's College London, London, UK

Simona Paladino
Dipartimento di Medicina Molecolare e Biotecnologie Mediche, Università Federico II; CEINGE Biotecnologie Avanzate, Napoli, Italy

Matthew Parent
Department of Physics and Astronomy, University of Maine, Orono, ME, USA

Sarah M. Plucinsky
Department of Chemistry, Lehigh University, Bethlehem, PA, USA

Kyle T. Root
Department of Chemistry, Lehigh University, Bethlehem, PA, USA

Saame Raza Shaikh
Department of Biochemistry & Molecular Biology, Department of Microbiology and Immunology and East Carolina Diabetes and Obesity Institute, East Carolina University, Greenville, NC, USA

Lukas K. Tamm
Department of Molecular Physiology and Biological Physics, Center for Membrane Biology, University of Virginia, Charlottesville, VA, USA

Alvaro M. Toledo
Department of Molecular Genetics and Microbiology, Stony Brook University, Stony Brook, NY, USA

Matthew Valles
Department of Physics and Astronomy, University of Maine, Orono, ME, USA

Stephen R. Wassall
Department of Physics, Indiana University-Purdue University Indianapolis, Indianapolis, IN, USA

Sung-Tae Yang
Department of Molecular Physiology and Biological Physics, Center for Membrane Biology, University of Virginia, Charlottesville, VA, USA

Chiara Zurzolo
Dipartimento di Medicina Molecolare e Biotecnologie Mediche, Università Federico II, Napoli, Italy; Unité de Trafic Membranaire et Pathogénèse, Institut Pasteur, Paris, France

PREFACE

How cell membranes are organized is a question that has fascinated researchers since the fluid mosaic model of membranes was put forward by Singer and Nicolson (Nicolson, 2014; Singer & Nicolson, 1972). In their original description of the mosaic nature of membranes, Singer and Nicolson predicted that proteins and lipids should be randomly distributed within membranes over distances from a few tenths of a micron or greater, but that short-range order could exist (Singer & Nicolson, 1972). In recent years, interest in membrane organization has become even more intense following the proposal by Kai Simons and colleagues that cell membranes contain submicron domains consisting of both proteins and lipids known as lipid rafts (Simons & Ikonen, 1997). The history of the lipid raft field has been a contentious one (Edidin, 2003; Hancock, 2006; Jacobson, Mouritsen, & Anderson, 2007; Kenworthy, 2008; Kraft, 2013; LaRocca et al., 2013; Lichtenberg, Goni, & Heerklotz, 2005; Lingwood & Simons, 2010; London, 2005; Munro, 2003; Owen, Williamson, Magenau, & Gaus, 2012; Pike, 2006; Simons & Gerl, 2010), and the purpose of this volume is not to take sides in the arguments for and against the lipid raft model. Rather, it is to highlight the breadth of methodologies and vast body of knowledge about biological membranes that have emerged in the search for lipid rafts.

In the literature, lipid rafts have often been equated with liquid-ordered domains (Ahmed, Brown, & London, 1997; Brown & London, 1998), and in vitro reconstitution systems have played an essential role in exploring the properties of membranes capable of forming coexisting liquid-ordered and liquid-disordered domains. One of the most widely studied model systems utilizes giant unilamellar vesicles (GUVs) consisting of ternary lipid mixtures (Veatch & Keller, 2002, 2003). These systems have been reviewed elsewhere (Morales-Penningston et al., 2010; Veatch, 2007, Veatch & Keller, 2005; Wesolowska, Michalak, Maniewska, & Hendrich, 2009) and will not be discussed here. Instead, we begin with a discussion of supported bilayer systems in the study of raft domains in Chapter 1 by Lukas Tamm and colleagues. This chapter highlights several approaches currently used to generate planar membranes, including both symmetric and asymmetric bilayers, and discusses how these have been used to investigate the partitioning of proteins and peptides between liquid-ordered and liquid-disordered domains. Next, the merits of giant plasma membrane-derived vesicles (GPMVs) as a model system for exploring raft domains are summarized by Ilya Levental and colleagues in Chapter 2. Derived from the plasma membrane of living cells, GPMVs contain a complex mixture of lipids as well as a high concentration of membrane proteins, but lack cytoskeletal underpinnings. They thus offer an intermediate level of complexity between reconstituted systems such as supported bilayers or GUVs and live cells.

Compared to model systems, raft-like domains in living cells remain much more challenging to detect. In recent years, super-resolution microscopy has opened up the possibility of visualizing lipid or protein clusters too small to be resolved using conventional approaches (Owen & Gaus, 2013; Owen, Magenau, Williamson, & Gaus, 2012). In Chapter 3, Samuel Hess and colleagues provide an update on how fluorescence photoactivation localization microscopy and other super-resolution techniques are being used to identify and characterize membrane domains not only at the cell surface, but also in intracellular compartments such as mitochondria. The use of high-resolution microscopy techniques to study domains is explored further in Chapter 4 by Dylan Owen and colleagues. Here, the focus is on signaling nanoclusters.

Lipid rafts are but one example of the types of domains that have been proposed to laterally organize cell membranes. The presence of domains that constrain the diffusion of proteins and lipids within the plasma membrane has long been inferred from approaches such as single molecule tracking and fluorescence recovery after photobleaching (Edidin, 1992; Jacobson, Sheets, & Simson, 1995; Sako & Kusumi, 1994). Interactions of proteins and lipids with such domains can cause them to undergo anomalous diffusion, where the mean-squared displacement of the diffusing molecules is not linear in time. In Chapter 5, Diego Krapf outlines mechanisms that contribute to anomalous diffusion in the plasma membrane. This comprehensive review starts off with a description of theoretical models to describe anomalous diffusion before summarizing examples of how a variety of membrane domains, including clathrin-coated pits and cytoskeletal corrals, impact the lateral diffusion of proteins and lipids at the cell surface.

Given that many classes of membrane domains rely on both protein–lipid and lipid–lipid interactions, it is not surprising that the exact lipid composition of membranes can heavily influence both the structure and function of domains. In Chapter 6, Saame Raza Shaikh describes the role of dietary fatty acids in modulating the structure and function of lipid rafts, with a special emphasis on the impact of polyunsaturated fatty acids. The works summarized here offers an interesting perspective on how dietary modulation of membrane lipids, through its effects on membrane domain organization, may directly impact human health.

The proposed functions of lipid rafts in mammalian cells have been extensively investigated, but only recently has the possibility that such domains play important roles in other classes of organisms been explored. Through the joint efforts of several authors including Erwin London and Amir Farnoud, our state of knowledge on the role of rafts in both pathogenic bacteria and yeast is reviewed in Chapter 7. In contrast, Chapter 8 revisits one of the earliest proposed functions of lipid rafts: the regulation of the trafficking of GPI-anchored proteins in polarized epithelial cells. As outlined by Simona Paladino, Chiara Zurzolo and colleagues, current models for how these proteins are handled by cells not only includes a potential role for rafts, but also for protein-driven oligomerization events and sorting steps at multiple intracellular compartments.

About the same time that the lipid raft model was being developed, another interesting class of membrane domains came to the forefront following the discovery of the protein caveolin-1 as a major structural protein of caveolae (Rothberg et al., 1992). Residing on the plasma membrane of many different cell types, these flask-shaped invaginations have been implicated in a wide range of cellular processes, including membrane trafficking and cell signaling (Parton & del Pozo, 2013). However, despite intense study, many questions about caveolae remain unanswered, including how the packing of caveolin-1 within membranes contributes to membrane bending and caveolar assembly. As described in Chapter 9, recent biophysical studies have now begun to shed light on this issue. Here, Kerney Jebrell Glover and colleagues describe how findings from his lab, as well as others in the field, are revealing the structure, topology, and oligomerization status of this integral membrane protein.

In closing, I would like to take this opportunity to thank my friend and former mentor Sid Simon for the invitation to put this volume together and Helene Kabes for editorial assistance. I would also like to acknowledge the sources of support for work in my own lab on membrane domain organization and function (NIH R01 GM106720, NIH R01 GM106672, and NIH R01 HL111259). And of course, none of this would be possible without the hard work of the authors, who graciously contributed their time and expertise to write the chapters, as well as those who provided feedback as reviewers. I hope that you the reader will benefit from the wealth of knowledge summarized in these chapters as much as I have.

Anne K. Kenworthy
Vanderbilt School of Medicine

REFERENCES

Ahmed, S. N., Brown, D. A., & London, E. (1997). On the origin of sphingolipid/cholesterol-rich detergent-insoluble cell membranes: physiological concentrations of cholesterol and sphingolipid induce formation of a detergent-insoluble, liquid-ordered lipid phase in model membranes. *Biochemistry, 36*, 10944–10953.

Brown, D. A., & London, E. (1998). Structure and origin of ordered lipid domains in biological membranes. *Journal of Membrane Biology, 164*(2), 103–114.

Edidin, M. (1992). Patches, posts and fences: proteins and plasma membrane domains. *Trends in Cell Biology, 2*(12), 376–380.

Edidin, M. (2003). The state of lipid rafts: from model membranes to cells. *Annual Review of Biophysics and Biomolecular Structure, 32*, 257–283.

Hancock, J. F. (2006). Lipid rafts: contentious only from simplistic standpoints. *Nature Reviews Molecular Cell Biology, 7*(6), 456–462.

Jacobson, K., Mouritsen, O. G., & Anderson, R. G. (2007). Lipid rafts: at a crossroad between cell biology and physics. *Nature Cell Biology, 9*(1), 7–14.

Jacobson, K., Sheets, E. D., & Simson, R. (1995). Revisiting the fluid mosaic model of membranes. *Science, 268*(5216), 1441–1442.

Kenworthy, A. K. (2008). Have we become overly reliant on lipid rafts? Talking point on the involvement of lipid rafts in T-cell activation. *EMBO Reports, 9*(6), 531–535.

Kraft, M. L. (2013). Plasma membrane organization and function: moving past lipid rafts. *Molecular Biology of the Cell, 24*(18), 2765–2768.

LaRocca, T. J., Pathak, P., Chiantia, S., Toledo, A., Silvius, J. R., Benach, J. L., et al. (2013). Proving lipid rafts exist: membrane domains in the prokaryote *Borrelia burgdorferi* have the same properties as eukaryotic lipid rafts. *PLoS Pathogens, 9*(5), e1003353.

Lichtenberg, D., Goni, F. M., & Heerklotz, H. (2005). Detergent-resistant membranes should not be identified with membrane rafts. *Trends in Biochemical Sciences, 30*(8), 430–436.

Lingwood, D., & Simons, K. (2010). Lipid rafts as a membrane-organizing principle. *Science, 327*(5961), 46–50.

London, E. (2005). How principles of domain formation in model membranes may explain ambiguities concerning lipid raft formation in cells. *Biochimica et Biophysica Acta, 1746*(3), 203–220.

Morales-Penningston, N. F., Wu, J., Farkas, E. R., Goh, S. L., Konyakhina, T. M., Zheng, J. Y., et al. (2010). GUV preparation and imaging: minimizing artifacts. *Biochimica et Biophysica Acta, 1798*(7), 1324–1332.

Munro, S. (2003). Lipid rafts: elusive or illusive? *Cell, 115*(4), 377–388.

Nicolson, G. L. (2014). The fluid-mosaic model of membrane structure: still relevant to understanding the structure, function and dynamics of biological membranes after more than 40 years. *Biochimica et Biophysica Acta, 1838*(6), 1451–1466.

Owen, D. M., & Gaus, K. (2013). Imaging lipid domains in cell membranes: the advent of super-resolution fluorescence microscopy. *Frontiers in Plant Science, 4*, 503.

Owen, D. M., Magenau, A., Williamson, D., & Gaus, K. (2012). The lipid raft hypothesis revisited–new insights on raft composition and function from super-resolution fluorescence microscopy. *Bioessays, 34*(9), 739–747.

Owen, D. M., Williamson, D. J., Magenau, A., & Gaus, K. (2012). Sub-resolution lipid domains exist in the plasma membrane and regulate protein diffusion and distribution. *Nature Communications, 3*, 1256.

Parton, R. G., & del Pozo, M. A. (2013). Caveolae as plasma membrane sensors, protectors and organizers. *Nature Reviews Molecular Cell Biology, 14*(2), 98–112.

Pike, L. J. (2006). Rafts defined: a report on the Keystone symposium on lipid rafts and cell function. *Journal of Lipid Research, 47*(7), 1597–1598.

Rothberg, K. G., Heuser, J. E., Donzell, W. C., Ying, Y. S., Glenney, J. R., & Anderson, R. G. (1992). Caveolin, a protein component of caveolae membrane coats. *Cell, 68*(4), 673–682.

Sako, Y., & Kusumi, A. (1994). Compartmentalized structure of the plasma membrane for receptor movements as revealed by a nanometer-level motion analysis. *Journal of Cell Biology, 125*(6), 1251–1264.

Simons, K., & Gerl, M. J. (2010). Revitalizing membrane rafts: new tools and insights. *Nature Reviews Molecular Cell Biology, 11*(10), 688–699.

Simons, K., & Ikonen, E. (1997). Functional rafts in cell membranes. *Nature, 387*, 569–572.

Singer, S. J., & Nicolson, G. L. (1972). The fluid mosaic model of the structure of cell membranes. *Science, 175*(23), 720–731.

Veatch, S. L. (2007). Electro-formation and fluorescence microscopy of giant vesicles with coexisting liquid phases. *Methods in Molecular Biology, 398*, 59–72.

Veatch, S. L., & Keller, S. L. (2002). Organization in lipid membranes containing cholesterol. *Physical Review Letters, 89*(26), 268101.

Veatch, S. L., & Keller, S. L. (2003). Separation of liquid phases in giant vesicles of ternary mixtures of phospholipids and cholesterol. *Biophysical Journal, 85*(5), 3074–3083.

Veatch, S. L., & Keller, S. L. (2005). Seeing spots: complex phase behavior in simple membranes. *Biochimica et Biophysica Acta, 1746*(3), 172–185.

Wesolowska, O., Michalak, K., Maniewska, J., & Hendrich, A. B. (2009). Giant unilamellar vesicles – a perfect tool to visualize phase separation and lipid rafts in model systems. *Acta Biochimica Polonica, 56*(1), 33–39.

PREVIOUS VOLUMES IN SERIES

Current Topics in Membranes and Transport

Volume 23 Genes and Membranes: Transport Proteins and Receptors* (1985)
Edited by Edward A. Adelberg and Carolyn W. Slayman

Volume 24 Membrane Protein Biosynthesis and Turnover (1985)
Edited by Philip A. Knauf and John S. Cook

Volume 25 Regulation of Calcium Transport across Muscle Membranes (1985)
Edited by Adil E. Shamoo

Volume 26 Na^+-H^+Exchange, Intracellular pH, and Cell Function* (1986)
Edited by Peter S. Aronson and Walter F. Boron

Volume 27 The Role of Membranes in Cell Growth and Differentiation (1986)
Edited by Lazaro J. Mandel and Dale J. Benos

Volume 28 Potassium Transport: Physiology and Pathophysiology* (1987)
Edited by Gerhard Giebisch

Volume 29 Membrane Structure and Function (1987)
Edited by Richard D. Klausner, Christoph Kempf, and Josvan Renswoude

Volume 30 Cell **Volume** Control: Fundamental and Comparative Aspects in Animal Cells (1987)
Edited by R. Gilles, Arnost Kleinzeller, and L. Bolis

Volume 31 Molecular Neurobiology: Endocrine Approaches (1987)
Edited by Jerome F. Strauss, III, and Donald W. Pfaff

Volume 32 Membrane Fusion in Fertilization, Cellular Transport, and Viral Infection (1988)
Edited by Nejat Düzgüneş and Felix Bronner

Volume 33 Molecular Biology of Ionic Channels* (1988)
Edited by William S. Agnew, Toni Claudio, and Frederick J. Sigworth

Volume 34 Cellular and Molecular Biology of Sodium Transport* (1989)
Edited by Stanley G. Schultz

Volume 35 Mechanisms of Leukocyte Activation (1990)
Edited by Sergio Grinstein and Ori D. Rotstein

*Part of the series from the Yale Department of Cellular and Molecular Physiology

Volume 36 Protein–Membrane Interactions* (1990)
Edited by Toni Claudio

Volume 37 Channels and Noise in Epithelial Tissues (1990)
Edited by Sandy I. Helman and Willy Van Driessche

Current Topics in Membranes

Volume 38 Ordering the Membrane Cytoskeleton Trilayer* (1991)
Edited by Mark S. Mooseker and Jon S. Morrow

Volume 39 Developmental Biology of Membrane Transport Systems (1991)
Edited by Dale J. Benos

Volume 40 Cell Lipids (1994)
Edited by Dick Hoekstra

Volume 41 Cell Biology and Membrane Transport Processes* (1994)
Edited by Michael Caplan

Volume 42 Chloride Channels (1994)
Edited by William B. Guggino

Volume 43 Membrane Protein–Cytoskeleton Interactions (1996)
Edited by W. James Nelson

Volume 44 Lipid Polymorphism and Membrane Properties (1997)
Edited by Richard Epand

Volume 45 The Eye's Aqueous Humor: From Secretion to Glaucoma (1998)
Edited by Mortimer M. Civan

Volume 46 Potassium Ion Channels: Molecular Structure, Function, and Diseases (1999)
Edited by Yoshihisa Kurachi, Lily Yeh Jan, and Michel Lazdunski

Volume 47 Amiloride-Sensitive Sodium Channels: Physiology and Functional Diversity (1999)
Edited by Dale J. Benos

Volume 48 Membrane Permeability: 100 Years since Ernest Overton (1999)
Edited by David W. Deamer, Arnost Kleinzeller, and Douglas M. Fambrough

Volume 49 Gap Junctions: Molecular Basis of Cell Communication in Health and Disease
Edited by Camillo Peracchia

Volume 50 Gastrointestinal Transport: Molecular Physiology
Edited by Kim E. Barrett and Mark Donowitz

CHAPTER ONE

Supported Lipid Bilayers as Models for Studying Membrane Domains

Volker Kiessling*, Sung-Tae Yang and Lukas K. Tamm
Department of Molecular Physiology and Biological Physics, Center for Membrane Biology, University of Virginia, Charlottesville, VA, USA
*Corresponding author: E-mail: vgk3c@virginia.edu

Contents

Abstract

Supported lipid bilayers have been in use for over 30 years. They have been employed to study the structure, composition, and dynamics of lipid bilayer phases, the binding and distribution of soluble, integral, and lipidated proteins in membranes, membrane fusion, and interactions of membranes with elements of the cytoskeleton. This review focuses on the unique ability of supported lipid bilayers to study liquid-ordered and liquid-disordered domains in membranes. We highlight methods to produce

Current Topics in Membranes, Volume 75
ISSN 1063-5823
http://dx.doi.org/10.1016/bs.ctm.2015.03.001

asymmetric lipid bilayers with lipid compositions that mimic those of the extracellular and cytoplasmic leaflets of cell membranes and the functional reconstitution of membrane proteins into such systems. Questions related to interleaflet domain coupling and membrane protein activation have been addressed and answered using advanced reconstitution and imaging procedures in symmetric and asymmetric supported membranes with and without coexisting lipid phase domains. Previously controversial topics regarding anomalous and anisotropic diffusion in membranes have been resolved by using supported membrane approaches showing that the propensity of certain lipid compositions to form "rafts" are important but overlaid with "picket-fence" interactions that are imposed by a subtended cytoskeletal network.

1. INTRODUCTION

A fascinating challenge of membrane biology over the last decade has been the study of lipid domains in cell membranes. Cellular membranes are composed of a bewildering variety of different lipids that form their lipid bilayer matrix into which membrane proteins of various topologies are embedded. The functional reasons for the enormous complexity of lipid compositions in biomembranes are only poorly understood. Some lipids activate enzymes while others have distinct roles in cell signaling. Yet other lipids and combinations of lipids are subject to thermodynamic forces that drive them into small or large assemblies in the plane of the lipid bilayer. A classic example is the combination of sphingomyelins (SMs), cholesterol, and unsaturated phospholipids that phase-separate into large domains in lipid model membranes (Crane & Tamm, 2003; Dietrich, Bagatolli, et al., 2001; Feigenson, 2009; Veatch & Keller, 2005). SMs and cholesterol are enriched in liquid-ordered (Lo) domains, whereas unsaturated phospholipids are enriched in liquid-disordered (Ld) domains. The lipids have more extended, i.e., more "ordered" alkyl chains and diffuse about an order of magnitude more slowly in Lo than in Ld phases. Despite these different properties, the lipids are still laterally mobile in both phases, i.e., they are considered to form a two-dimensional "liquid" as opposed to a gel phase. Gel phases with very slowly diffusing lipids are formed, e.g., by SMs or ceramides, in the absence of cholesterol. Whether Lo-phase lipid domains, often referred to as "lipid rafts," also occur in cell membranes has been the subject of intense debates over the past 15 years or so (Edidin, 2003; Kusumi et al., 2012; Lingwood & Simons, 2010; Pike, 2006).

Lipid rafts were originally defined by cell biologists based on the observation that the extraction of cell membranes with cold detergent yielded fractions that were enriched in SMs and cholesterol (Brown & London, 1998; Simons & Ikonen, 1997). These fractions also frequently contained proteins that are known to cooperate in cell signaling. Although it was later shown that the aggregation of sphingomyelin and cholesterol can be artifactually enhanced by the action of detergent extraction (Lichtenberg, Goni, & Heerklotz, 2005), higher resolution optical techniques on cell membranes have revealed that these lipids retain special properties in cellular membranes that organize them in nanoscale domains even in the presence of the high densities of membrane proteins that are encountered in cell membranes (Honigmann, Mueller, et al., 2014). These domains are not only very small (10−100 nm), but may also be relatively short-lived and undergoing continuous lipid and protein exchange, which makes them quite difficult to detect in cell membranes. However, lipid domains in pure lipid model systems are much larger (1−10 μm) and long-lived, which renders their study much easier in these systems. Despite these clear differences between lipid domains in cellular and model membranes, we believe that studies of lipid domains and their interactions with membrane proteins in model systems yield highly valuable insight into their nature since the model systems amplify and reveal underlying physical−chemical principles that must also occur in cell membranes, but are much more difficult to uncover in live cells.

There have been two principal model systems to study the coexistence of Lo and Ld phases in lipid bilayers. They are giant unilamellar vesicles (GUVs) and supported lipid bilayers (SLBs). This article reviews pertinent results on lipid domain formation, lipid asymmetry, and protein insertion in SLBs. Although we focus on SLBs, many of the same phenomena can also be observed in GUVs. Despite the high complementarity between these two model systems, they both have their own advantages and disadvantages. In GUVs, one does not need to worry about potential adverse interactions with the substrate that may occur in supported bilayers, but SLBs are flat and extend over many millimeters and therefore offer much better in-focus imaging capabilities than GUVs.

SLBs have been introduced more than 30 years ago as model systems for mimicking cellular membranes (Tamm, 1984; Tamm & McConnell, 1985). They are particularly well suited to study the lateral mobility and phase behavior of lipids (Groves, Boxer, & McConnell, 1997; McConnell, Watts, Weis, & Brian, 1986; Sackmann, 1996; Tamm, 1988; Wright, Palmer, & Thompson, 1988). It has been recognized early on that integral membrane proteins with significant protein domains on both sides of the membrane cannot be reconstituted into supported bilayers in a laterally mobile form.

Therefore, multiple strategies have been devised to lift SLBs from their solid supports by introduction of an intervening polymer cushion. Several such designs improve the lateral mobility of integral membrane proteins (Floyd, Ragains, Skehel, Harrison, & van Oijen, 2008; Goennenwein, Tanaka, Hu, Moroder, & Sackmann, 2003; Naumann et al., 2000; Wagner & Tamm, 2000). As will be reviewed below, the same kinds of polymer cushions are also advantageous to preserve symmetric and asymmetric lipid domain structures in supported membranes.

2. METHODS FOR SUPPORTED LIPID BILAYER PREPARATION

To set the stage, we will first briefly review methods for supported bilayer preparation and their applications for intra- and interleaflet lipid interactions. In particular, we want to emphasize the utility of the different techniques for the study of domain structure, bilayer asymmetry, and reconstitution of integral membrane proteins.

2.1 Direct vesicle fusion

The most common technique to form supported bilayers is the direct vesicle fusion (VF) method (Brian & McConnell, 1984; Kalb, Frey, & Tamm, 1992) (Figure 1(A)), which was developed shortly after the original Langmuir–Blodgett/Langmuir–Schäfer (LB/LS) method (see below). Small unilamellar vesicles (SUVs, diameter <50 nm) or large unilamellar vesicles (LUVs, diameter >50 nm) of a specific lipid composition are prepared and added to a clean hydrophilic substrate. Typical substrates are glass or quartz for fluorescence microscopy applications and mica for atomic force microscopy (AFM) applications. After a 30–60-min incubation time, during which vesicles adhere to the substrate and rupture to form a continuous membrane, the nonadhering vesicles are washed out with buffer.

The simplicity of the procedure makes this a very attractive approach. However, for the study of domains in membranes, this method has several disadvantages. First, the domains formed from phase-separating lipid compositions are too small to be visualized by normal fluorescence microscopy. Yet, this problem can be circumvented by using AFM or optical superresolution detection techniques. Second, integral membrane proteins usually end up mostly immobile in the supported bilayer. The water cleft between the substrate and lipid bilayer is only 1–2 nm wide (Fromherz, Kiessling, Kottig, & Zeck, 1999; Kiessling & Tamm, 2003). Proteins with large

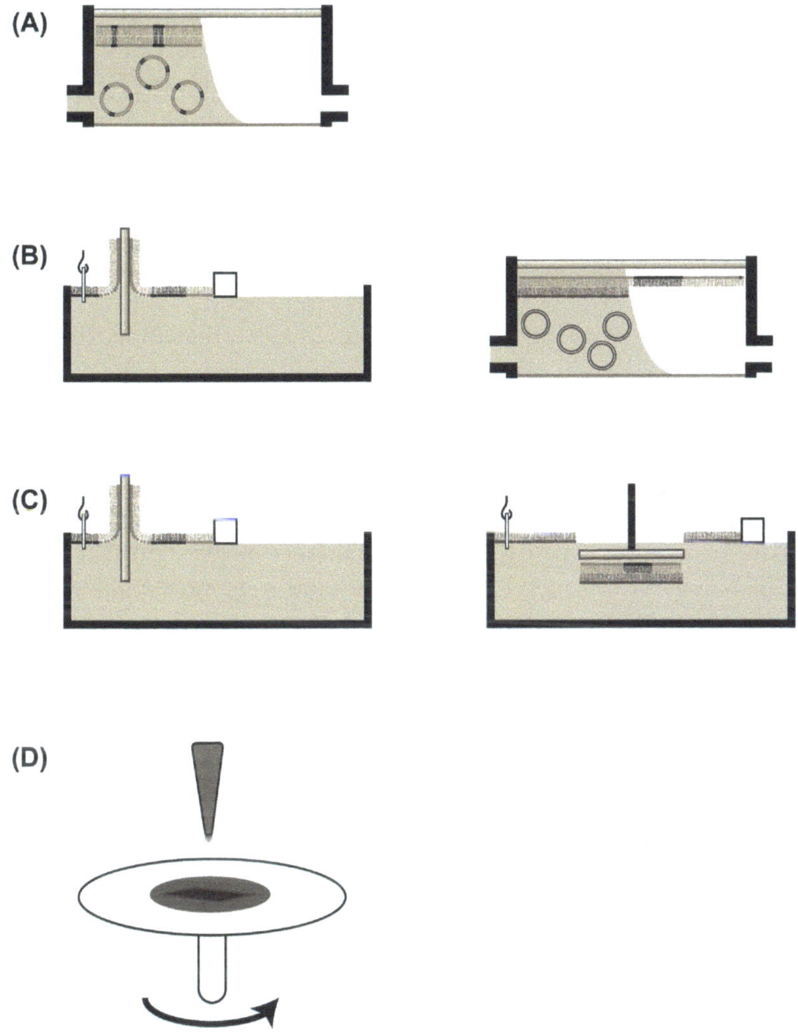

Figure 1 ***Methods for supported lipid bilayer preparation.*** (A) Direct vesicle fusion (VF). (B) Langmuir–Blodgett/vesicle fusion (LB/VF). (C) Langmuir–Blodgett/Langmuir–Schäfer (LB/LS). (D) Spin coating. (See color plate)

cytosolic or extracellular domains that are facing outward in the vesicle get stuck to the substrate upon vesicle adhesion. Protein mobility can be improved by cushioning the bilayer with polymers like cellulose from the solid substrate (Goennenwein et al., 2003). A different protein reconstitution approach that results in highly oriented protein insertion is the so-called direct incorporation method (Milhiet et al., 2006). A small volume of protein stabilized in detergent is added to an already formed SLB. The detergent is later removed by extensive rinsing with buffer and/or by incubating with biobeads. Third, studying bilayer asymmetry with different defined lipid compositions in both leaflets is not as straightforward as with the methods discussed below. However, Visco, Chiantia, and Schwille (2014) created bilayer asymmetry and triggered domain formation by using a methyl-β-cyclodextrin-mediated lipid exchange mechanism.

2.2 Langmuir–Blodgett/vesicle fusion

To overcome some of the limitations of the direct VF method, the preparation of the two leaflets of the SLB is accomplished in two subsequent steps (Kalb et al., 1992) (Figure 1(B)). First, a lipid monolayer is transferred from the water–air interface of a Langmuir trough to the substrate by quickly immersing a carefully cleaned hydrophilic slide into the water subphase and slowly retracting it through the interface. Before transfer, the surface pressure of the monolayer is set to the value (~32 mN/m) that matches that of biological cell membranes. In a second step, vesicles are added to the dry supported monolayer to form the second leaflet of the bilayer (Kalb et al., 1992).

Lipid compositions with coexisting lipid phases separate into micrometer-size distinguishable domains when spread at the water–air interface. These domains are maintained when the monolayer is transferred to the substrate (Crane & Tamm, 2004; Dietrich, Bagatolli, et al., 2001).

Reconstitution of integral membrane proteins that are laterally mobile is achieved by incorporating them into the vesicles that form the second leaflet of the supported bilayer (Kalb & Tamm, 1992). Soluble *N*-ethylmaleimide-sensitive-factor-activating protein receptors (SNARE) reconstituted by this technique are preferentially oriented with their large cytosolic domains facing away from the substrate (Liang, Kiessling, & Tamm, 2013; Wagner & Tamm, 2001) and FRAP measurements reveal that most protein is laterally mobile.

Protein mobility can be further enhanced especially for larger proteins by incorporating a polymer cushion between the substrate and the bilayer (Wagner & Tamm, 2000). Adding 3 mol% of lipid that is covalently linked to 77 subunits of polyethylene glycol (PEG) and a silane group at the end of the PEG chain to the first monolayer stabilizes the proximal leaflet of the SLB and increases the distance between the bilayer and the support to ~4 nm (Kiessling & Tamm, 2003). The increased stability of the proximal leaflet turns out to be especially relevant for studies on membrane domains.

The Langmuir–Blodgett/Vesicle Fusion (LB/VF) technique allows preparation of supported bilayers that maintain an asymmetric lipid composition for several hours (Crane, Kiessling, & Tamm, 2005) if the membrane is composed of lipids that are in the Ld phase. However, to achieve bilayer asymmetry with lipids that are in the Lo phase or that form membranes with coexisting Lo and Ld phase domains, using the PEG cushion becomes crucial to maintain stability of the domains in the proximal leaflet during distal leaflet lipid assembly via VF (Kiessling, Crane, & Tamm, 2006).

2.3 Langmuir–Blodgett/Langmuir–Schäfer

After a dry supported lipid monolayer has been prepared by the LB method as described above, transferring a second monolayer from the LB trough to the substrate can also complete bilayer formation (Tamm, 1984; Tamm & McConnell, 1985). For this, the substrate is first held horizontally above the subphase orienting the hydrophobic lipid chains of the already supported monolayer and the lipid chains at the water interface toward each other before passing the substrate through the interface (Figure 1(C)).

In principle, this technique, like the LB/VF method described above, also allows one to generate SLBs with asymmetric lipid compositions. However, in contrast to the LB/VF technique, SLB preparations with fluorescently labeled lipids added to only one monolayer did not result in fully asymmetric dye distributions (Crane et al., 2005).

As described above for the LB/VF technique, the bilayer can be cushioned and lifted from the surface by adding polymer conjugated lipids to the LB layer. Utilizing the lipopolymers dioctadecylamine(poly(ethyloxazoline)-8988) and octadecyl(poly(ethyloxazoline)-5822) and the LB/LS technique, the Naumann group was able to prepare asymmetric bilayers with coexisting Lo and Ld domains in registration in both leaflets (Garg, Ruhe, Ludtke, Jordan, & Naumann, 2007). Integral membrane proteins can be reconstituted by the direct incorporation method after the SLB has been prepared (Siegel, Kimble-Hill, Garg, Jordan, & Naumann, 2011).

2.4 Spin coating

Spin-coating substrates with lipids in an organic solvent is a relatively new technique to prepare stacked SLBs to study membrane domains (Mennicke & Salditt, 2002). Lipid mixtures are dissolved in an organic solvent and spread over the surface of a cleaned substrate on top of a spin coater. The substrate is then rotated at 1000–3000 rpm for ~30 s (Figure 1(D)). After residual solvent is removed under vacuum, the lipids are hydrated in the desired buffer solution. Depending on the rinsing procedure and buffer conditions, this results in one to three stacked SLBs (Honigmann, van den Bogaart, et al., 2013; Honigmann, Mueller, Hell, & Eggeling, 2013). So far, only peptides or truncated integral membrane proteins have been incorporated into these model membranes by mixing them with the lipids in the organic solvent (Honigmann, van den Bogaart, et al., 2013; Mennicke & Salditt, 2002). The method does not work for larger proteins that denature in organic solvent. Coexisting Lo and Ld domains in symmetric spin-coated SLBs have been observed by confocal and STED microscopy (Honigmann, van den Bogaart, et al., 2013; Honigmann, Mueller, et al., 2013).

2.5 Domains in supported bilayers with symmetric lipid leaflets

Model membrane systems continue to be excellent tools to investigate the molecular interactions that determine the highly dynamic structure of biological membranes. In pure lipid bilayers, we need to distinguish lipid–lipid interactions within the plane of the membrane and transverse interactions between the two leaflets of the bilayer. The first type of interactions may lead to clusters of particular lipid species or to cholesterol-dependent Lo phase domains ("lipid rafts"). The second type of interactions plays a role in lipid interdigitation and domain coupling between the two leaflets. SLBs are uniquely useful for investigating these latter interactions.

Supported bilayers prepared by the VF method are usually considered symmetric, although depending on the support material, lipid composition, buffer conditions, and temperature, asymmetry can be induced (Lin, Blanchette, Ratto, & Longo, 2006; Richter, Maury, & Brisson, 2005; Rossetti, Textor, & Reviakine, 2006; Wacklin & Thomas, 2007). Domains in these bilayers are formed in situ and range from tens of nanometers to a few micrometers in size. Due to the small size, AFM proved to be the technique of choice for studying their structure and dynamics (Dufrene, Barger, Green, & Lee, 1997). Image contrast between domain and nondomain areas can be achieved by height differences, e.g., Lo domains in ternary "lipid-raft" mixtures are $\sim$0.8 nm higher than the surrounding Ld-phase areas. Additional information can be gained by performing force measurements to probe the mechanical stability of the bilayer (Chiantia, Ries, Kahya, & Schwille, 2006). For further reading about the different aspects of AFM studies and domains in supported membranes, we refer to a number of excellent review articles (Giocondi et al., 2010; Janshoff & Steinem, 2001; Lin, Blanchette, Ratto, & Longo, 2007).

While the domains formed on mica substrates for AFM studies are dynamic in the sense that they can still change their shape and size, the much larger (tens of micrometer) domains transferred by the LB process are, as a whole, immobile. The lipids that constitute Lo domains are mobile and the partitioning of fluorescently labeled lipids is consistent with results obtained from LUVs and GUVs (Crane & Tamm, 2004; Dietrich, Bagatolli, et al., 2001). Although these domains do not move in the plane of the membrane or fuse with each other like their counterparts do in GUVs, they can be dissolved by the extraction of cholesterol with methyl-β-cyclodextrin (Yang, Kiessling, Simmons, White, & Tamm, in press).

2.6 Domains in supported bilayers with asymmetric lipid leaflets

The LB/VF and LB/LS techniques can be used to prepare asymmetric bilayers with defined and distinct lipid compositions in each leaflet. Lo phase domains formed at the water—air interface of a Langmuir trough and transferred to the substrate in two steps by the LB/LS method register when the membrane—surface interaction is reduced by lipopolymers (Garg et al., 2007). Our group found that lipid compositions that by themselves do not show any phase separation will form Lo domains on top of already existing Lo domains of a phase-separated monolayer (Kiessling et al., 2006). Prerequisite for this effect is the presence of populations of inner leaflet lipids with low chain melting temperatures ($T_m < T_{environment}$, most phosphatidylcholines (PCs)) and inner leaflet lipids with high chain melting temperatures ($T_m > T_{environment}$, most phosphatidylethanolamines (PEs) and phosphatidylserines (PSs)) and cholesterol (Wan, Kiessling, & Tamm, 2008). More details on the lipid requirements of transbilayer coupling of Lo phase domains can be found in a relatively recent review (Kiessling, Wan, & Tamm, 2009).

Recently, a new method to assemble symmetric and asymmetric supported bilayers with coexisting Lo and Ld phases by VF has been introduced. In this method, SLBs are prepared from SUVs consisting of PC and SM lipids before the distal leaflet is subsequently enriched with cholesterol by a methyl-β-cyclodextrin-mediated lipid exchange (Visco et al., 2014). With PC and SM present in both leaflets, this method produces micrometer-size bilayer-spanning domains. Interestingly, the presence of a single-phase proximal leaflet suppresses the domain formation in the distal leaflet of otherwise phase-separating lipid compositions. This correlates well with the observation that the success rate (i.e., forming a bilayer without destroying the first monolayer) of forming a single-phase monolayer on top of phase-separated monolayers by the LB/VF technique was reduced although, of course, such bilayers can be formed (Wan et al., 2008). Together these observations confirm results from free-standing lipid bilayers that the phase behavior of the two leaflets are not independent, but that the lipid composition in one leaflet can induce or suppress phase separation in the other leaflet (Collins & Keller, 2008).

3. TARGETING MEMBRANE DOMAINS WITH SOLUBLE PROTEINS

3.1 Cholera toxin-GM1 targeting

The ubiquitous ganglioside GM1 has been a traditional marker for lipid rafts in cell and model membranes. Owing to its mostly saturated hydrocarbon chain ceramide moiety, GM1 preferentially partitions into Lo phases. Its oligosaccharide polar moiety constitutes the primary cellular receptor for subunit B of cholera toxin. Fluorescently labeled cholera toxin B subunits (Fl-CTB) have therefore frequently been used to image lipid rafts. Indeed, early reconstitutions of lipid rafts in SLBs have used Fl-CTB to image Lo phases in model membranes composed of SM, 1-palmitoyl-2-oleoyl-*sn*-glycero-3-phosphocholine (POPC), and cholesterol (Dietrich, Bagatolli, et al., 2001), and lipid extracts from mouse kidney cells (Dietrich, Volovyk, Levi, Thompson, & Jacobson, 2001). In both cases, raft-targeting of GM1-CTB critically depends on the cholesterol content and parallels the appearance of Lo phases as visualized by fluorescent lipid analogs.

In the absence of cholesterol, GM1 still clusters in SLBs as visualized by Fl-CTB. GM1 clusters ranging from 7 nm at 0.1 mol% GM1 to 20 nm at 10 mol% GM1 were observed by AFM (Shi et al., 2007). Recent high-resolution lateral diffusion measurements of CTB bound to GM1 similarly concluded that GM1 undergoes anomalous subdiffusion due to interleaflet coupling between GM1s in the two leaflets of a supported bilayer (Spillane et al., 2014). When the saturated hydrocarbon chains of GM1 were replaced with oleic acids, the anomalous subdiffusion was abolished indicating that GM1 clusters had dissolved and were no longer coupled to GM1 molecules in the other leaflet of the bilayer.

3.2 Glycosylphosphatidylinositol-anchored protein

Glycosylphosphatidylinositol-anchored proteins (GPI-APs) are located in the extracellular leaflet of the plasma membrane and were the focus of the original raft hypothesis (Simons & Ikonen, 1997). Simons and coworkers postulated that GPI-APs are incorporated into sphingolipid—cholesterol rafts in the Golgi complex and carried along with them to the apical membrane of MDCK cells. The function of GPI-APs within signaling pathways involves different dimerization states that are stabilized by their association with lipid rafts (Suzuki et al., 2012).

The urokinase plasminogen activator receptor (uPAR), a GPI-AP, was reconstituted into polymer-tethered SLBs to study how ligand binding alters the dimerization state and to quantify the partitioning into Lo domains (Ge, Siegel, Jordan, & Naumann, 2014). It was found that binding of urokinase plasminogen activator (uPA) suppresses uPAR dimerization while binding of vitronectin increases the dimerization levels. As expected, uPAR partitioned preferentially into Lo domains of bilayers containing 1,2-dioleoyl-*sn*-glycero-3-phosphocholine (DOPC), 1,2-dipalmitoyl-*sn*-glycero-3-phosphocholine (DPPC), and cholesterol. However, upon binding to uPA the affinity to the Lo phase was reduced while upon binding to vitronectin the affinity to the Lo phase was further increased.

3.3 Calcium- and noncalcium-mediated C2 domain targeting

C2 domains are part of many signaling molecules including phospholipases and synaptotagmins. They are usually translocated to membranes by an elevation of the surrounding calcium concentration. They typically contain two calcium binding sites and membrane translocation occurs through calcium coordination with acidic lipids that are prevalent in the inner leaflet of cell membranes. Since inner leaflet lipids do not on their own form ordered domains (see Section 2.6), C2 domain binding could indicate the distribution of acidic lipids in nonhomogenous membranes. Indeed, when C2A domains of synaptotagmin are bound to asymmetric supported lipid bilayers with an SM-PC cholesterol outer-leaflet raft-mimicking lipid mixture and a PC-PE-PS-cholesterol inner leaflet-mimicking mixture, the protein and hence PS are found enriched in the induced Ld domains of the inner leaflet (Wan, Kiessling, Cafiso, & Tamm, 2011). The enhancement of C2A binding to Ld over Lo domains is two- to threefold. Similar results are observed for tandem C2A-C2B domain binding to inner leaflet lipid mixtures that in addition to the above lipids also contain up to 5 mol% of the signaling lipid phosphatidylinositol 4,5 biphosphate (PIP2). Contrary to C2A, the synaptotagmin C2B domain not only binds via calcium coordination to acidic lipids, but also binds specifically to the doubly phosphorylated headgroup of PIP2. Even in the absence of calcium, C2A-C2B binds to acidic lipids via a polybasic motive on the opposite side from the calcium binding loops of C2B. This calcium-independent binding interaction also reveals an enrichment of acidic lipids in the Ld over the Lo phase in asymmetric supported bilayers (Wan et al., 2011). Interestingly, and perhaps reminiscent of the CTB-detected clustering of GM1 described above, C2A-C2B and C2B, but not C2A demix PS in single-phase cholesterol-free membranes (Lai, Tamm, Ellena, & Cafiso, 2011).

3.4 Targeting membrane domains with Ras and Rac1

Ras proteins are small G-proteins that are involved in numerous growth-factor-regulated signaling events. Many cancers can be traced to disturbed Ras protein regulation. All Ras proteins are C-terminally farnesylated, but N- and H-Ras have additional palmitylation sites. K-Ras has no further lipid modification, but contains a polybasic amino acid sequence just before the terminal farnesylation. All three types of Ras proteins have been reported to associate with lipid rafts in cells and the argument that raft association is important for cell signaling has been made frequently (Hancock, 2003; Kholodenko, Hancock, & Kolch, 2010). However, raft associations of Ras proteins have not been uncontroversial in the cell biological literature. To shed more light on this problem and to distinguish between different models of lipid association of Ras proteins, Winter and coworkers have investigated the distribution of N-Ras and K-Ras in GUV and SLB model membranes. Farnesylation alone is not sufficient to stably anchor Ras in a cell or model membrane. When additionally palmitylated, N-Ras partitions into the Ld phase of two-phase lipid bilayers consisting of DPPC, DOPC, and cholesterol (Weise, Triola, Brunsveld, Waldmann, & Winter, 2009). Subsequently the proteins move by slow diffusion to Lo—Ld domain boundaries, where they also form small clusters. This behavior is independent of whether N-Ras is in its GDP- or GTP-bound form.

A different behavior is observed in similar studies with K-Ras (Weise et al., 2011). This protein is translocated to membranes with acidic lipids through electrostatic coupling of eight lysine residues right next to the farnesyl tail. To model this electrostatic interaction, the authors included equimolar DPPG and DOPG (5—10 mol% each) into their DPPC/DOPC/cholesterol phase-separated model membranes. Like N-Ras, K-Ras partitions into the Ld phase, but unlike N-Ras it does not eventually accumulate at phase boundaries. However, after prolonged incubation, K-Ras formed quite extended protein-rich fluid-phase domains within the Ld phase domains. The protein-rich fluid domains are very likely also enriched in fluid anionic lipids although this was not formally shown in this study. Like with N-Ras, the protein-induced fluid—fluid phase-separation is again independent of GDP or GTP loading of K-Ras.

Rac1 is another small G-protein belonging to the Rho family that has been claimed to be raft-associated in cell membranes (Jaffe & Hall, 2005; del Pozo et al., 2004). Rho-family GTPases regulate cytoskeletal organization and cell adhesion and hence are important modulators in development, angiogenesis, immune responses, and cell cycle control. Rac1 is geranylgeranylated at its extreme C-terminus. The geranylgeranyl modification has a longer hydrophobic isoprenyl tail than the farnesyl modification in Ras and therefore, provides stable membrane translocation of Rac1 upon GTP binding. (Rac1 binds to a GDP dissociation inhibitor (RhoGDI) in its GDP-bound form, which renders this complex soluble and its translocation to the membranes is mediated through the action of a GDP—GTP nucleotide exchange factor (GEF).) When Rac1 is bound to phase-separated SLB composed of brain SM, brain PC, and cholesterol, it translocates primarily to the boundary region between the Lo and Ld phases (Moissoglu et al., 2014). FRET analysis in cell membranes shows that Rac1 requires the presence of lipid rafts, but is still predominantly found in Ld phase lipids, which is consistent with its corrugated geranylgeranyl tail not fitting well into the highly ordered hydrophobic tails of Lo phase lipids. Single-molecule fluorescence observations that Rac1 inserts into heterogenous membranes at the edges of Lo phase lipid domains resolve the apparent discrepancy that cholesterol and rafts are required, but that Rac1 is preferentially found in nonraft regions of cell membranes (Moissoglu et al., 2014).

3.5 Fusion peptide targeting of membrane domains

Influenza, human immunodeficiency, and other enveloped viruses are enveloped by membranes that are highly enriched in cholesterol and SM, therefore, are often considered to be raft-like. These viral membrane envelopes acquire their cholesterol-rich character by budding during their biogenesis from cholesterol rich nanodomains in the cell membrane. An important physiological question is whether this raft-like character is also required for virus entry. We have recently investigated this question using a single particle approach. We found that liposomes that were decorated with HIV gp41 fusion peptides were targeted to Lo/Ld phase boundaries in SLBs (Yang et al., in press; Figure 2). The boundaries of the reconstituted lipid rafts were not only the sites of HIV liposome binding, but also the sites of fusion

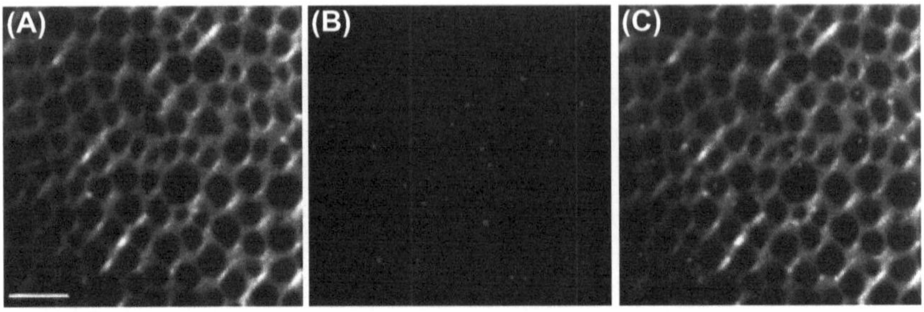

Figure 2 ***HIV-fusion peptide-mediated vesicle binding to Lo/Ld phase boundary regions.*** (A) Fluorescence micrograph of a supported lipid bilayer composed of brain SM/brain PS/cholesterol (50:25:25) and labeled with rhodamine-PE, which partitions into the Ld phase. (B) Total internal reflection fluorescence micrograph of membrane-bound liposomes composed of brain PC/brain PS (75:25) and labeled with the fluorescent lipid analog DiD. The liposomes were preincubated with HIV-fusion peptide. (C) The overlay of the images in (A) and (B) shows that most liposomes target the domain boundaries. Scale bar is 10 μm. (See color plate)

of these particles with the supported membrane. It is likely that fusion peptide insertion at the phase boundary reduces the line tension that is associated with raft domains that are surrounded by nonraft regions of the bilayer. This reduction in line tension may provide energy that supports the otherwise energetically unfavorable fusion process.

4. DOMAIN PARTITIONING OF INTEGRAL MEMBRANE PROTEINS IN SUPPORTED LIPID BILAYERS

The functional significance of a heterogenously organized membrane arises from the fact that functional units assemble or cluster within the same membrane domain. Investigating the interactions that drive the formation of these clusters or domains must necessarily include integral membrane proteins. In particular, the option to prepare asymmetric SLBs makes the supported membrane system an ideal tool to study these interactions in a context that mimics the plasma membrane.

4.1 SNARE proteins

SNARE proteins form a very stable four helical bundle ("SNARE complex") during membrane fusion in exocytosis, with their folding providing the necessary energy for fusion and thereby constituting the minimal fusion machinery (Weber et al., 1998). In neuronal synapses, two SNAREs, syntaxin 1a and SNAP-25, are found in the postsynaptic plasma membrane while a third SNARE, synaptobrevin 2, resides in the synaptic vesicle membrane. Both plasma membrane SNAREs, syntaxin 1a and SNAP-25, are clustered in different domains of the membrane (van den Bogaart, Lang, & Jahn, 2013; Lang et al., 2001). Cholesterol depletion causes dispersion of the SNARE clusters in PC12 cells. However, it has also been shown that these clusters are distinct from detergent resistant membrane (DRM) fractions. These characteristics of the SNARE clusters nicely illustrate the complexity and diversity of protein—protein and protein—lipid interactions that drive domain formation in membranes, of which the classical "lipid rafts" are just one example.

It has been confirmed that syntaxin 1a prefers "nonraft" Ld phase domains over Lo phase domains in experiments with GUVs consisting of DOPC, SM, and cholesterol (Bacia, Schuette, Kahya, Jahn, & Schwille, 2004). The symmetric lipid composition of these GUVs represents only the lipids of the outer leaflet of the plasma membrane. To complement these findings with studies in a more physiological asymmetric lipid environment, we showed recently that full-length syntaxin 1a still partitions into the Ld phase domain when reconstituted into asymmetric SLBs containing anionic and PE inner leaflet lipids where domain formation in the distal leaflet is induced by the proximal first monolayer representing the outer leaflet of the plasma membrane (Yang et al., in press; Figure 3).

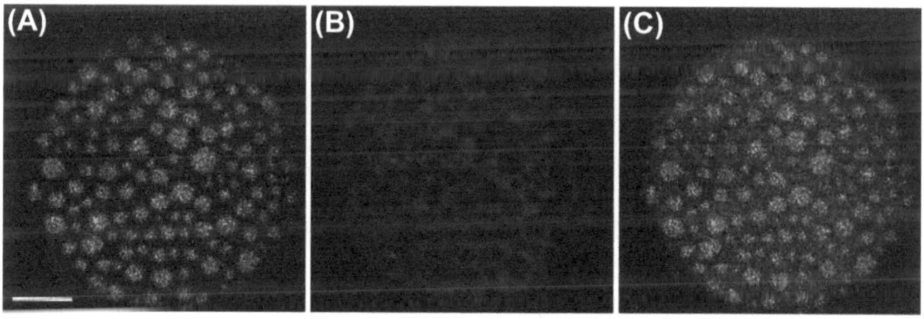

Figure 3 ***Lateral organization of syntaxin 1a in asymmetric supported lipid bilayers with Lo/Ld phase separation.*** (A) Fluorescence micrograph of nitro-benzoxadiazol-aminoethyl-(NBD)-labeled DPPE in an asymmetric supported lipid bilayer formed by the LB/VF technique. The proximal leaflet was composed of brain SM/brain PS/cholesterol (40:40:20) and labeled with NBD-DPPE which partitions into the Lo phase. (B) Fluorescence micrograph of Alexa-546-labeled syntaxin 1a of the supported lipid bilayer in (A). Syntaxin 1a labeled with Alexa-546 was reconstituted into vesicles composed of brain PC/brain PE/brain PS/brain PIP_2/cholesterol (32:30:15:3:20). The vesicles were added to the LB monolayer to form the distal leaflet of the supported bilayer and integrate syntaxin 1a into the membrane. (C) The overlay image shows that syntaxin 1a partitions preferentially into the Ld phase. Scale bar is 10 μm. (See color plate)

The SM-independent, but cholesterol-dependent clustering of syntaxin 1a was first shown in LUV preparations (Murray & Tamm, 2009). Later, it was shown in GUVs that PIP2 can also promote clustering of truncated syntaxin 1a consisting of just the transmembrane domain and a polybasic linker region (SxTM+) (van den Bogaart et al., 2011). In a follow-up study, this result was confirmed in stacked supported membranes prepared by the spin-coating method (Honigmann, van den Bogaart, et al., 2013). Interestingly, these syntaxin clusters are highly enriched in PIP2. Soluble C2A-C2B domains target these clusters with and without calcium present while they bind the regions between the clusters only in the presence of calcium. When PS-containing liposomes were added to these membranes in the presence of calcium, they bound preferentially to the SxTM+/PIP2/C2A-C2B clusters. From this, the authors conclude that these clusters act as molecular beacons for synaptic vesicle recruitment.

4.2 Integrins

Integrins are transmembrane receptors that attach to the extracellular (ECM) matrix and are involved in signal transduction from the ECM to the cell interior. It has been shown that integrins in complicity with lipid rafts recruit the G-protein Rac1 to the plasma membrane (Guan, 2004; Moissoglu et al., 2014; Palazzo, Eng, Schlaepfer, Marcantonio, & Gundersen, 2004; del Pozo et al., 2004). The Naumann group was able to functionally reconstitute integrins into their lipopolymer stabilized supported membrane system and quantify the partitioning into Lo domains and the oligomerization states by photon counting histogram analysis (Hussain, Siegel, Ge, Jordan, & Naumann, 2013; Siegel et al., 2011). When the integrin isoforms ($\alpha_v\beta_3$ and $\alpha_5\beta_1$) were incorporated into symmetric bilayers composed of DOPC, DPPC, and cholesterol, they strongly partitioned into the ("nonraft") Ld phase. However, upon binding of the ECM ligands vitronectin and fibronectin, both integrins translocated into Lo phase domains. Surprisingly, the observed change in phase preference did not correlate with a change of the oligomerization state; the protein stayed mostly monomeric. Independent experiments in single-phase bilayers showed that cholesterol concentration and ligand binding had only subtle influences on integrin oligomerization. Although the exact mechanisms for the ligand-dependent phase partitioning are not known, it seems plausible that ligand binding induces conformational changes in integrins that modify the hydrophobic matching between protein and lipid environment (Siegel et al., 2011). Interestingly, when integrin was reconstituted into asymmetric bilayers, where only one leaflet showed phase separation, the protein preferentially partitioned into the ("raft like") Lo phase. No changes upon ECM ligand binding were observed (Hussain et al., 2013). This work highlights the importance of bilayer asymmetry in the lateral organization of the membrane and the unique role that asymmetric supported bilayers can play in elucidating the relevant molecular interactions.

5. INTERACTIONS OF CYTOSKELETAL FILAMENTS WITH SUPPORTED LIPID BILAYERS

It has long been known that in addition to lipid heterogeneity, elements of the cytoskeleton couple to membranes and play an important role in membrane domain formation. In its original conception, the "picket-fence" model of membrane compartmentalization by the cytoskeleton was seen as an alternative to the "raft hypothesis," but more recent incarnations of these models feature cooperating roles of the cytoskeleton and membrane lipids (Kusumi et al., 2012). The lateral diffusion of lipids is significantly affected when actin is coupled to SLBs with even a single fluid lipid species, i.e., without any potential to form phase-separated ordered domains. For example, when a network of actin was coupled to an SLB with GM1-bound cholera toxin B, the lipid diffusion coefficient decreased about twofold and the protein diffusion coefficient decreased about ninefold (Heinemann, Vogel, & Schwille, 2013). When myosin filaments were added and activated by ATP, the actin filaments were rearranged, shortened, broken, and gradually concentrated in actomyosin foci on the reconstituted SLBs (Vogel, Petrasek, Heinemann, & Schwille, 2013). Actin polymerization on the surface of GUVs induced by a PIP2-binding regulatory protein induced fluid—fluid phase separation and domain formation under conditions (37 °C) where the same membranes were uniform and single phase in the absence of actin (Liu & Fletcher, 2006). Temperature-dependent studies further showed that the actin network elevates the miscibility transition temperature (Liu & Fletcher, 2006).

Two theoretical studies predicted stable mosaics of Lo and Ld domains under the influence of a mesh of a mechanically coupled cytoskeleton. In the first study, Monte Carlo simulations showed that two-component membranes near critical points undergo anomalous subdiffusion and that coupling to the cytoskeleton shifts the miscibility transition temperature (Ehrig, Petrov, & Schwille, 2011). The second study applied a conserved order parameter Ising model to membranes with lipids near a miscibility critical point and concluded that domains are broken up into much smaller units when they are pinned to a mesh of cytoskeleton (Machta, Papanikolaou, Sethna, & Veatch, 2011). These predictions have been verified and refined recently by visualizing lipid domain structures and dynamics by stimulated emission depletion (STED) microscopy and FCS experiments, respectively, with nanometer resolution (Honigmann, Mueller, et al., 2014; Honigmann, Sadeghi, et al., 2014). When an actin meshwork was pinned to Ld components in a three-component Lo/Ld two-phase SLB, Ld phases aligned parallel to the actin fibers with the Lo phases occupying the interstitial areas. Diffusion of an Ld lipid marker was anisotropic with a faster component along the actin fibers and a slower component perpendicular to it. However, when the actin meshwork was coupled to Lo components in the same model membranes, the Lo phase domains aligned with the cytoskeleton and Ld phase domains occupied the interstitial areas. As expected from the above mentioned previous studies, demixing phase transitions shifted to higher temperatures, but the new STED results confirm that this happens also when observed at the nanoscopic level. These experiments nicely prove that the cytoskeleton can have a profound influence on the phase behavior of an underlying membrane, but that the intrinsic physical phase behavior of the lipids in that membrane is still required to produce the observed anomalous and anisotropic diffusion results.

6. CONCLUSIONS

Supported lipid bilayers have offered a lot of new insight into the molecular interactions that give rise to domains in complex membrane systems. These include in-plane lipid interactions that form raft-like Lo domains in Ld lipid bilayers as well as protein interactions with Lo and Ld domains and interfaces between Lo and Ld domains. They also include lateral and transverse protein—protein interactions and the modification of all of these interactions by elements of the cytoskeleton. One aspect that can be particularly well studied in supported lipid bilayers are transverse lipid interactions and the coupling of lipids and lipid domains across the two leaflets in asymmetric supported membranes. The sequential build-up of asymmetric SLBs, their flat geometry coupled with high resolution optical and force microscopy, and increasingly sophisticated functional membrane protein reconstitution techniques offer a wide-open playing field to investigate almost all aspects of cellular membrane complexity that are hard to achieve with any other model system. This field is poised to grow and the described approaches will doubtless continue to answer many open questions of the biology of cell membranes that benefit from the ability to rebuild increasingly complex membranes from its components and thus approach and reconstitute the many physiological functions of cellular, vesicular, and viral membranes.

ACKNOWLEDGMENTS

This work was supported by NIH grants P01 GM072694 and R01 AI30577. We thank members of the Tamm lab, past and present, for many stimulating discussions and the work from our group cited in this review.

REFERENCES

Bacia, K., Schuette, C. G., Kahya, N., Jahn, R., & Schwille, P. (2004). SNAREs prefer liquid-disordered over "raft" (liquid-ordered) domains when reconstituted into giant unilamellar vesicles. *Journal of Biological Chemistry, 279*(36), 37951–37955.

van den Bogaart, G., Lang, T., & Jahn, R. (2013). Microdomains of SNARE proteins in the plasma membrane. *Current Topics in Membrane, 72*, 193–230.

van den Bogaart, G., Meyenberg, K., Risselada, H. J., Amin, H., Willig, K. I., Hubrich, B. E., et al. (2011). Membrane protein sequestering by ionic protein-lipid interactions. *Nature, 479*(7374), 552–555.

Brian, A. A., & McConnell, H. M. (1984). Allogeneic stimulation of cytotoxic T cells by supported planar membranes. *Proceedings of the National Academy of Sciences of the United States of America, 81*(19), 6159–6163.

Brown, D. A., & London, E. (1998). Functions of lipid rafts in biological membranes. *Annual Reviews of Cell and Developmental Biology, 14*, 111–136.

Chiantia, S., Ries, J., Kahya, N., & Schwille, P. (2006). Combined AFM and two-focus SFCS study of raft exhibiting model membranes. *ChemPhysChem*, 7(11), 2409–2418.

Collins, M. D., & Keller, S. L. (2008). Tuning lipid mixtures to induce or suppress domain formation across leaflets of unsupported asymmetric bilayers. *Proceedings of the National Academy of Sciences of the United States of America, 105*(1), 124–128.

Crane, J. M., Kiessling, V., & Tamm, L. K. (2005). Measuring lipid asymmetry in planar supported bilayers by fluorescence interference contrast microscopy. *Langmuir, 21*(4), 1377–1388.

Crane, J. M., & Tamm, L. K. (2003). Cholesterol content and phospholipid asymmetry: effects on raft formation in planar model membranes. *Biophysical Journal, 84*(2), 372A.

Crane, J. M., & Tamm, L. K. (2004). Role of cholesterol in the formation and nature of lipid rafts in planar and spherical model membranes. *Biophysical Journal, 86*(5), 2965–2979.

Dietrich, C., Bagatolli, L. A., Volovyk, Z. N., Thompson, N. L., Levi, M., Jacobson, K., et al. (2001). Lipid rafts reconstituted in model membranes. *Biophysical Journal, 80*(3), 1417–1428.

Dietrich, C., Volovyk, Z. N., Levi, M., Thompson, N. L., & Jacobson, K. (2001). Partitioning of Thy-1, GM1, and cross-linked phospholipid analogs into lipid rafts reconstituted in supported model membrane monolayers. *Proceedings of the National Academy of Sciences of the United States of America, 98*(19), 10642–10647.

Dufrene, Y. F., Barger, W. R., Green, J. B. D., & Lee, G. U. (1997). Nanometer-scale surface properties of mixed phospholipid monolayers and bilayers. *Langmuir, 13*(18), 4779–4784.

Edidin, M. (2003). The state of lipid rafts: from model membranes to cells. *Annual Reviews of Biophysics and Biomolecular Structure, 32*, 257–283.

Ehrig, J., Petrov, E. P., & Schwille, P. (2011). Near-critical fluctuations and cytoskeleton-assisted phase separation lead to subdiffusion in cell membranes. *Biophysical Journal, 100*(1), 80–89.

Feigenson, G. W. (2009). Phase diagrams and lipid domains in multicomponent lipid bilayer mixtures. *Biochimica et Biophysica Acta, 1788*(1), 47–52.

Floyd, D. L., Ragains, J. R., Skehel, J. J., Harrison, S. C., & van Oijen, A. M. (2008). Single-particle kinetics of influenza virus membrane fusion. *Proceedings of the National Academy of Sciences of the United States of America, 105*(40), 15382–15387.

Fromherz, P., Kiessling, V., Kottig, K., & Zeck, G. (1999). Membrane transistor with giant lipid vesicle touching a silicon chip. *Applied Physics A, 69*(5), 571–576.

Garg, S., Ruhe, J., Ludtke, K., Jordan, R., & Naumann, C. A. (2007). Domain registration in raft-mimicking lipid mixtures studied using polymer-tethered lipid bilayers. *Biophysical Journal, 92*(4), 1263–1270.

Ge, Y., Siegel, A. P., Jordan, R., & Naumann, C. A. (2014). Ligand binding alters dimerization and sequestering of urokinase receptors in raft-mimicking lipid mixtures. *Biophysical Journal, 107*(9), 2101–2111.

Giocondi, M. C., Yamamoto, D., Lesniewska, E., Milhiet, P. E., Ando, T., & Le Grimellec, C. (2010). Surface topography of membrane domains. *Biochimica et Biophysica Acta, 1798*(4), 703–718.

Goennenwein, S., Tanaka, M., Hu, B., Moroder, L., & Sackmann, E. (2003). Functional incorporation of integrins into solid supported membranes on ultrathin films of cellulose: impact on adhesion. *Biophysical Journal, 85*(1), 646–655.

Groves, J. T., Boxer, S. G., & McConnell, H. M. (1997). Electric field-induced reorganization of two-component supported bilayer membranes. *Proceedings of the National Academy of Sciences of the United States of America, 94*(25), 13390–13395.

Guan, J. L. (2004). Cell biology. Integrins, rafts, Rac, and Rho. *Science, 303*(5659), 773–774.

Hancock, J. F. (2003). Ras proteins: different signals from different locations. *Nature Reviews Molecular Cell Biology, 4*(5), 373–384.

Heinemann, F., Vogel, S. K., & Schwille, P. (2013). Lateral membrane diffusion modulated by a minimal actin cortex. *Biophysical Journal, 104*(7), 1465–1475.

Honigmann, A., van den Bogaart, G., Iraheta, E., Risselada, H. J., Milovanovic, D., Mueller, V., et al. (2013). Phosphatidylinositol 4,5-bisphosphate clusters act as molecular beacons for vesicle recruitment. *Nature Structural and Molecular Biology, 20*(6), 679–686.

Honigmann, A., Mueller, V., Hell, S. W., & Eggeling, C. (2013). STED microscopy detects and quantifies liquid phase separation in lipid membranes using a new far-red emitting fluorescent phosphoglycerolipid analogue. *Faraday Discussions, 161*, 77–89.

Honigmann, A., Mueller, V., Ta, H., Schoenle, A., Sezgin, E., Hell, S. W., et al. (2014). Scanning STED-FCS reveals spatiotemporal heterogeneity of lipid interaction in the plasma membrane of living cells. *Nature Communications, 5*, 5412.

Honigmann, A., Sadeghi, S., Keller, J., Hell, S. W., Eggeling, C., & Vink, R. (2014). A lipid bound actin meshwork organizes liquid phase separation in model membranes. *eLife, 3*, e01671.

Hussain, N. F., Siegel, A. P., Ge, Y., Jordan, R., & Naumann, C. A. (2013). Bilayer asymmetry influences integrin sequestering in raft-mimicking lipid mixtures. *Biophysical Journal, 104*(10), 2212–2221.

Jaffe, A. B., & Hall, A. (2005). Rho GTPases: biochemistry and biology. *Annual Reviews of Cell and Developmental Biology, 21*, 247–269.

Janshoff, A., & Steinem, C. (2001). Scanning force microscopy of artificial membranes. *ChemBioChem, 2*(11), 799–808.

Kalb, E., Frey, S., & Tamm, L. K. (1992). Formation of supported planar bilayers by fusion of vesicles to supported phospholipid monolayers. *Biochimica et Biophysica Acta, 1103*(2), 307–316.

Kalb, E., & Tamm, L. K. (1992). Incorporation of cytochrome b5 into supported phospholipid-bilayers by vesicle fusion to supported monolayers. *Thin Solid Films, 210*(1–2), 763–765.

Kholodenko, B. N., Hancock, J. F., & Kolch, W. (2010). Signalling ballet in space and time. *Nature Reviews Molecular Cell Biology, 11*(6), 414–426.

Kiessling, V., Crane, J. M., & Tamm, L. K. (2006). Transbilayer effects of raft-like lipid domains in asymmetric planar bilayers measured by single molecule tracking. *Biophysical Journal, 91*(9), 3313–3326.

Kiessling, V., & Tamm, L. K. (2003). Measuring distances in supported bilayers by fluorescence interference-contrast microscopy: polymer supports and SNARE proteins. *Biophysical Journal, 84*(1), 408–418.

Kiessling, V., Wan, C., & Tamm, L. K. (2009). Domain coupling in asymmetric lipid bilayers. *Biochimica et Biophysica Acta, 1788*(1), 64–71.

Kusumi, A., Fujiwara, T. K., Chadda, R., Xie, M., Tsunoyama, T. A., Kalay, Z., et al. (2012). Dynamic organizing principles of the plasma membrane that regulate signal transduction: commemorating the fortieth anniversary of Singer and Nicolson's fluid-mosaic model. *Annual Reviews of Cell and Developmental Biology, 28*, 215–250.

Lai, A. L., Tamm, L. K., Ellena, J. F., & Cafiso, D. S. (2011). Synaptotagmin 1 modulates lipid acyl chain order in lipid bilayers by demixing phosphatidylserine. *Journal of Biological Chemistry, 286*(28), 25291–25300.

Lang, T., Bruns, D., Wenzel, D., Riedel, D., Holroyd, P., Thiele, C., et al. (2001). SNAREs are concentrated in cholesterol-dependent clusters that define docking and fusion sites for exocytosis. *EMBO Journal, 20*(9), 2202–2213.

Liang, B., Kiessling, V., & Tamm, L. K. (2013). Prefusion structure of syntaxin-1A suggests pathway for folding into neuronal trans-SNARE complex fusion intermediate. *Proceedings of the National Academy of Sciences of the United States of America, 110*(48), 19384–19389.

Lichtenberg, D., Goni, F. M., & Heerklotz, H. (2005). Detergent-resistant membranes should not be identified with membrane rafts. *Trends in Biochemical Sciences, 30*(8), 430–436.

Lin, W. C., Blanchette, C. D., Ratto, T. V., & Longo, M. L. (2006). Lipid asymmetry in DLPC/DSPC-supported lipid bilayers: a combined AFM and fluorescence microscopy study. *Biophysical Journal, 90*(1), 228–237.

Lin, W. C., Blanchette, C. D., Ratto, T. V., & Longo, M. L. (2007). Lipid domains in supported lipid bilayer for atomic force microscopy. *Methods in Molecular Biology, 400*, 503–513.

Lingwood, D., & Simons, K. (2010). Lipid rafts as a membrane-organizing principle. *Science, 327*(5961), 46–50.

Liu, A. P., & Fletcher, D. A. (2006). Actin polymerization serves as a membrane domain switch in model lipid bilayers. *Biophysical Journal, 91*(11), 4064–4070.

Machta, B. B., Papanikolaou, S., Sethna, J. P., & Veatch, S. L. (2011). Minimal model of plasma membrane heterogeneity requires coupling cortical actin to criticality. *Biophysical Journal, 100*(7), 1668–1677.

McConnell, H. M., Watts, T. H., Weis, R. M., & Brian, A. A. (1986). Supported planar membranes in studies of cell-cell recognition in the immune system. *Biochimica et Biophysica Acta, 864*(1), 95–106.

Mennicke, U., & Salditt, T. (2002). Preparation of solid-supported lipid bilayers by spin-coating. *Langmuir, 18*(21), 8172–8177.

Milhiet, P. E., Gubellini, F., Berquand, A., Dosset, P., Rigaud, J. L., Le Grimellec, C., et al. (2006). High-resolution AFM of membrane proteins directly incorporated at high density in planar lipid bilayer. *Biophysical Journal, 91*(9), 3268–3275.

Moissoglu, K., Kiessling, V., Wan, C., Hoffman, B. D., Norambuena, A., Tamm, L. K., et al. (2014). Regulation of Rac1 translocation and activation by membrane domains and their boundaries. *Journal of Cell Science, 127*(Pt 11), 2565–2576.

Murray, D. H., & Tamm, L. K. (2009). Clustering of syntaxin-1a in model membranes is modulated by phosphatidylinositol 4,5-bisphosphate and cholesterol. *Biochemistry, 48*(21), 4617–4625.

Naumann, C. A., Frank, C. W., Prucker, O., Lehmann, T., Ruehe, J., & Knoll, W. (2000). Lateral mobility of phospholipids in tethered polymer-supported membranes. *Biophysical Journal, 78*(1), 273A.

Palazzo, A. F., Eng, C. H., Schlaepfer, D. D., Marcantonio, E. E., & Gundersen, G. G. (2004). Localized stabilization of microtubules by integrin- and FAK-facilitated Rho signaling. *Science, 303*(5659), 836–839.

Pike, L. J. (2006). Rafts defined: a report on the Keystone symposium on lipid rafts and cell function. *Journal of Lipid Research, 47*(7), 1597–1598.

del Pozo, M. A., Alderson, N. B., Kiosses, W. B., Chiang, H. H., Anderson, R. G., & Schwartz, M. A. (2004). Integrins regulate Rac targeting by internalization of membrane domains. *Science, 303*(5659), 839–842.

Richter, R. P., Maury, N., & Brisson, A. R. (2005). On the effect of the solid support on the interleaflet distribution of lipids in supported lipid bilayers. *Langmuir, 21*(1), 299–304.

Rossetti, F. F., Textor, M., & Reviakine, I. (2006). Asymmetric distribution of phosphatidyl serine in supported phospholipid bilayers on titanium dioxide. *Langmuir, 22*(8), 3467–3473.

Sackmann, E. (1996). Supported membranes: Scientific and practical applications. *Science, 271*(5245), 43–48.

Shi, J., Yang, T., Kataoka, S., Zhang, Y., Diaz, A. J., & Cremer, P. S. (2007). GM1 clustering inhibits cholera toxin binding in supported phospholipid membranes. *Journal of the American Chemical Society, 129*(18), 5954–5961.

Siegel, A. P., Kimble-Hill, A., Garg, S., Jordan, R., & Naumann, C. A. (2011). Native ligands change integrin sequestering but not oligomerization in raft-mimicking lipid mixtures. *Biophysical Journal, 101*(7), 1642–1650.

Simons, K., & Ikonen, E. (1997). Functional rafts in cell membranes. *Nature, 387*(6633), 569–572.

Spillane, K. M., Ortega-Arroyo, J., de Wit, G., Eggeling, C., Ewers, H., Wallace, M. I., et al. (2014). High-speed single-particle tracking of GM1 in model membranes reveals anomalous diffusion due to interleaflet coupling and molecular pinning. *Nano Letters, 14*(9), 5390–5397.

Suzuki, K. G., Kasai, R. S., Hirosawa, K. M., Nemoto, Y. L., Ishibashi, M., Miwa, Y., et al. (2012). Transient GPI-anchored protein homodimers are units for raft organization and function. *Nature Chemical Biology, 8*(9), 774–783.

Tamm, L. K. (1984). The substrate supported lipid bilayer-a new model membrane system. *Klinische Wochenschrift, 62*, 502–503.

Tamm, L. K. (1988). Lateral diffusion and fluorescence microscope studies on a monoclonal-antibody specifically bound to supported phospholipid bilayers. *Biochemistry, 27*(5), 1450–1457.

Tamm, L. K., & McConnell, H. M. (1985). Supported phospholipid bilayers. *Biophysical Journal, 47*(1), 105–113.

Veatch, S. L., & Keller, S. L. (2005). Seeing spots: complex phase behavior in simple membranes. *Biochimica et Biophysica Acta, 1746*(3), 172–185.

Visco, I., Chiantia, S., & Schwille, P. (2014). Asymmetric supported lipid bilayer formation via methyl-beta-cyclodextrin mediated lipid exchange: Influence of asymmetry on lipid dynamics and phase behavior. *Langmuir, 30*(25), 7475–7484.

Vogel, S. K., Petrasek, Z., Heinemann, F., & Schwille, P. (2013). Myosin motors fragment and compact membrane-bound actin filaments. *eLife, 2*, e00116.

Wacklin, H. P., & Thomas, R. K. (2007). Spontaneous formation of asymmetric lipid bilayers by adsorption of vesicles. *Langmuir, 23*(14), 7644–7651.

Wagner, M. L., & Tamm, L. K. (2000). Tethered polymer-supported planar lipid bilayers for reconstitution of integral membrane proteins: silane-polyethyleneglycol-lipid as a cushion and covalent linker. *Biophysical Journal, 79*(3), 1400–1414.

Wagner, M. L., & Tamm, L. K. (2001). Reconstituted syntaxin1A/SNAP25 interacts with negatively charged lipids as measured by lateral diffusion in planar supported bilayers. *Biophysical Journal, 81*(1), 266–275.

Wan, C., Kiessling, V., Cafiso, D. S., & Tamm, L. K. (2011). Partitioning of synaptotagmin I C2 domains between liquid-ordered and liquid-disordered inner leaflet lipid phases. *Biochemistry, 50*(13), 2478–2485.

Wan, C., Kiessling, V., & Tamm, L. K. (2008). Coupling of cholesterol-rich lipid phases in asymmetric bilayers. *Biochemistry, 47*(7), 2190–2198.

Weber, T., Zemelman, B. V., McNew, J. A., Westermann, B., Gmachl, M., Parlati, F., et al. (1998). SNAREpins: minimal machinery for membrane fusion. *Cell, 92*(6), 759–772.

Weise, K., Kapoor, S., Denter, C., Nikolaus, J., Opitz, N., Koch, S., et al. (2011). Membrane-mediated induction and sorting of K-Ras microdomain signaling platforms. *Journal of American Chemical Society, 133*(4), 880–887.

Weise, K., Triola, G., Brunsveld, L., Waldmann, H., & Winter, R. (2009). Influence of the lipidation motif on the partitioning and association of N-Ras in model membrane subdomains. *Journal of American Chemical Society, 131*(4), 1557–1564.

Wright, L. L., Palmer, A. G., 3rd, & Thompson, N. L. (1988). Inhomogeneous translational diffusion of monoclonal antibodies on phospholipid Langmuir–Blodgett films. *Biophysical Journal, 54*(3), 463–470.

Yang, S.-T., Kiessling, V., Simmons, J. A., White, J. M., & Tamm, L. K. HIV gp41-mediated membrane fusion occurs at edges of cholesterol-rich lipid domains. *Nature Chemical Biology*, in press.

CHAPTER TWO

Giant Plasma Membrane Vesicles: Models for Understanding Membrane Organization

Kandice R. Levental and Ilya Levental*
Department of Integrative Biology and Pharmacology, University of Texas Health Science Center at Houston – Medical School, Houston, TX, USA
*Corresponding author: E-mail: ilya.levental@uth.tmc.edu

Contents

Current Topics in Membranes, Volume 75
ISSN 1063-5823
http://dx.doi.org/10.1016/bs.ctm.2015.03.009

Abstract

The organization of eukaryotic membranes into functional domains continues to fascinate and puzzle cell biologists and biophysicists. The lipid raft hypothesis proposes that collective lipid interactions compartmentalize the membrane into coexisting liquid domains that are central to membrane physiology. This hypothesis has proven controversial because such structures cannot be directly visualized in live cells by light microscopy. The recent observations of liquid–liquid phase separation in biological membranes are an important validation of the raft hypothesis and enable application of the experimental toolbox of membrane physics to a biologically complex phase-separated membrane. This review addresses the role of giant plasma membrane vesicles (GPMVs) in refining the raft hypothesis and expands on the application of GPMVs as an experimental model to answer some of key outstanding problems in membrane biology.

1. INTRODUCTION

Lipid rafts have received a tremendous amount of research attention, with a PubMed search for "lipid raft" returning an average of ~220 papers published per year over the last decade. This intense interest can be attributed to a perfect storm of potential biological importance and scientific controversy. Recent estimates suggest that far from being small, isolated, specialized membrane domains, lipid rafts make up a major fraction—if not the majority—of the plasma membrane (PM) of mammalian cells (Levental et al., 2009; Meder, Moreno, Verkade, Vaz, & Simons, 2006; Owen, Williamson, Magenau, & Gaus, 2012; Sanchez, Tricerri, & Gratton, 2012). Thus, as a central mechanism for compartmentalizing and organizing biological membranes, rafts could conceivably be involved in a broad variety of signal transduction and membrane trafficking mechanisms, and have been implicated in nearly as many. However, nearly 20 years after their initial postulation, the properties, biological relevance, and even existence of raft domains still remain hotly debated.

Much of this debate is predicated on the adage that "seeing is believing." Shortly after the introduction of the raft hypothesis (Simons & Ikonen, 1997), it became clear that these domains, if present, are not visible by diffraction-limited light microscopy. Since then, a variety of approaches including fluorescence spectroscopy (Engel et al., 2010; Kenworthy, 2007; LaRocca et al., 2013), electron spin resonance spectroscopy (Ge et al., 2003), electron microscopy (LaRocca et al., 2013; Prior & Hancock, 2012), and superresolution microscopy (Eggeling et al., 2009; Hess et al., 2007) have been applied in attempts to provide definitive evidence for the existence of ordered, lipid-driven membrane domains in biological membranes. Into this context, the direct microscopic observation of ordered membrane domains in intact PMs isolated as giant plasma membrane vesicles (GPMVs) (Baumgart et al., 2007) was a critical validation of the raft hypothesis.

Because of this conceptual importance and their unique experimental versatility, GPMVs are poised to become a key tool in the methodological arsenal of membrane biology and biophysics. Because detailed protocols are available for GPMV isolation (Levental & Levental, 2015; Sezgin et al., 2012), this review will instead begin with a historical perspective on rafts and the role of GPMV experiments in the current understanding of membrane organization. It will also cover similar/alternative intact PM preparations and focus on important caveats and limitations of GPMVs as a coherent model for the live cell PM. Finally, we will attempt an exhaustive overview of the many uses of GPMVs for measuring the properties of isolated membranes and their domains, and end on a discussion of the future outlook for this methodology in membrane research.

2. THE CONTINUING STORY OF LIPID RAFTS

Observations from the past decade of research (Simons & Gerl, 2010) have been synthesized into a conceptual model of lipid rafts as a set of interactions between membrane components, which collectively contribute to membrane organization (Lingwood & Simons, 2010). Specifically, sterols, sphingolipids, glycolipids, and lipids with saturated acyl chains preferentially interact with each other to the exclusion of lipids with highly unsaturated (or branched) acyl chains. Lipid-anchored and transmembrane proteins also have preferential affinity for specific lipids and/or membrane environments. The net result of these interactions is lateral membrane domains of distinct protein and lipid compositions that can mediate biological functions, for example, by concentrating specific proteins to promote interactions. The specific properties of those domains—i.e., their size, lifetime, abundance, and composition—are strictly dependent on the nature of a given membrane. That is, the domains can manifest as very small (a few molecules) and highly dynamic (microseconds–milliseconds) lipid–protein complexes, which can be stabilized by more specific protein–protein, protein–lipid, and protein–glycan interactions to form larger (tens to hundreds of nanometers) and more stable signaling platforms. These can then be further coalesced (usually by nonbiological perturbations) into the microscopically observable liquid-ordered raft phases of GPMVs. Importantly, the driving forces for these organizational states are the same: preferential interactions between raft lipids and proteins inducing the formation of a liquid-ordered domain. Thus the composition and organization of the macroscopic, coalesced ordered domains in GPMVs are assumed to be reflective of smaller, more dynamic in vivo domains (Figure 1).

This conceptual model has coalesced from three disparate, but ultimately convergent, lines of evidence: (1) detergent resistance of cellular membranes; (2) sensitivity to cholesterol depletion; (3) liquid–liquid phase separation in synthetic lipid model membranes. The general aspects and important caveats of these approaches are discussed below.

2.1 Detergent resistance

The original and still most widely used technique for the analysis of membrane rafts is the isolation of detergent-resistant membranes (DRMs—alternative names for the same phenomenon include detergent-insoluble glycolipid-enriched fractions, DIG or GEMs). DRMs are intact membranes remaining after cellular solubilization by nonionic detergents—most commonly Triton X-100—at cold temperatures (4 °C). These DRMs are consistently enriched in specific cellular lipids (the raft lipids noted above—cholesterol, glycosphingolipids, saturated lipids) and lipid-anchored proteins, and depleted of unsaturated lipids, and certain PM proteins, most notably the transferrin receptor and the low-density lipoprotein receptor. The same lipids and proteins enriched in DRMs are preferentially sorted to apical PMs in epithelial cells (Brown & Rose, 1992), and cluster together following antibody-mediated cross-linking of the cell surface (Harder, Scheiffele, Verkade, & Simons, 1998), leading to the hypothesis that detergent-resistant membrane fractions represent lateral membrane domains that serve to segregate membrane components for polarized membrane traffic in the late secretory pathway (Schuck & Simons, 2004) and signaling at the PM (Simons & Toomre, 2000). Because of their ease of isolation and apparent relationship to cellular structures, DRMs became the operational definition of lipid rafts, with the crucial distinction between physiological domains and detergent isolated residue often being overlooked.

2.2 Caveats of DRMs

Much of the controversy around lipid rafts arose because of methodological issues associated with their isolation and identification. By their very nature,

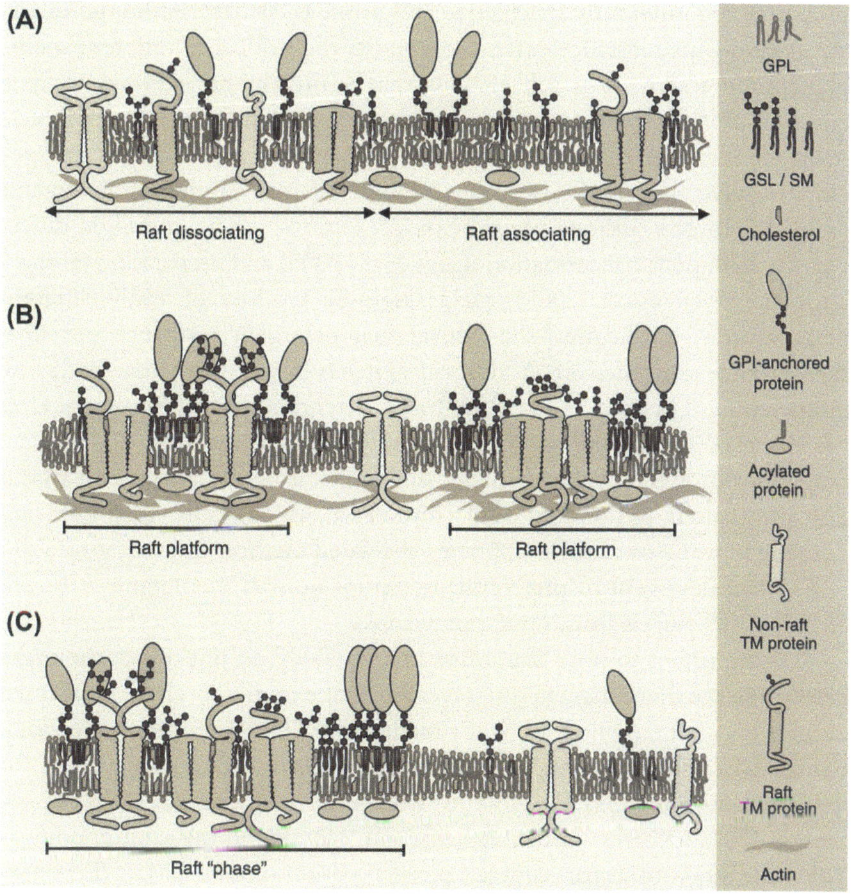

Figure 1 ***Lipid rafts as functional organizers of cellular membranes.*** This figure *(adapted from Lingwood & Simons, 2010)* shows the various possible organization states of cholesterol-rich mammalian membranes. (A) Preferential interactions between raft lipids and proteins (including cholesterol, sphingolipids, GPI-anchored proteins, acylated proteins) manifest themselves as domains of various sizes and stabilities. (B) Specific interactions between proteins can stabilize smaller, more dynamic domains into larger, functional raft platforms. (C) Under certain conditions, e.g., release from constraint by the actin cytoskeleton, domains coalesce into microscopic phases. *From Lingwood, D., & Simons, K. (2010). Lipid rafts as a membrane-organizing principle. Science, 327(5961), 46–50. Reprinted with permission from AAAS.* (See color plate)

raft and nonraft regions of the same cellular membrane are not dramatically different. They are composed of similar lipids and proteins, with small physicochemical differences driving their separation. Thus, methods to isolate raft from nonraft membrane regions based on differences in physical properties are inherently speculative and prone to artifact. Although differential detergent resistance certainly supports the notion of different membrane subdomains, it is self-evident that a membrane exposed to lytic concentrations of cold detergent is unlikely to remain in its native physical state, and therefore that DRMs should not be directly equated with any specific in vivo structure (Brown, 2006). Different nonionic detergents (Schuck, Honsho, Ekroos, Shevchenko, & Simons, 2003), and slight differences in detergent concentration (Gaus et al., 2005) and preparation temperature (Levental et al., 2009), yield insoluble residues of vastly different compositions, invalidating the notion that detergent resistance per se is the defining raft criterion. Additionally, it has been shown that both low temperature (Levental et al., 2009) and detergents (Hao, Mukherjee, & Maxfield, 2001; Heerklotz, 2002) can induce the formation of membrane domains, although it is also possible that these treatments stabilize preexisting domains (Pathak & London, 2011; Zhou et al., 2013) rather than inducing domains in nonseparating membranes. A related method that uses sonication rather than detergent to preferentially disrupt nonraft membranes (Ostrom & Liu, 2007) suffers from similar drawbacks.

It is important to note that while DRMs likely do not reflect the organization of the membrane in situ, several recent experiments have confirmed insights into raft targeting originally made using DRMs. For example, post-translational palmitoylation has been clearly shown to be a requirement for the association of certain proteins with DRMs (Levental, Grzybek, & Simons, 2010; Zhang, Trible, & Samelson, 1998), and the same modification was shown to impart ordered phase partitioning in GPMVs (Levental, Lingwood, Grzybek, Coskun, & Simons, 2010). Similarly, temperatures at which ordered domains are observed in GPMVs correlate to those at which DRMs could be efficiently isolated (Levental et al., 2009). It is likely that DRMs do selectively isolate membranes with raft-like characteristics, and thus that DRM association can be used to suggest raft residence; however, DRM association alone is far from a definitive criterion.

2.3 Cholesterol depletion

A classical method for confirming raft association of a protein or cellular process is cholesterol depletion, usually by the cholesterol-chelating polysaccharide cyclodextrin (CD). The rationale behind this paradigm is that cholesterol is an essential component for raft formation, and thus its depletion should interfere with raft assembly and function (Zidovetzki & Levitan, 2007). Concordantly, cholesterol depletion reduces the detergent resistance of a number of otherwise detergent-resistant proteins (Scheiffele, Roth, & Simons, 1997). Possibly resulting from this effect, perturbation of cellular cholesterol also affects polarized membrane traffic, the original cellular function for which rafts were postulated. Indeed, by the criterion of cholesterol sensitivity, nearly every cellular process can be judged to be raft-dependent, as cholesterol depletion has been shown to affect everything from growth factor signaling (Chen & Resh, 2002) to immune cell activation (Sheets, Holowka, & Baird, 1999), apoptosis (Calleros, Lasa, Rodriguez-Alvarez, Toro, & Chiloeches, 2006), cytoskeletal assembly, cell adhesion, and motility (Norman et al., 2010), among dozens of others. In intact cells, acute cholesterol depletion can even induce the formation of micrometric membrane domains that are selective for certain lipid and protein components (Hao et al., 2001). However, such an inference must be viewed with a skeptical eye because of the inherently pleiotropic nature of cholesterol modulation, as discussed below.

2.4 Caveats of cholesterol depletion

Cholesterol is the most abundant lipid of the PM (up to 40 mol% (Gerl et al., 2012)) and therefore a major structural and functional component thereof. Thus, its acute depletion is inherently pleiotropic. One reason is due to the essential role of cholesterol for the barrier function of the PM, leading to increased ion leakage across the PM upon cholesterol depletion (Haines, 2001). Such a perturbation may have wide-ranging consequences, including cytotoxicity (Mahammad, Dinic, Adler, & Parmryd, 2010). Similarly, cholesterol is a central determinant of nearly all bulk membrane physical properties (e.g., fluidity, bending modulus, etc.), which certainly contribute to membrane physiology and are dramatically altered by acute depletion. Because of these effects, it is nearly impossible to definitively implicate raft disruption as the sole mechanism for the responses of cholesterol depleted cells. Thus, neither detergent resistance nor CD sensitivity alone (nor indeed, the two combined) is definitive evidence for rafts.

2.5 Phase separation in model membranes

In parallel with the biochemical data showing that certain membrane components are resistant to detergent solubilization, and that resistance is mediated by lipid—lipid interactions involving cholesterol and saturated sphingolipids, bio physical studies of model membranes began to elucidate the physicochemical basis for these phenomena (McConnell & Vrljic, 2003). As early as the 1980s, Harden McConnell and colleagues visualized coexisting fluid domains in Langmuir monolayers of cholesterol and phospholipids (Rice & McConnell, 1989) (Figure 2(A)). This behavior was theoretically interpreted by invoking the presence of a novel two-dimensional state—the liquid-ordered (L_o) phase—arising from the interactions between cholesterol and extended conformations of phospholipid acyl chains (Ipsen, Karlstrom, Mouritsen, Wennerstrom, & Zuckermann, 1987). This L_o phase is an intermediate between the highly ordered but conformationally and translationally locked crystalline/gel state of pure phospholipids below their phase transition temperature and the fluid, disordered $L\alpha$ phase (which became known as the L_d phase in relation to L_o). Approximately a decade later, technology for producing free-standing bilayers as either giant unilamellar vesicles (GUVs) or supported lipid bilayers allowed direct visualization and characterization of the same liquid–liquid phase separation phenomenon in more biomimetic bilayers containing sphingolipids in addition to phosphatidylcholine and cholesterol (Dietrich, Bagatolli, et al., 2001; Korlach, Schwille, Webb, & Feigenson, 1999; Samsonov, Mihalyov, & Cohen, 2001; Veatch & Keller, 2002) (Figure 2(B)). These methods allow careful mapping of phase diagrams in these systems, showing that liquid–liquid phase coexistence is observable through a wide range of biomimetic compositions (Bezlyepkina, Gracia, Shchelokovskyy, Lipowsky, & Dimova, 2013; Feigenson & Buboltz, 2001; Veatch & Keller, 2003). Significantly, similar behavior could be observed in lipid mixtures isolated directly from human samples (including red blood cells (Keller, Pitcher, Huestis, & McConnell, 1998) and brush border

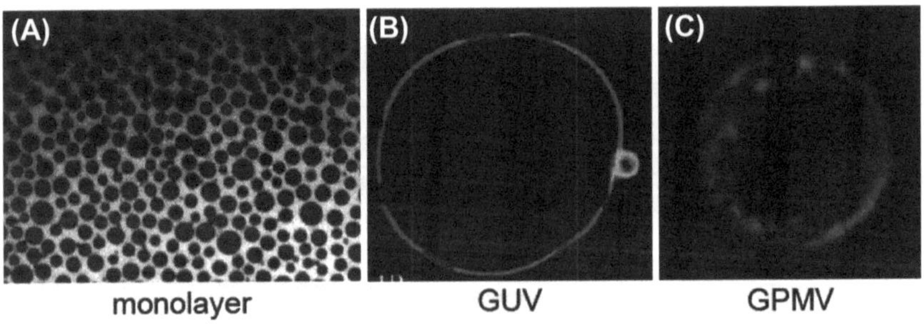

Figure 2 ***Liquid–liquid phase separation in biomimetic membranes.*** The coexistence of two liquid phases can be observed across various model membrane modalities in the presence of cholesterol. Shown are (A) Langmuir monolayer, (B) GUV composed of synthetic lipids *(from Korlach et al., 1999; copyright © by the National Academy of Sciences)*, and (C) GPMV isolated from a mammalian cell *(from Levental et al., 2010b)*, imaged by inclusion of minor fluorescent lipid components with preferential partitioning for either ordered or disordered domains. (See color plate)

membranes (Dietrich, Volovyk, Levi, Thompson, & Jacobson, 2001)), confirming that the capacity for fluid—fluid phase separation exists in mammalian membranes. This was later expanded to include fungal PMs with the reconstitution of yeast PM lipids (Klose et al., 2010). A key finding was that macroscopic phase separation could be triggered by a biological stimulus. Specifically, the addition of a glycolipid-binding bacterial protein (the B subunit of cholera toxin) to supported bilayers containing its receptor glycolipid (GM1) was sufficient to induce phase separation in a membrane containing as little as 0.2 mol% GM1 (Hammond et al., 2005). This phase separation could then sort "bystander" components of the membrane, i.e., those not directly bound by the ligand, including dissolved membrane peptides as models for transmembrane proteins.

Altogether, the necessary role of cholesterol in the formation of the liquid-ordered phase, the relative detergent insolubility of liquid-ordered domains (Ahmed, Brown, & London, 1997), and their preferential recruitment of sphingolipids, glycolipids, and GPI-anchored proteins make a strong case for the L_o/L_d paradigm as the physicochemical underpinning for the raft hypothesis (Simons & Vaz, 2004). This case has been recently bolstered by the observation of what appears to be ordered and disordered lipid domains in vivo, in inflated yeast vacuoles (Toulmay & Prinz, 2013).

2.6 Caveats of membrane model systems

Although phase separation in model membranes provided key physical evidence to explain the plausibility of raft formation in cells, important caveats prohibit a direct connection between these phenomena: (1) limited lipid compositional complexity of model membranes; (2) paucity or lack of membrane proteins; and (3) absence of other cellular structures. The classical synthetic "raft mixture," designed to represent mammalian PMs (Simons & van Meer, 1988), includes an unsaturated lipid component (usually doubly unsaturated dioleoyl phosphatidylcholine; DOPC), a saturated lipid component (dipalmitoyl PC (DPPC) or sphingomyelin (SM)), and cholesterol, at roughly equimolar quantities. This is a necessary and reasonable simplification of the more complex biological membrane, but only recently has the true compositional complexity of biological membranes become clear. Mass spectrometric studies of cellular lipidomes have revealed several hundreds (Gerl et al., 2012; Sampaio et al., 2011), if not thousands (Atilla-Gokcumen et al., 2014), of different lipid species in biological membranes. Moreover, it seems that lipid species with two unsaturated acyl chains are not highly abundant, rather glycerophospholipids usually contain a saturated acyl chain together with either a monounsaturated or polyunsaturated fatty acid. Besides bulk compositional complexity, plasmalemmal membranes are often asymmetric with respect to the two leaflets, with the exoplasmic leaflet generally excluding charged (anionic) lipids, most notably phosphatidylserine (PS). In contrast, model membranes are symmetric, with some notable exceptions (Cheng, Megha, & London, 2009; Collins & Keller, 2008). Another defining property of biological membranes is their high protein content. Membranes are estimated to be ~50% protein by weight, and ~20% of the cross-sectional area at the membrane core is occupied by polypeptides rather than lipids (Dupuy & Engelman, 2008). These estimates are in stark contrast to synthetic membranes, which usually exclude transmembrane proteins. Even when proteins are included, technical limitations limit their levels to <5 mol%, far below the load of native membranes. Finally, the structure and physiology of native membranes is likely strongly coupled to other cellular components, most notably the cortical cytoskeleton and the extracellular matrix. Because of the intimate physical lamination and biochemical interaction between these layers, it is likely that their dynamics and organization are tightly coupled. These interactions are difficult to reconstitute in vitro, although both theoretical (Machta, Papanikolaou, Sethna, & Veatch, 2011) and experimental (Honigmann et al., 2014; Zhao, Wu, & Veatch, 2013) attempts have recently borne fruit.

3. PHASE SEPARATION IN GPMVs

The above-mentioned caveats have prevented an unequivocal convergence on a working model for the functional partitioning of biological membranes. In this context, the discovery of liquid–liquid phase separation in isolated PMs provided unambiguous proof of the feasibility of lipid-driven membrane domains in cellular membranes (Figure 2(C)).

3.1 Discovery of detachable membrane blebs and use as isolated PMs

PM blebbing during apoptosis or cell motility has been known from the early days of cytology (reviewed in (Charras, 2008)). The mechanism typically involves a local delamination of the PM from the underlying cytoskeleton (e.g., by dissociation of protein–protein interactions), coupled with contraction of the actomyosin cortex, which leads to flow of fluid (cytoplasm or extracellular water) to fill the expanding area. The final step involves reassembly of the cytoskeletal cortex and retraction of the bleb.

Similar steps likely occur in blebs induced by chemical necrosis, although these do not require actin contraction for formation, and usually are not retracted because of the death of the cell (Barros et al., 2003).

GPMVs are induced by chemical treatment, and the first observations of blebs induced by cysteine-alkylating chemicals were made on isolated cells and tumor tissue in 1961 (Belkin & Hardy, 1961). A variety of chemicals have since been shown to produce these structures, sharing in common the capacity to passively traverse the PM in live cells and covalently modify free sulfhydryls (Scott, Perkins, Zschunke, Hoerl, & Maercklein, 1979). Initially, the utility of these cell-detached membrane blebs was realized as a method for isolation of highly enriched PM fractions (Scott, 1976), though the method was relatively sparsely used, with notable exceptions for in situ structural studies of the IgE receptor (Holowka & Baird, 1983), the lipid composition of the PM in mast cells (Fridriksson et al., 1999), and physical characterization of biological membranes separated from the cellular milieu (Dai & Sheetz, 1999; Ge et al., 2003; Tank, Wu, & Webb, 1982).

3.2 Phase separation in GPMVs as validation of the raft hypothesis

The breakthrough observation of liquid—liquid phase separation in GPMVs was made in 2007 (Baumgart et al., 2007). Vesicles from cultured mammalian cells robustly separated into liquid domains that sorted both exogenous and native lipids and proteins (Figure 3). This characteristic was the most important in the context of the raft hypothesis: the components that had been identified as canonical raft lipids and proteins (e.g., GPI-anchored proteins and the glycosphingolipid GM1) cosegregated together into one of these phases, away from several predicted nonraft components such as prenylated proteins and unsaturated lipids. It is important to note that several components did not partition as expected, and this caveat is expanded on in Section 5.2 (Figure 3(F)).

This phase separation was soon shown to be critically dependent on membrane cholesterol (Levental et al., 2009), validating another key prediction of the raft hypothesis. Distinct diffusivity (Levental et al., 2009) and order (Kaiser et al., 2009) values for the coexisting domains confirmed that the phenomenon observed in these biological membranes was indeed liquid-ordered/liquid-disordered phase immiscibility. These observations provided a critical conceptual validation for the central tenet of the raft hypothesis by showing that biological membranes have the capacity for separation into coexisting fluid phases of distinct compositions and physical properties.

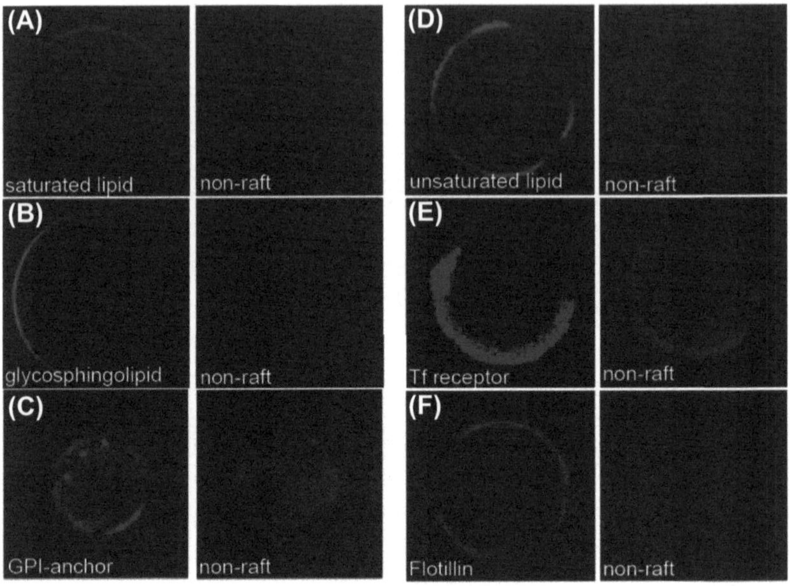

Figure 3 ***Partitioning of lipids and proteins in GPMVs.*** Phase separation in GPMVs sorts native and exogenous components of the membranes, often in accordance with predictions from the raft hypothesis. (A) Saturated lipids, (B) glycosphingolipids, and (C) GPI-anchored proteins partition strongly to the raft phase (in this figure, the nonraft phase is marked by the fluorescent unsaturated phospholipid rhodamine-DOPE). (D) Unsaturated lipids and (E) the transferrin receptor partition to the nonraft phase, consistent with their predicted exclusion from lipid rafts. However, (F) flotillin and some other predicted raft proteins are sometimes excluded from the raft phase. *Figures adapted from Sengupta et al. (2008) and Levental et al. (2010b). Reprinted from* Biochimica et Biophysica Acta, 1778*(1), Sengupta P, Hammond A, Holowka D, Baird B, Structural determinants for partitioning of lipids and proteins between coexisting fluid phases in giant plasma membrane vesicles, 20–32, Copyright (2008), with permission from Elsevier.* (See color plate)

3.3 Critical fluctuations in GPMVs

An important question remained: why do PMs isolated as GPMVs show stable, microscopic raft domains whereas live cell PMs are laterally homogeneous even at the highest resolution of light microscopy? The riddle remains unsolved, but an important clue was provided by characterization of the miscibility phase transition in GPMVs (Veatch et al., 2008). Detailed microscopic analysis of domain edge fluctuations coupled to rigorous theoretical predictions revealed critical phase transitions in GPMVs, in analogy to similar behavior previously observed in biomimetic monolayers (Keller, Anderson, & McConnell, 2000) and bilayers (Veatch, Soubias, Keller, & Gawrisch, 2007). This insight was both surprising and revealing. Critical behavior has been extensively studied, with the conclusion that many features of critical systems are "universal," i.e., their properties are often entirely independent of the specific features of the system (Honerkamp-Smith, Veatch, & Keller, 2009). Thus, insights from ferromagnets or liquid—vapor coexistence in simple fluids can be extrapolated to make bold predictions in isolated PMs. One such prediction is the lifetime and size of critical fluctuations persisting beyond the dissolution of macroscopic domains. The theory predicts that domain size (formally fluctuation correlation length) falls off inversely with temperature above a critical point ($T-T_c$). Thus, for a vesicle with T_c (i.e., critical miscibility transition temperature) around room temperature (as observed by Veatch et al. and others), "domains" of a few tens of nanometers would be expected at physiological temperatures. This prediction is a compelling bridge between GPMVs and live cell PMs, but remains to be experimentally confirmed. An important sidenote is that the predicted criticality of the PM would lead to some unexpected properties when coupled to relatively immobile cellular components (Zhao et al., 2013) like the cytoskeleton (Machta et al., 2011) or extracellular matrix (Groves, 2013). Namely, a small fraction of immobile membrane molecules can lead to compositional inhomogeneity persisting well above the macroscopic phase transition temperature. For the case of the cytoskeleton, this theory has recently received some compelling experimental support in model membranes (Honigmann et al., 2014).

4. CAVEATS AND ALTERNATIVES TO GPMVs AS PLASMA MEMBRANE MODELS

A number of important caveats must be considered before inferences from GPMVs are applied to in vivo settings. Among these are (1) unknown enzymatic activity and effects of chemical treatments during GPMV formation; (2) loss of strict membrane asymmetry; (3) lack of assembled cytoskeleton; and (4) chemical equilibrium. These caveats do not invalidate the utility of GPMVs, but rather define the parameters for them to serve as an intermediate model system with the experimental versatility of synthetic membranes coupled with the compositional complexity of native ones.

4.1 Unknown mechanism of formation and enzymatic activity

Some features of the well-characterized blebbing phenomena associated with apoptosis or cell motility likely extend to chemically induced GPMVs (discussed in Section 3.1 above). However, there are important differences. Chemical blebbing is unlike apoptotic blebbing in that the former shows no obvious evidence of nuclear fragmentation and generally produces blebs free of electron—dense lumenal components (Scott et al., 1979). Further, unlike blebs for motility, GPMVs are not retracted by actin assembly, but instead "pinch off" the surface of the cells. The mechanisms involved in the formation and scission of these blebs are unknown. It has been suggested that degradation of the lipidic second messenger phosphoinositide bisphosphate (PIP_2) is responsible for GPMV formation (Keller, Lorizate, & Schwille, 2009), consistent with the key role for PIP_2 in mediating PM-to-actin attachment (Janmey & Lindberg, 2004). It is not clear why chemical treatments should lead to PIP_2 degradation, though a speculation is that cellular ATP quickly becomes depleted by interference of the blebbing chemicals with the oxidative phosphorylation machinery, leading to an absence of "fuel" for the kinase reactions required to regenerate phosphorylated inositides. Although the lipid composition of GPMVs has been characterized (Fridriksson et al., 1999) and is generally in line with predictions for the PM, a detailed lipidomic comparison between GPMVs and a separate, well-validated PM prep remains an important step to confirming the composition of GPMVs as representative of intact PMs.

Other important considerations are the possible effects of the alkylating chemicals themselves. For example, the most common isolation protocol uses the highly reactive formaldehyde, which nonspecifically cross-links proteins and lipids primarily by reacting with primary amines. Although the concentration used for GPMV isolation is ~50-fold less than is used for fixation, significant cross-linking is still likely, and its effects on the organization of the membrane are unclear. In addition, there are more specific effects of the isolation chemicals that clearly have direct effects on proteins and membrane domains. For example, it appears that the reducing agent dithiothreitol (DTT; the other ingredient of the most common GPMV isolation protocol) leads to a reduction in protein palmitoylation, which in turn reduces the raft phase partitioning of proteins reliant on this modification for order phase affinity (see Section 5.1) (Levental, Lingwood, et al., 2010). Additionally, formaldehyde and DTT appear to cooperatively cross-link membrane proteins to the amine-containing glycerophospholipids phosphatidylethanolamine (PE), resulting in extraction of proteins from the ordered domains (Levental, Grzybek, & Simons, 2011). Alternative preparations based on N-ethylmaleimide do not cross-link proteins or reduce disulfide bridges (Levental, Lingwood, et al., 2010), but they alkylate free cysteines, with probable effects on protein structure and function.

4.2 Loss of membrane asymmetry

Lipid asymmetry between apposing bilayer leaflets is a fundamental property of nearly all PMs, with mammalian cells expending significant energetic resources to prevent negatively charged lipids (mainly PS) from flipping to the extracellular leaflet. Regulation of leaflet asymmetry is clinically relevant, as exposure of negatively charged lipids is required for apoptosis, blood coagulation, and macrophage activation (Fadeel & Xue, 2009). In contrast to native PMs, GPMVs are not strictly asymmetric with respect to PS exposure, as evidenced by the binding of Annexin V (Baumgart et al., 2007; Keller et al., 2009); an alternate, though unlikely, explanation for these observations is that GPMVs are somehow permeable to large polypeptides and therefore that Annexin V staining actually stains endoplasmic leaflet PS. PS exposure in GPMVs is likely an active process, as the Ca^{2+} influx associated with GPMV formation probably induces scramblase activity (Sahu, Gummadi, Manoj, & Aradhyam, 2007). It is useful to note that while PM asymmetry is classically defined by the absence of exofacial PS, the membranes are asymmetric in a number of other ways that are likely unaffected by induction of GPMVs. Examples include large glycolipids, which most likely remain confined to the extracellular leaflet, as well as lipid anchored and transmembrane proteins which almost certainly maintain their appropriate topology in GPMVs.

4.3 Lack of assembled cytoskeleton and chemical equilibrium

While the ingredients for cytoskeletal assembly (actin/tubulin monomers, accessory proteins, etc.) are present in GPMV lumens, no assembled cytoskeleton is visible by TEM (Scott et al., 1979). While this is not unassailable evidence for the absence of polymerized cytoskeleton, it is highly unlikely that a membrane tightly coupled to the actin framework would expand with the dynamics observed in GPMV formation. Because the interplay between the actin meshwork and the PM is such an integral part of membrane physiology (Sheetz, Sable, & Dobereiner, 2006), it must be emphasized that GPMVs are a model for the PM in isolation and that the coupling between it and other cellular elements remains poorly understood.

Another important consideration in comparing GPMVs to native membranes is that they are in a state of chemical equilibrium, in stark contrast to living cells. It is possible that cells input energy to reorganize the PM or regulate partitioning of proteins between coexisting domains. A recently proposed example is the dynamic remodeling of GPI-anchored proteins by the activity of putative actin microfilament bundles, named "actin asters" (Goswami et al., 2008; Gowrishankar et al., 2012).

4.4 Alternative PM model systems

Synthetic GUVs are the most direct alternative to GPMVs, providing complete experimental control of composition in exchange for reduced compositional complexity and protein content. However, this section will focus on the alternative protocols for preparation of intact PMs from live cells. Nearly concomitant with the discovery of phase separation in GPMVs, Lingwood and Simons discovered a similar phenomenon in blebs called plasma membrane spheres (PMS) (Lingwood, Ries, Schwille, & Simons, 2008). The advantage of PMS is that reactive chemicals are not required for bleb isolation, rather the cells bleb after prolonged incubation in glucose-supplemented phosphate-buffered saline. Large, selective, lateral domains are observed following cross-linking of the raft glycolipid GM1 by CTxB, or cooling below 10 °C (Levental et al., 2011). The cholesterol dependence of domain formation, and the difference in order between coexisting domains (Kaiser et al., 2009), confirmed raft phase separation. The major disadvantage of PMS is that both isolation and domain induction by CTxB are somewhat cell-type specific (mostly done in A431 epithelial carcinoma cells), in contrast to GPMVs, which are readily derived from essentially all cell types tested thus far (~20, including several primary cell fractions; personal observations). Several other nonchemical blebbing protocols have been developed, based on hypotonic swelling (Cohen, Ushiro, Stoscheck, & Chinkers, 1982), boron salts (Del Piccolo, Placone, He, Agudelo, & Hristova, 2012), and even laser irradiation (Kelly et al., 2009); however, phase separation in these has not been evaluated. Notably, several of the caveats associated with GPMVs (e.g., asymmetry and enzymatic activity) are likely also present in these alternative systems.

5. APPLICATIONS OF GPMVs

GPMVs are a methodological breakthrough for the lipid raft field because they allow application of the experimental toolbox of synthetic membrane biophysics to an intact membrane with the lipid and protein complexity/diversity of a mammalian PM. This section will focus on the uses of GPMVs for investigation of both raft-associated phenomena and as a model for the PM. The most common of these are (1) to measure component partitioning between coexisting domains; and (2) to quantify the properties of the domains.

5.1 Partitioning between domains

Most of the thousands of papers on the role of lipid rafts in cellular physiology focus on a specific raft-associated protein, with an overwhelming majority defining this affinity by detergent resistance. However, because of the widely acknowledged caveats of DRMs described above, the protein composition of membrane rafts and the structural determinants of protein raft association remain vague. In this context, perhaps the greatest advantage of GPMVs is their use for simple and accurate quantification of lipid and protein partitioning between raft and nonraft phases. For proteins, it is usually necessary to express a fluorescent chimera, although native receptors have also been measured by labeling of endogenous ligands (Baumgart et al., 2007). Labeled lipids are even easier because they partition passively into the cellular PMs or isolated GPMVs and can be quantified directly therein. An example of this quantification is shown in Figure 4. Here, a fluorescently labeled protein of interest (Figure 4(B)) is imaged simultaneously with a reference marker whose raft preference is known (Figure 4(A)). The relative concentration of the protein of interest in the raft phase is then quantified by a fluorescence intensity line scan (Figure 4(D)) through the two phases (white arrow in Figure 4(C)). Raft partitioning is defined as the ratio of raft to nonraft intensity, expressed as the equilibrium partition coefficient ($K_{p,raft} = Int_{raft}/Int_{nonraft}$).

The first publications employing GPMVs focused on selective protein partitioning between phases. Baumgart et al. reported that while GPI anchored proteins and the GM1-binding cholera toxin (and the polycyclic aromatic fluorophore naphthopyrene) preferentially partitioned into the more ordered (i.e., raft) phase, all other components tested were excluded from the more highly packed domain, including some with predicted raft affinity, like the dually palmitoylated Src-family kinase Lyn, glycosylated proteins, and the IgE receptor (FcεRI) (Baumgart et al., 2007). Sengupta, Hammond, Holowka, & Baird (2008) further explored the structural determinant of partitioning in GPMVs for both lipids and proteins, observing that a saturated lipid (dipalmitoyl phosphatidylethanolamine; DPPE) labeled with a common hydrophobic fluorophore (nitrobenzodiazole) strongly

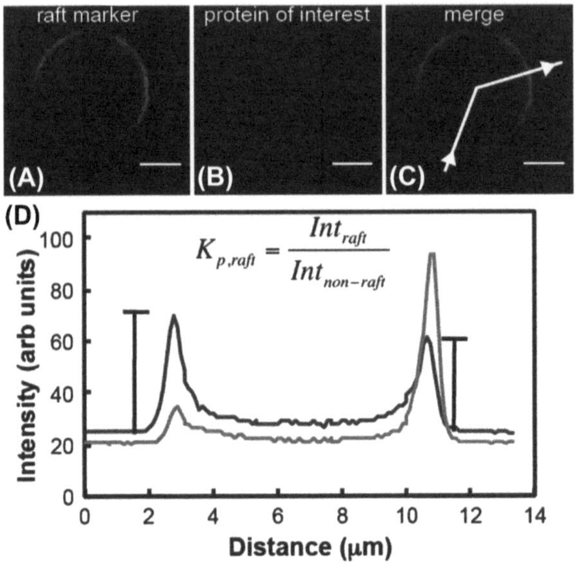

Figure 4 ***Quantification of protein partitioning in GPMVs.*** Component partitioning in GPMVs can be quantified by standard fluorescent microscopy approaches. GPMVs stained with (A) a marker for the raft domain can be isolated from cells expressing (B) a fluorescently tagged protein of interest. (C) The merged image can be used to (D) calculate the ratio of fluorescent intensity of the protein in the raft and nonraft phases, which is a direct quantification of the raft partition coefficient ($K_{p,raft}$) of the protein. *Figure adapted from Levental et al. (2010b).* (See color plate)

preferred the raft phase, with the opposite preference for an unsaturated version of the same lipid with the same tag, fully in line with the predicted affinity of saturated lipids for raft domains (Figure 3(A)–(C)). However, the same lipid labeled with fluorescein (DPPE-FITC) was excluded from raft domains, highlighting a ubiquitous potential artifact of lipid analogs: lipids are relatively small molecules and their partitioning behavior (and likely cellular function) is exquisitely dependent on their specific structure, amphiphilicity, and dipole moment. Fluorophores and other chemical tags are potentially major perturbations of these properties because they are often bulky and hydrophilic relative to the completely hydrophobic membrane core. This effect was recently highlighted in a survey of phase partitioning of a number of fluorescent lipid probes (Sezgin et al., 2012). In this study, the majority of labeled "raft lipids" (including sphingomyelin, glycosphingolipids, and cholesterol) showed little ordered phase affinity in either synthetic GUVs or biological GPMVs. Moreover, there was a clear difference between synthetic and biological membrane models, with all probes showing greater ordered phase preference in GPMVs—the hypothesized origin of this effect is discussed in Section 5.3.

In contrast to lipids, tagging proteins with fluorophores is not expected to affect their partitioning, because proteins are relatively large and the probes (most often fluorescent protein chimeras) are far away from the membrane-interacting portions of the protein. An exception is when the probe changes the structure of the protein, either directly or by inducing oligomerization, as is the case for the green fluorescent protein (Sengupta et al., 2008). Observations of protein partitioning in GPMVs have yielded confusing results inconsistent with predictions from previous characterizations of raft composition. Specifically, many raft proteins—such as flotillin (Sengupta et al., 2008), caveolin (Johnson et al., 2010), influenza hemagglutinin (Nikolaus et al., 2010), and acylated intracellular proteins like Lyn, Lck, Fyn, and Ras (Johnson et al., 2010) are either completely or largely excluded from the raft phase in GPMVs. This discrepancy was partially addressed by the observation that DTT in the GPMV isolation medium may depalmitoylate membrane proteins (Levental, Lingwood, et al., 2010). Palmitoylation is a posttranslational lipid modification in which a saturated fatty acid (16-carbon palmitate) is enzymatically added to reduced sulfhydryl residues in protein cysteines. For some otherwise-soluble proteins, this modification mediates stable membrane attachment (Resh, 2006). For many others, including those already anchored into the membrane via a transmembrane polypeptide, palmitoylation has been shown to impart raft affinity (Levental, Grzybek, et al., 2010). Because palmitoylation (in contrast to other lipid modifications) is regulated dynamically and reversibly in cells, it is postulated that this modification acts as a primary mechanism for switchable raft affinity in membrane proteins. This importance of palmitoylation in raft targeting was shown directly in GPMVs for the single-pass transmembrane adaptor protein linker for activation of T-cells (LAT) by mutation of its palmitoylation sites, reduction of palmitoylation by pharmacological inhibitors, and direct depalmitoylation by either DTT or hydroxylamine. This study estimated that approximately one-third of all raft proteins require palmitoylation for raft residence, with another third of raft proteins relying on GPI anchors (Levental, Lingwood, et al., 2010).

The most detailed study to date using GPMVs to probe the structural determinants of protein partitioning to raft domains set out to experimentally validate the long-standing theoretical prediction that rafts selectively recruit proteins based on the length of their transmembrane domains (TMDs) (Diaz-Rohrer, Levental, Simons, & Levental, 2014). Because ordered domains have been shown to be thicker than disordered domains in model membranes (Garcia-Saez, Chiantia, & Schwille, 2007; Heberle et al., 2013), the selection of longer TMDs by raft domains should be favored to minimize the hydrophobic mismatch between lipids and polypeptide (Mouritsen & Bloom, 1984). The experiments focused on mutations of the LAT transmembrane domain that systematically varied its length from the wild-type 24 amino acids down to 18. Quantification of partitioning in GPMVs unequivocally demonstrated a direct relationship between raft association and TMD length in a biological membrane, confirming results from synthetic model membranes (Lin & London, 2013). Importantly and unsurprisingly, TMD length was not the sole determining feature of raft partitioning, as mutations of specific residues in the TMD also reduced raft phase association in GPMVs. Intriguingly, there was a direct correlation between raft partitioning in GPMVs and subcellular localization. Variants with strong raft association were localized to the PM, as expected from the distribution of the native LAT protein. In contrast, variants with truncated/mutated TMDs and disrupted raft association were missorted to degradative endosomes including late endosomes and lysosomes. The key sorting step appears to be in the early endosomes, where proteins with raft affinity were recycled back to the PM (Diaz-Rohrer et al., 2014).

5.2 Caveats of protein partitioning in GPMVs

Although chemical depalmitoylation by DTT explains some of the unexpected lack of raft partitioning for predicted raft proteins, it is certainly not a complete explanation, as GPMVs derived by DTT-free preparations often show similar results (personal observations). Thus, there remains an important, unresolved discrepancy between raft composition assayed by GPMVs and DRMs. There are three possible explanations as follows:

1. GPMVs reflect native partitioning while DRMs are artifactually permissive, recruiting a number of nonraft proteins. This could occur by binding of nonraft proteins to intact membranes remaining after detergent lysis, or by the detergent changing the nature of the raft and nonraft domains during extraction.
2. DRMs accurately reflect raft composition, while GPMV raft phases are too exclusive because GPMV isolation disrupts protein-protein interactions, membrane asymmetry, or specific lipids required for proper raft partitioning.
3. GPMVs and DRMs probe different aspects of raft composition. For example, the raft phase in GPMVs may only recruit proteins with high

inherent affinity for ordered domains, excluding any that require specific protein–protein interactions or have relatively weak affinity for raft domains, whereas DRMs may include all proteins with any raft affinity. Our opinion is that a combination of (1) and (3) is most likely correct, though continued optimization of the GPMV system may reveal artifacts that lead to exclusion of native raft proteins from the raft phase.

This discussion also highlights the importance of avoiding the conceptual false dichotomy behind the classification of proteins as either "raft" or "non-raft." Raft partitioning is a continuum from complete raft phase preference to complete exclusion. Practically, most lipids and proteins partition to both phases in GPMVs and to both detergent-resistant and detergent-labile membrane fractions. This is not surprising from the standpoint of solution theory because of the similarity between the coexisting phases, and emphasizes that it is more correct to conceptualize rafts in terms of enrichments of certain proteins and lipids, rather than absolute selection.

5.3 Probing order and other physical properties of raft and nonraft domains in GPMVs

Microscopic phase separation in GPMVs allows access to the standard experimental toolbox of membrane biophysics in a biologically complex membrane. In the context of rafts, one of the most important aspects of membrane organization is lipid order. Though lipid acyl chain order can be quantified directly by nuclear magnetic resonance spectroscopy, it is more commonly estimated by fluorescent probes whose spectral properties are sensitive to membrane environment. Although numerous such probes are available (Sezgin, Sadowski, & Simons, 2014), the most commonly used is Laurdan (Parasassi, De Stasio, Ravagnan, Rusch, & Gratton, 1991), or its carboxylated analog C-Laurdan (Kim et al., 2007). Laurdan exhibits a red-shift in its emission spectrum because of the dipolar relaxation induced by aqueous solvent, with the extent of this shift dependent on the water exclusion of the membrane environment, which is in turn dependent on the packing/order of the membrane (Figure 5(A)). Thus, Laurdan is an indirect measure of membrane order (Owen, Rentero, Magenau, Abu-Siniyeh, & Gaus, 2011), typically reported by the normalized metric, generalized polarization (GP).

Comparison of Laurdan GP in various phase-separated model membranes revealed that while both synthetic and natural membranes form coexisting high and low order domains (representative of raft and nonraft domains, respectively), the absolute order of the two phases was highly

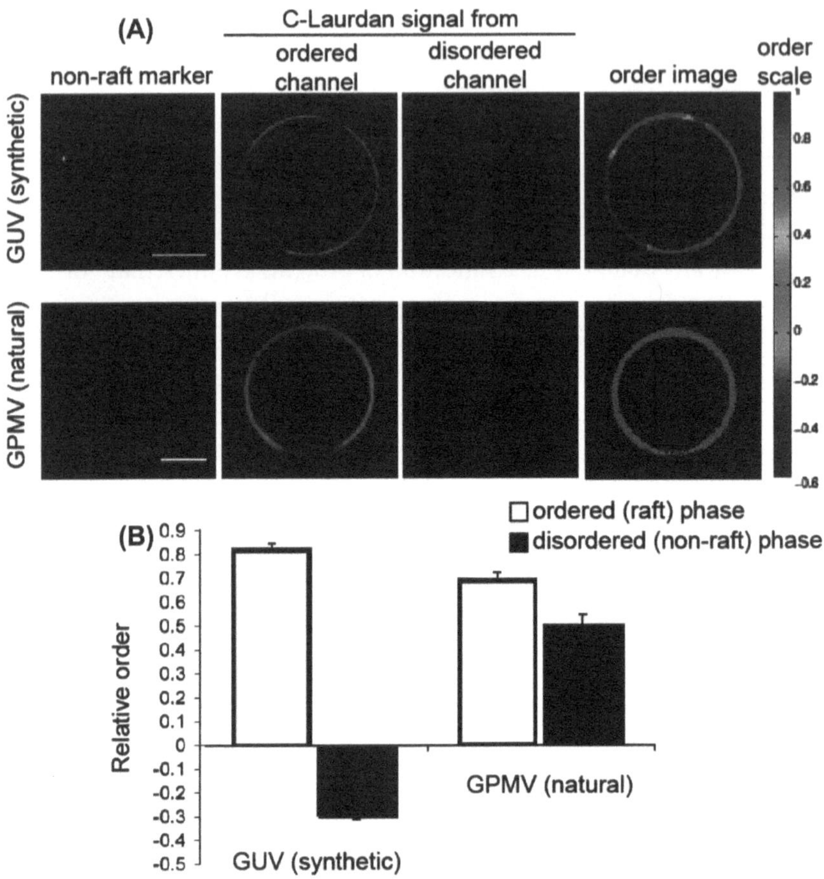

Figure 5 ***Quantification of domain order in model membranes.*** Microscopy of the environment-sensitive dye C-Laurdan can be used to measure the relative order differences between phases. (A) Images of the ordered (~440 nm) and disordered (~490 nm) emission channels are processed to yield pixel-by-pixel order maps. (B) The disparity between ordered and disordered domains is much greater in synthetic GUVs than in cell-derived GPMVs, with the relatively disordered, nonraft phase responsible for most of the difference. *Figure adapted from Kaiser et al. (2009).* (See color plate)

system dependent (Kaiser et al., 2009). The most profound differences between systems were (1) the order of the relative disordered (nonraft) phase and (2) the difference in order (ΔGP) between coexisting domains. GUVs composed of the "standard raft mixture" of DOPC:SM:Chol (2:2:1) had highly disordered (GP ~ −0.3) nonraft domains and very ordered raft domain (GP ~ 0.8) leading to an exaggerated interdomain differential (ΔGP > 1). In contrast, the nonraft domain was much more ordered in both biological membrane systems assayed (GPMVs and PMS) resulting in a small interdomain difference of $\Delta GP < 0.2$ (Figure 5(B)). This effect can probably be attributed to the peculiar lipid composition commonly used for synthetic model "raft" experiments. As described in Section 3.3, doubly unsaturated lipids like DOPC are not highly abundant in biological membranes (Gerl et al., 2012; Sampaio et al., 2011), and likely contribute to the high disorder of the nonraft phase in GUVs. Additionally, the high protein content of the biological membranes probably contributes significantly to the order of the raft phase.

A related point concerns the discrepancy in component partitioning between GUVs and GPMVs. As noted in Section 5.1, nearly all fluorescent lipid analogs tested by Sezgin et al. (2012) showed greater raft partitioning in GPMVs than in GUVs. A similar effect has been observed for proteins, though in less direct comparisons. The GPI-anchored protein placental alkaline phosphatase was shown to preferentially partition into the nonraft phase when reconstituted into GUVs (Kahya, Brown, & Schwille, 2005), whereas GPI-anchored proteins in general are highly raft phase enriched in GPMVs (Baumgart et al., 2007; Levental, Lingwood, et al., 2010; Sengupta et al., 2008). Similarly, while LAT partitions into the raft phase of GPMVs (Diaz-Rohrer et al., 2014; Levental, Lingwood, et al., 2010), a palmitoylated peptide mimicking the transmembrane domain of LAT is entirely excluded from the ordered domain of GUVs (Shogomori et al., 2005). It is likely that these discrepancies are directly attributable to the differences in domain order described above: the unnaturally disordered nonraft domains and tightly packed raft domains of GUVs combine to prevent partitioning of all but the most raft-integral components (i.e., saturated lipids) to the ordered domain. An exciting result is that the specific order values and interdomain differences may contribute to the bioactivity of membrane molecules. A fluorescent analog of GM1 bound its ligand CTxB avidly in the disordered domain of GUVs, but was completely incapable of binding when dissolved in the more ordered phase (Sezgin et al., in press). This effect is dependent on the interdomain difference, with smaller, more biomimetic ΔGP values promoting ordered phase binding (Sezgin et al., in press).

5.4 Miscibility transition temperature and its relationship to raft domains in cells

The criticality observed in GPMVs suggests that differences in the critical miscibility transition temperature (T_{misc} or T_c) predict changes in the size and lifetime of domains at physiological temperatures, and thus that perturbations affecting the temperature at which microscopic domain formation can be observed in GPMVs are potentially regulators of domain properties in live cells. The most obvious example of such a perturbation is cholesterol modulation, which was shown to dramatically affect T_{misc}; however, the direction of the effects was surprising. Cholesterol depletion of cells prior to GPMV isolation actually increased the miscibility transition temperature, effectively stabilizing separated domains, while the opposite effect was induced by cholesterol loading cells (Levental et al., 2009). This effect can be explained by the potentially nonlinear effects of cholesterol depletion on the coexisting domains, previously observed in synthetic membrane systems (Veatch & Keller, 2003). For example, cholesterol may be preferentially depleted from nonraft domains, thus making those domains more disordered and more likely to separate from the raft domains. Such specific modulation of the properties of the disordered phase was recently observed by treating GPMVs with bile acids (Zhou et al., 2013). These biological detergents were shown to preferentially partition into nonraft domains of GPMVs, reduce the order of those domains, and thereby dramatically stabilize phase separation (similar behavior was observed in synthetic membranes treated with Triton X-100 (Pathak & London, 2011)). Most importantly, the effects on the biophysical properties of GPMVs could be directly correlated to the organization and signaling of the PM in vivo. The same treatments that stabilized phase separation in GPMVs also enhanced the lateral clustering of the oncogene Ras in intact PMs, as evidenced by electron microscopy of basal PM sheets. Significantly, rearrangement of Ras affected signal transduction downstream of growth factor receptors. Stimulation of live cells by bile acids potentiated the activation of the mitogen-activated protein kinase pathway by the stimulating ligand epidermal growth factor. Similar stabilization of microdomain formation was also observed for two different lipid binding protein ligands, namely cholera toxin (GM1) and Annexin V (PS) (Johnson et al., 2010). To date, we are not aware of any observations of protein—protein binding affecting phase separation in GPMVs, but very few of these have been tested and it would be surprising if such interactions were unimportant for regulating membrane properties.

Finally, an intriguing case of exogenous pharmaceuticals affecting phase separation in GPMVs was recently shown for liquid general anesthetics. Several anesthetic alcohols, but not structurally similar nonanesthetics, decrease the critical miscibility transition temperature in a concentration dependent manner, with the magnitude of the effect directly scaling with anesthetic efficacy (Gray, Karslake, Machta, & Veatch, 2013). This finding is intriguing because of the well-known correlation between anesthetic potency and hydrophobicity, which has led to speculation that the still unknown mechanism of action of the broad variety of general anesthetics may have its origins in direct effects on the structure of membranes (Cantor, 1997). The observations in GPMVs support this idea, although much work remains to be done to confirm it.

5.5 Other uses of GPMVs

GPMVs are an excellent model for the composition, properties, and functions of the cellular PM in isolation. In this role, they have been used to measure the biophysical properties (Tank et al., 1982) and composition of the PM (Fridriksson et al., 1999), and the penetration of the PM by quantum dots (Dubavik et al., 2012), cell-penetrating peptides (Pae et al., 2014), and bacterial toxins (Manni, Sot, & Goni, 2014). Other potential uses include biotechnological applications, such as introducing cellular and viral proteins to synthetic supported bilayers (Costello, Hsia, Millet, Porri, & Daniel, 2013) and the formation of nanotube networks (Bauer, Davidson, & Orwar, 2006).

6. CONCLUSIONS AND PERSPECTIVES

The discovery of liquid–liquid phase separation in GPMVs was a vital breakthrough for the lipid raft field, definitively confirming the capacity of compositionally complex, protein-rich, mammalian membranes to undergo lipid-driven ordered domain formation. Although occurring in isolation from other cellular components and usually under nonphysiological temperatures, this capacity is unlikely to be an artifact. Without straying too far into teleology, it seems unlikely that an energy-free mechanism for structural organization of the membrane would have gone unutilized by evolution in the billions of years since the invention of the molecule responsible for the liquid-ordered phase, cholesterol. Moreover, there is increasing evidence that even in nonsterol producing prokaryotes, polycyclic molecules like hopanoids may serve a similar order-promoting role, suggesting that the liquid-ordered phase and potentially liquid–liquid coexistence in the membrane is an ancient evolutionary adaptation, predating the oxygenation of the Earth's atmosphere (Saenz, Sezgin, Schwille, & Simons, 2012). Nanometric domains extrapolated from critical fluctuations in GPMVs and the observations of such domains formed by modulated phases in biomimetic membranes (Goh, Amazon, & Feigenson, 2013) support the notion that the phase separation observed under nonphysiological conditions likely reflects certain aspects of organization in live cells. However, convincing evidence of existence of lipid-mediated membrane domains does little to answer the crucial mechanistic question of how these structures influence cellular physiology. Hypotheses are easy to conjure—regulation of protein–protein interactions (Sheets et al., 1999), local concentration effects on reactions between signaling molecules (Coskun, Grzybek, Drechsel, & Simons, 2010; Suzuki, 2012), conformational/activity changes induced by the membrane environment (Coskun et al., 2010; Sezgin et al., in press), etc. (Brown & London, 1998)—but specific molecular mechanisms remain to be clearly delineated.

Another open question concerns the state of membrane domains in GPMVs above the miscibility transition temperature, and the relationship of those to the nanometric organization of the PM in live cells. The exciting developments in superresolution microscopy of signaling domains (see chapters "Dances with Membranes: Breakthroughs from Super-Resolution Imaging" and "The Nanoscale Organization of Signaling Domains at the Plasma Membrane" by Curthoys et al., 2015; Griffié, Burn, & Owen, 2015) may yield insights into this key question, although the putatively rapid dynamics of raft domains present a significant hurdle. An important prediction from critical fluctuations is that the critical miscibility temperature in GPMVs is related to the size and lifetime of domains at physiological conditions. If true, this would provide a simple and coherent readout for changes to raft properties induced by extrinsic perturbations. However, this hypothesis remains experimentally validated. Finally, a key unresolved issue is whether the protein composition of the raft phase in GPMVs reflects the same in rafts of living cells. The discrepancy between the relative inclusivity of DRMs and exclusivity of GPMV raft domains suggests that one or both of these must be critically reevaluated.

Ultimately, the ease of GPMV isolation and visualization, their quantitative and repeatable nature, and their amenability to the biophysical toolbox combine to make an ideal model system for experimental investigation at the intersection of membrane biophysics and cell biology.

ACKNOWLEDGMENTS

We would like to acknowledge funding support from the Cancer Prevention and Research Institute of Texas (CPRIT) grant R1215.

REFERENCES

Ahmed, S. N., Brown, D. A., & London, E. (1997). On the origin of sphingolipid/cholesterol-rich detergent-insoluble cell membranes: physiological concentrations of cholesterol and sphingolipid induce formation of a detergent-insoluble, liquid-ordered lipid phase in model membranes. *Biochemistry, 36*(36), 10944–10953.

Atilla-Gokcumen, G. E., Muro, E., Relat-Goberna, J., Sasse, S., Bedigian, A., Coughlin, M. L., et al. (2014). Dividing cells regulate their lipid composition and localization. *Cell, 156*(3), 428–439.

Barros, L. F., Kanaseki, T., Sabirov, R., Morishima, S., Castro, J., Bittner, C. X., et al. (2003). Apoptotic and necrotic blebs in epithelial cells display similar neck diameters but different kinase dependency. *Cell Death & Differentiation, 10*(6), 687–697.

Bauer, B., Davidson, M., & Orwar, O. (2006). Direct reconstitution of plasma membrane lipids and proteins in nanotube-vesicle networks. *Langmuir, 22*(22), 9329–9332.

Baumgart, T., Hammond, A. T., Sengupta, P., Hess, S. T., Holowka, D. A., Baird, B. A., et al. (2007). Large-scale fluid/fluid phase separation of proteins and lipids in giant plasma membrane vesicles. *Proceedings of the National Academy of Sciences of the United States of America, 104*(9), 3165–3170.

Belkin, M., & Hardy, W. G. (1961). Relation between water permeability and integrity of sulfhydryl groups in malignant and normal cells. *Journal of Biophysical and Biochemical Cytology, 9*(4), 733–745.

Bezlyepkina, N., Gracia, R. S., Shchelokovskyy, P., Lipowsky, R., & Dimova, R. (2013). Phase diagram and tie-line determination for the ternary mixture DOPC/eSM/cholesterol. *Biophysical Journal, 104*(7), 1456–1464.

Brown, D. A. (2006). Lipid rafts, detergent-resistant membranes, and raft targeting signals. *Physiology (Bethesda), 21*, 430–439.

Brown, D. A., & London, E. (1998). Functions of lipid rafts in biological membranes. *Annual Review of Cell and Developmental Biology, 14*, 111–136.

Brown, D. A., & Rose, J. K. (1992). Sorting of GPI-anchored proteins to glycolipid-enriched membrane subdomains during transport to the apical cell surface. *Cell, 68*(3), 533–544.

Calleros, L., Lasa, M., Rodriguez-Alvarez, F. J., Toro, M. J., & Chiloeches, A. (2006). RhoA and p38 MAPK mediate apoptosis induced by cellular cholesterol depletion. *Apoptosis, 11*(7), 1161–1173.

Cantor, R. S. (1997). The lateral pressure profile in membranes: a physical mechanism of general anesthesia. *Biochemistry, 36*(9), 2339–2344.

Charras, G. T. (2008). A short history of blebbing. *Journal of Microscopy, 231*(3), 466–478.

Cheng, H. T., Megha, & London, E. (2009). Preparation and properties of asymmetric vesicles that mimic cell membranes: effect upon lipid raft formation and transmembrane helix orientation. *Journal of Biological Chemistry, 284*(10), 6079–6092.

Chen, X., & Resh, M. D. (2002). Cholesterol depletion from the plasma membrane triggers ligand-independent activation of the epidermal growth factor receptor. *Journal of Biological Chemistry, 277*(51), 49631–49637.

Cohen, S., Ushiro, H., Stoscheck, C., & Chinkers, M. (1982). A native 170,000 epidermal growth factor receptor-kinase complex from shed plasma membrane vesicles. *Journal of Biological Chemistry, 257*(3), 1523–1531.

Collins, M. D., & Keller, S. L. (2008). Tuning lipid mixtures to induce or suppress domain formation across leaflets of unsupported asymmetric bilayers. *Proceedings of the National Academy of Sciences of the United States of America, 105*(1), 124–128.

Coskun, U., Grzybek, M., Drechsel, D., & Simons, K. (2011). Regulation of human EGF receptor by lipids. *Proceedings of the National Academy of Sciences of the United States of America, 108*(22), 9044–9048.

Costello, D. A., Hsia, C. Y., Millet, J. K., Porri, T., & Daniel, S. (2013). Membrane fusion-competent virus-like proteoliposomes and proteinaceous supported bilayers made directly from cell plasma membranes. *Langmuir, 29*(21), 6409–6419.

Curthoys, N. M., Parent, M., Mlodzianoski, M., Nelson, A. J., Lilieholm, J., Butler, M. B., et al. (2015). Dances with Membranes: Breakthroughs from Super-resolution Imaging. *Current Topics in Membranes, 75*, 59–124.

Dai, J., & Sheetz, M. P. (1999). Membrane tether formation from blebbing cells. *Biophysical Journal, 77*(6), 3363–3370.

Del Piccolo, N., Placone, J., He, L., Agudelo, S. C., & Hristova, K. (2012). Production of plasma membrane vesicles with chloride salts and their utility as a cell membrane mimetic for biophysical characterization of membrane protein interactions. *Analytical Chemistry, 84*(20), 8650–8655.

Diaz-Rohrer, B., Levental, K., Simons, K., & Levental, I. (2014). Membrane raft association is a determinant of plasma membrane localization. *Proceedings of the National Academy of Sciences of the United States of America, 111*(23).

Dietrich, C., Bagatolli, L. A., Volovyk, Z. N., Thompson, N. L., Levi, M., Jacobson, K., et al. (2001). Lipid rafts reconstituted in model membranes. *Biophysical Journal, 80*(3), 1417–1428.

Dietrich, C., Volovyk, Z. N., Levi, M., Thompson, N. L., & Jacobson, K. (2001). Partitioning of Thy-1, GM1, and cross-linked phospholipid analogs into lipid rafts reconstituted in supported model membrane monolayers. *Proceedings of the National Academy of Sciences of the United States of America, 98*(19), 10642–10647.

Dubavik, A., Sezgin, E., Lesnyak, V., Gaponik, N., Schwille, P., & Eychmuller, A. (2012). Penetration of amphiphilic quantum dots through model and cellular plasma membranes. *ACS Nano, 6*(3), 2150–2156.

Dupuy, A. D., & Engelman, D. M. (2008). Protein area occupancy at the center of the red blood cell membrane. *Proceedings of the National Academy of Sciences of the United States of America, 105*(8), 2848–2852.

Eggeling, C., Ringemann, C., Medda, R., Schwarzmann, G., Sandhoff, K., Polyakova, S., et al. (2009). Direct observation of the nanoscale dynamics of membrane lipids in a living cell. *Nature, 457*(7233), 1159–1162.

Engel, S., Scolari, S., Thaa, B., Krebs, N., Korte, T., Herrmann, A., et al. (2010). FLIM-FRET and FRAP reveal association of influenza virus haemagglutinin with membrane rafts. *Biochemical Journal, 425*(3), 567–573.

Fadeel, B., & Xue, D. (2009). The ins and outs of phospholipid asymmetry in the plasma membrane: roles in health and disease. *Critical Reviews in Biochemistry and Molecular Biology, 44*(5), 264–277.

Feigenson, G. W., & Buboltz, J. T. (2001). Ternary phase diagram of dipalmitoyl-PC/dilauroyl-PC/cholesterol: nanoscopic domain formation driven by cholesterol. *Biophysical Journal, 80*(6), 2775–2788.

Fridriksson, E. K., Shipkova, P. A., Sheets, E. D., Holowka, D., Baird, B., & McLafferty, F. W. (1999). Quantitative analysis of phospholipids in functionally important membrane domains from RBL-2H3 mast cells using tandem high-resolution mass spectrometry. *Biochemistry, 38*(25), 8056–8063.

Garcia-Saez, A. J., Chiantia, S., & Schwille, P. (2007). Effect of line tension on the lateral organization of lipid membranes. *Journal of Biological Chemistry, 282*(46), 33537–33544.

Gaus, K., Rodriguez, M., Ruberu, K. R., Gelissen, I., Sloane, T. M., Kritharides, L., et al. (2005). Domain-specific lipid distribution in macrophage plasma membranes. *Journal of Lipid Research, 46*(7), 1526–1538.

Ge, M., Gidwani, A., Brown, H. A., Holowka, D., Baird, B., & Freed, J. H. (2003). Ordered and disordered phases coexist in plasma membrane vesicles of RBL-2H3 mast cells. An ESR study. *Biophysical Journal, 85*(2), 1278–1288.

Gerl, M. J., Sampaio, J. L., Urban, S., Kalvodova, L., Verbavatz, J. M., Binnington, B., et al. (2012). Quantitative analysis of the lipidomes of the influenza virus envelope and MDCK cell apical membrane. *Journal of Cell Biology, 196*(2), 213–221.

Goh, S. L., Amazon, J. J., & Feigenson, G. W. (2013). Toward a better raft model: modulated phases in the four-component bilayer, DSPC/DOPC/POPC/CHOL. *Biophysical Journal, 104*(4), 853–862.

Goswami, D., Gowrishankar, K., Bilgrami, S., Ghosh, S., Raghupathy, R., Chadda, R., et al. (2008). Nanoclusters of GPI-anchored proteins are formed by cortical actin-driven activity. *Cell, 135*(6), 1085–1097.

Gowrishankar, K., Ghosh, S., Saha, S., R., C., Mayor, S., & Rao, M. (2012). Active remodeling of cortical actin regulates spatiotemporal organization of cell surface molecules. *Cell, 149*(6), 1353–1367.

Gray, E., Karslake, J., Machta, B. B., & Veatch, S. L. (2013). Liquid general anesthetics lower critical temperatures in plasma membrane vesicles. *Biophysical Journal, 105*(12), 2751–2759.

Griffié, J., Burn, G., & Owen, D. M. (2015). The Nanoscale Organization of Signaling Domains at the Plasma Membrane. *Current Topics in Membranes, 75*, 125–166.

Groves, J. T. (2013). Cell membranes: glycans' imprints. *Nat Mater, 12*(2), 96–97.

Haines, T. H. (2001). Do sterols reduce proton and sodium leaks through lipid bilayers? *Progress in Lipid Research, 40*(4), 299–324.

Hammond, A. T., Heberle, F. A., Baumgart, T., Holowka, D., Baird, B., & Feigenson, G. W. (2005). Crosslinking a lipid raft component triggers liquid ordered-liquid disordered phase separation in model plasma membranes. *Proceedings of the National Academy of Sciences of the United States of America, 102*(18), 6320–6325.

Hao, M., Mukherjee, S., & Maxfield, F. R. (2001). Cholesterol depletion induces large scale domain segregation in living cell membranes. *Proceedings of the National Academy of Sciences of the United States of America, 98*(23), 13072.

Harder, T., Scheiffele, P., Verkade, P., & Simons, K. (1998). Lipid domain structure of the plasma membrane revealed by patching of membrane components. *Journal of Cell Biology, 141*(4), 929–942.

Heberle, F. A., Petruzielo, R. S., Pan, J., Drazba, P., Kucerka, N., Standaert, R. F., et al. (2013). Bilayer thickness mismatch controls domain size in model membranes. *Journal of the American Chemical Society, 135*(18), 6853–6859.

Heerklotz, H. (2002). Triton promotes domain formation in lipid raft mixtures. *Biophysical Journal, 83*(5), 2693–2701.

Hess, S. T., Gould, T. J., Gudheti, M. V., Maas, S. A., Mills, K. D., & Zimmerberg, J. (2007). Dynamic clustered distribution of hemagglutinin resolved at 40 nm in living cell membranes discriminates between raft theories. *Proceedings of the National Academy of Sciences of the United States of America, 104*(44), 17370–17375.

Holowka, D., & Baird, B. (1983). Structural studies on the membrane-bound immunoglobulin E-receptor complex. 1. Characterization of large plasma membrane vesicles from rat basophilic leukemia cells and insertion of amphipathic fluorescent probes. *Biochemistry, 22*(14), 3466–3474.

Honerkamp-Smith, A. R., Veatch, S. L., & Keller, S. L. (2009). An introduction to critical points for biophysicists; observations of compositional heterogeneity in lipid membranes. *Biochimica et Biophysica Acta, 1788*(1), 53–63.

Honigmann, A., Sadeghi, S., Keller, J., Hell, S. W., Eggeling, C., & Vink, R. (2014). A lipid bound actin meshwork organizes liquid phase separation in model membranes. *eLife, 3*, e01671.

Ipsen, J. H., Karlstrom, G., Mouritsen, O. G., Wennerstrom, H., & Zuckermann, M. J. (1987). Phase equilibria in the phosphatidylcholine-cholesterol system. *Biochimica et Biophysica Acta, 905*(1), 162–172.

Janmey, P. A., & Lindberg, U. (2004). Cytoskeletal regulation: rich in lipids. *Nature Reviews Molecular Cell Biology, 5*(8), 658–666.

Johnson, S. A., Stinson, B. M., Go, M. S., Carmona, L. M., Reminick, J. I., Fang, X., et al. (2010). Temperature-dependent phase behavior and protein partitioning in giant plasma membrane vesicles. *Biochimica et Biophysica Acta, 1798*(7), 1427–1435.

Kahya, N., Brown, D. A., & Schwille, P. (2005). Raft partitioning and dynamic behavior of human placental alkaline phosphatase in giant unilamellar vesicles. *Biochemistry, 44*(20), 7479–7489.

Kaiser, H. J., Lingwood, D., Levental, I., Sampaio, J. L., Kalvodova, L., Rajendran, L., et al. (2009). Order of lipid phases in model and plasma membranes. *Proceedings of the National Academy of Sciences of the United States of America, 106*(39), 16645–16650.

Keller, S. L., Anderson, T. G., & McConnell, H. M. (2000). Miscibility critical pressures in monolayers of ternary lipid mixtures. *Biophysical Journal, 79*, 2033–2042.

Keller, H., Lorizate, M., & Schwille, P. (2009). PI(4,5)P2 degradation promotes the formation of cytoskeleton-free model membrane systems. *ChemPhysChem, 10*(16), 2805–2812.

Keller, S. L., Pitcher, W. H., Huestis, W. H., & McConnell, H. M. (1998). Red blood cell lipids form immiscible liquids. *Physical Review Letters, 81*(22), 5019.

Kelly, C. V., Kober, M. M., Kinnunen, P., Reis, D. A., Orr, B. G., & Banaszak Holl, M. M. (2009). Pulsed-laser creation and characterization of giant plasma membrane vesicles from cells. *Journal of Biological Physics, 35*(3), 279–295.

Kenworthy, A. K. (2007). Fluorescence recovery after photobleaching studies of lipid rafts. *Methods in Molecular Biology, 398*, 179–192.

Kim, H. M., Choo, H. J., Jung, S. Y., Ko, Y. G., Park, W. H., Jeon, S. J., et al. (2007). A two-photon fluorescent probe for lipid raft imaging: C-laurdan. *Chembiochem, 8*(5), 553–559.

Klose, C., Ejsing, C. S., Garcia-Saez, A. J., Kaiser, H. J., Sampaio, J. L., Surma, M. A., et al. (2010). Yeast lipids can phase-separate into micrometer-scale membrane domains. *Journal of Biological Chemistry, 285*(39), 30224–30232.

Korlach, J., Schwille, P., Webb, W. W., & Feigenson, G. W. (1999). Characterization of lipid bilayer phases by confocal microscopy and fluorescence correlation spectroscopy. *Proceedings of the National Academy of Sciences of the United States of America, 96*(15), 8461.

LaRocca, T. J., Pathak, P., Chiantia, S., Toledo, A., Silvius, J. R., Benach, J. L., et al. (2013). Proving lipid rafts exist: membrane domains in the prokaryote *Borrelia burgdorferi* have the same properties as eukaryotic lipid rafts. *PLoS Pathogens, 9*(5), e1003353.

Levental, I., Byfield, F. J., Chowdhury, P., Gai, F., Baumgart, T., & Janmey, P. A. (2009). Cholesterol-dependent phase separation in cell-derived giant plasma-membrane vesicles. *Biochemical Journal, 424*(2), 163–167.

Levental, I., Grzybek, M., & Simons, K. (2010). Greasing their way: lipid modifications determine protein association with membrane rafts. *Biochemistry, 49*(30), 6305–6316.

Levental, I., Grzybek, M., & Simons, K. (2011). Raft domains of variable properties and compositions in plasma membrane vesicles. *Proceedings of the National Academy of Sciences of the United States of America, 108*(28), 11411–11416.

Levental, K. R., & Levental, I. (2015). Isolation of giant plasma membrane vesicles for evaluation of plasma membrane structure and protein partitioning. *Methods in Molecular Biology, 1232*, 65–77.

Levental, I., Lingwood, D., Grzybek, M., Coskun, U., & Simons, K. (2010). Palmitoylation regulates raft affinity for the majority of integral raft proteins. *Proceedings of the National Academy of Sciences of the United States of America, 107*(51), 22050–22054.

Lin, Q., & London, E. (2013). Altering hydrophobic sequence lengths shows that hydrophobic mismatch controls affinity for ordered lipid domains (rafts) in the multitransmembrane strand protein perfringolysin O. *Journal of Biological Chemistry, 288*(2), 1340–1352.

Lingwood, D., Ries, J., Schwille, P., & Simons, K. (2008). Plasma membranes are poised for activation of raft phase coalescence at physiological temperature. *Proceedings of the National Academy of Sciences of the United States of America, 105*(29), 10005–10010.

Lingwood, D., & Simons, K. (2010). Lipid rafts as a membrane-organizing principle. *Science, 327*(5961), 46–50.

Machta, B. B., Papanikolaou, S., Sethna, J. P., & Veatch, S. L. (2011). Minimal model of plasma membrane heterogeneity requires coupling cortical actin to criticality. *Biophysical Journal, 100*(7), 1668–1677.

Mahammad, S., Dinic, J., Adler, J., & Parmryd, I. (2010). Limited cholesterol depletion causes aggregation of plasma membrane lipid rafts inducing T cell activation. *Biochimica et Biophysica Acta, 1801*(6), 625–634.

Manni, M. M., Sot, J., & Goni, F. M. (2015). Interaction of *C. perfringens* epsilon-toxin with biological and model membranes: a putative protein receptor in cells. *Biochimica et Biophysica Acta, 1848*(3), 797–804.

McConnell, H. M., & Vrljic, M. (2003). Liquid-liquid immiscibility in membranes. *Annual Review of Biophysics and Biomolecular Structure, 32*, 469–492.

Meder, D., Moreno, M. J., Verkade, P., Vaz, W. L., & Simons, K. (2006). Phase coexistence and connectivity in the apical membrane of polarized epithelial cells. *Proceedings of the National Academy of Sciences of the United States of America, 103*(2), 329–334.

Mouritsen, O. G., & Bloom, M. (1984). Mattress model of lipid-protein interactions in membranes. *Biophysical Journal, 46*(2), 141–153.

Nikolaus, J., Scolari, S., Bayraktarov, E., Jungnick, N., Engel, S., Pia Plazzo, A., et al. (2010). Hemagglutinin of influenza virus partitions into the nonraft domain of model membranes. *Biophysical Journal, 99*(2), 489–498.

Norman, L. L., Oetama, R. J., Dembo, M., Byfield, F., Hammer, D. A., Levitan, I., et al. (2010). Modification of cellular cholesterol content affects traction force, adhesion and cell spreading. *Cellular and Molecular Bioengineering, 3*(2), 151–162.

Ostrom, R. S., & Liu, X. (2007). Detergent and detergent-free methods to define lipid rafts and caveolae. *Methods in Molecular Biology, 400*, 459–468.

Owen, D. M., Rentero, C., Magenau, A., Abu-Siniyeh, A., & Gaus, K. (2011). Quantitative imaging of membrane lipid order in cells and organisms. *Nature Protocols, 7*(1), 24–35.

Owen, D. M., Williamson, D. J., Magenau, A., & Gaus, K. (2012). Sub-resolution lipid domains exist in the plasma membrane and regulate protein diffusion and distribution. *Nature Communications, 3*, 1256.

Pae, J., Saalik, P., Liivamagi, L., Lubenets, D., Arukuusk, P., Langel, U., et al. (2014). Translocation of cell-penetrating peptides across the plasma membrane is controlled by cholesterol and microenvironment created by membranous proteins. *Journal of Controlled Release, 192*, 103–113.

Parasassi, T., De Stasio, G., Ravagnan, G., Rusch, R. M., & Gratton, E. (1991). Quantitation of lipid phases in phospholipid vesicles by the generalized polarization of Laurdan fluorescence. *Biophysical Journal, 60*(1), 179–189.

Pathak, P., & London, E. (2011). Measurement of lipid nanodomain (raft) formation and size in sphingomyelin/POPC/cholesterol vesicles shows TX-100 and transmembrane helices increase domain size by coalescing preexisting nanodomains but do not induce domain formation. *Biophysical Journal, 101*(10), 2417–2425.

Prior, I. A., & Hancock, J. F. (2012). Ras trafficking, localization and compartmentalized signalling. *Seminars in Cell & Developmental Biology*, 145–153.

Resh, M. D. (2006). Palmitoylation of ligands, receptors, and intracellular signaling molecules. *Science's STKE, 2006*(359), re14.

Rice, P. A., & McConnell, H. M. (1989). Critical shape transitions of monolayer lipid domains. *Proceedings of the National Academy of Sciences of the United States of America, 86*(17), 6445–6448.

Saenz, J. P., Sezgin, E., Schwille, P., & Simons, K. (2012). Functional convergence of hopanoids and sterols in membrane ordering. *Proceedings of the National Academy of Sciences of the United States of America, 109*(35), 14236–14240.

Sahu, S. K., Gummadi, S. N., Manoj, N., & Aradhyam, G. K. (2007). Phospholipid scramblases: an overview. *Archives of Biochemistry and Biophysics, 462*(1), 103–114.

Sampaio, J. L., Gerl, M. J., Klose, C., Ejsing, C. S., Beug, H., Simons, K., et al. (2011). Membrane lipidome of an epithelial cell line. *Proceedings of the National Academy of Sciences of the United States of America, 108*(5), 1903–1907.

Samsonov, A. V., Mihalyov, I., & Cohen, F. S. (2001). Characterization of cholesterol-sphingomyelin domains and their dynamics in bilayer membranes. *Biophysical Journal, 81*(3), 1486–1500.

Sanchez, S. A., Tricerri, M. A., & Gratton, E. (2012). Laurdan generalized polarization fluctuations measures membrane packing micro-heterogeneity in vivo. *Proceedings of the National Academy of Sciences of the United States of America, 109*(19), 7314–7319.

Scheiffele, P., Roth, M. G., & Simons, K. (1997). Interaction of influenza virus haemagglutinin with sphingolipid-cholesterol membrane domains via its transmembrane domain. *EMBO Journal, 16*(18), 5501–5508.

Schuck, S., Honsho, M., Ekroos, K., Shevchenko, A., & Simons, K. (2003). Resistance of cell membranes to different detergents. *Proceedings of the National Academy of Sciences of the United States of America, 100*(10), 5795–5800.

Schuck, S., & Simons, K. (2004). Polarized sorting in epithelial cells: raft clustering and the biogenesis of the apical membrane. *Journal of Cell Science, 117*(Pt 25), 5955–5964.

Scott, R. E. (1976). Plasma membrane vesiculation: a new technique for isolation of plasma membranes. *Science, 194*(4266), 743–745.

Scott, R. E., Perkins, R. G., Zschunke, M. A., Hoerl, B. J., & Maercklein, P. B. (1979). Plasma membrane vesiculation in 3T3 and SV3T3 cells. I. Morphological and biochemical characterization. *Journal of Cell Science, 35*, 229–243.

Sengupta, P., Hammond, A., Holowka, D., & Baird, B. (2008). Structural determinants for partitioning of lipids and proteins between coexisting fluid phases in giant plasma membrane vesicles. *Biochimica et Biophysica Acta, 1778*(1), 20–32.

Sezgin, E., Grzybek, M., Buhl, T., Dirkx, R., Gutmann, T., Coskun, U., et al. Adaptive lipid packing and bioactivity in membrane domains. *PLoS One*, in press.

Sezgin, E., Levental, I., Grzybek, M., Schwarzmann, G., Mueller, V., Honigmann, A., et al. (2012). Partitioning, diffusion, and ligand binding of raft lipid analogs in model and cellular plasma membranes. *Biochimica et Biophysica Acta, 1818*(7), 1777–1784.

Sezgin, E., Sadowski, T., & Simons, K. (2014). Measuring lipid packing of model and cellular membranes with environment sensitive probes. *Langmuir, 30*(27), 8160–8166.

Sheets, E. D., Holowka, D., & Baird, B. (1999). Critical role for cholesterol in Lyn-mediated tyrosine phosphorylation of FcepsilonRI and their association with detergent-resistant membranes. *Journal of Cell Biology, 145*(4), 877–887.

Sheetz, M. P., Sable, J. E., & Dobereiner, H. G. (2006). Continuous membrane-cytoskeleton adhesion requires continuous accommodation to lipid and cytoskeleton dynamics. *Annual Review of Biophysics and Biomolecular Structure, 35*, 417–434.

Shogomori, H., Hammond, A. T., Ostermeyer-Fay, A. G., Barr, D. J., Feigenson, G. W., London, E., et al. (2005). Palmitoylation and intracellular domain interactions both contribute to raft targeting of linker for activation of T cells. *Journal of Biological Chemistry, 280*(19), 18931–18942.

Simons, K., & Gerl, M. J. (2010). Revitalizing membrane rafts: new tools and insights. *Nature Reviews Molecular Cell Biology, 11*(10), 688–699.

Simons, K., & Ikonen, E. (1997). Functional rafts in cell membranes. *Nature, 387*(6633), 569–572.

Simons, K., & Toomre, D. (2000). Lipid rafts and signal transduction. *Nature Reviews Molecular Cell Biology, 1*(1), 31–39.

Simons, K., & van Meer, G. (1988). Lipid sorting in epithelial cells. *Biochemistry, 27*(17), 6197–6202.

Simons, K., & Vaz, W. L. (2004). Model systems, lipid rafts, and cell membranes. *Annual Review of Biophysics and Biomolecular Structure, 33*, 269–295.

Suzuki, K. G. (2012). Lipid rafts generate digital-like signal transduction in cell plasma membranes. *Biotechnology Journal, 7*(6), 753–761.

Tank, D. W., Wu, E. S., & Webb, W. W. (1982). Enhanced molecular diffusibility in muscle membrane blebs: release of lateral constraints. *Journal of Cell Biology, 92*(1), 207–212.

Toulmay, A., & Prinz, W. A. (2013). Direct imaging reveals stable, micrometer-scale lipid domains that segregate proteins in live cells. *Journal of Cell Biology, 202*(1), 35–44.

Veatch, S. L., Cicuta, P., Sengupta, P., Honerkamp-Smith, A., Holowka, D., & Baird, B. (2008). Critical fluctuations in plasma membrane vesicles. *ACS Chemical Biology, 3*(5), 287–293.

Veatch, S. L., & Keller, S. L. (2002). Organization in lipid membranes containing cholesterol. *Physical Review Letters, 89*(26), 268101.

Veatch, S. L., & Keller, S. L. (2003). Separation of liquid phases in giant vesicles of ternary mixtures of phospholipids and cholesterol. *Biophysical Journal, 85*(5), 3074–3083.

Veatch, S. L., Soubias, O., Keller, S. L., & Gawrisch, K. (2007). Critical fluctuations in domain-forming lipid mixtures. *PNAS, 104*(45), 17650–17655.

Zhang, W., Trible, R. P., & Samelson, L. E. (1998). LAT palmitoylation: its essential role in membrane microdomain targeting and tyrosine phosphorylation during T cell activation. *Immunity, 9*(2), 239–246.

Zhao, J., Wu, J., & Veatch, S. L. (2013). Adhesion stabilizes robust lipid heterogeneity in supercritical membranes at physiological temperature. *Biophysical Journal, 104*(4), 825–834.

Zhou, Y., Maxwell, K. N., Sezgin, E., Lu, M., Liang, H., Hancock, J. F., et al. (2013). Bile acids modulate signaling by functional perturbation of plasma membrane domains. *Journal of Biological Chemistry, 288*(50), 35660–35670.

Zidovetzki, R., & Levitan, I. (2007). Use of cyclodextrins to manipulate plasma membrane cholesterol content: evidence, misconceptions and control strategies. *Biochimica et Biophysica Acta, 1768*(6), 1311–1324.

CHAPTER THREE

Dances with Membranes: Breakthroughs from Super-resolution Imaging

Nikki M. Curthoys, Matthew Parent, Michael Mlodzianoski, Andrew J. Nelson, Jennifer Lilieholm, Michael B. Butler, Matthew Valles and Samuel T. Hess*

Department of Physics and Astronomy, University of Maine, Orono, ME, USA

*Corresponding author: E-mail: sam.hess@umit.maine.edu

Contents

Current Topics in Membranes, Volume 75
ISSN 1063-5823
http://dx.doi.org/10.1016/bs.ctm.2015.03.008

Abstract

Biological membrane organization mediates numerous cellular functions and has also been connected with an immense number of human diseases. However, until recently, experimental methodologies have been unable to directly visualize the nanoscale details of biological membranes, particularly in intact living cells. Numerous models explaining membrane organization have been proposed, but testing those models has required indirect methods; the desire to directly image proteins and lipids in living cell membranes is a strong motivation for the advancement of technology. The development of super-resolution microscopy has provided powerful tools for quantification of membrane organization at the level of individual proteins and lipids, and many of these tools are compatible with living cells. Previously inaccessible questions are now being addressed, and the field of membrane biology is developing rapidly. This chapter discusses how the development of super-resolution microscopy has led to fundamental advances in the field of biological membrane organization. We summarize the history and some models explaining how proteins are organized in cell membranes, and give an overview of various super-resolution techniques and methods of quantifying super-resolution data. We discuss the application of super-resolution techniques to membrane biology in general, and also with specific reference to the fields of actin and actin-binding proteins, virus infection, mitochondria, immune cell biology, and phosphoinositide signaling. Finally, we present our hopes and expectations for the future of super-resolution microscopy in the field of membrane biology.

1. INTRODUCTION

1.1 Fundamentals of membrane organization

Biological membranes mediate vast numbers of cellular functions, serve as the fundamental barrier between cell exterior and interior, and spatially define most cellular organelles (Alberts, 2002). Biological membranes are composed of proteins, lipids, and other small molecules, typically arranged in two opposing monolayers (i.e., a bilayer) (Alberts, 2002). The bilayer arrangement allows hydrogen bonding between the aqueous phase (i.e., the cytoplasm on one side, and extracellular medium on the other) and hydrophilic lipid head groups, while restricting interactions between hydrophobic lipid tails and the hydrophilic cytoplasm and extracellular medium (Tanford, 1991). The lipids found in membranes include a large number (thousands) of molecular species, which can be subdivided into several general classes: phospholipids (including sphingolipids and glycerophospholipids with saturated and/or unsaturated fatty acid chains), glycolipids, and sterols (including cholesterol) (Alberts, 2002; Lehninger, Nelson, & Cox, 2013).

1.1.1 Membrane lateral organization

1.1.1.1 Singer–Nicolson fluid mosaic model

The idea of a cell membrane as a mosaic structure of globular proteins within a phospholipid bilayer was proposed in 1971 by S.J. Singer (Singer, 1971) and popularized the following year (Singer & Nicolson, 1972). This "fluid mosaic" model proposes the free lateral diffusion of membrane proteins, which assume a long-scale random distribution in the two dimensional homogenous lipid fluid phase (see Figure 1(A)). One fundamental departure from this theory has been pivotal in shaping membrane research from the 1980s. Rather than being strewn randomly throughout a homogenous cell plasma membrane, proteins and lipids were proposed to be laterally heterogeneous, distributed in discrete patches. While researchers worldwide were embracing this view of cell membranes, to this day there remains deep division in the community: what are the driving forces behind plasma membrane heterogeneity?

In the 10 years after the popular Singer–Nicolson (Singer & Nicolson, 1972) paper was published, researchers were theorizing that cell plasma membranes were organized into discrete lipid domains, and already proposing lipid–protein interactions (Moore, Lentz, & Meissner, 1978) analogous to the present day lipid shell model and boundary lipid theories (Anderson & Jacobson, 2002); researchers were also beginning to theorize that cytoskeletons could modulate lateral mobility of membrane molecules (Karnovsky, Kleinfeld, Hoover, & Klausner, 1982). The ability of glycosphingolipids to self-associate and form discrete patches (reviewed in (Thompson & Tillack, 1985)), was then understood to also encompass protein distributions. Glycosphingolipid self-association in the Golgi could form patches with which

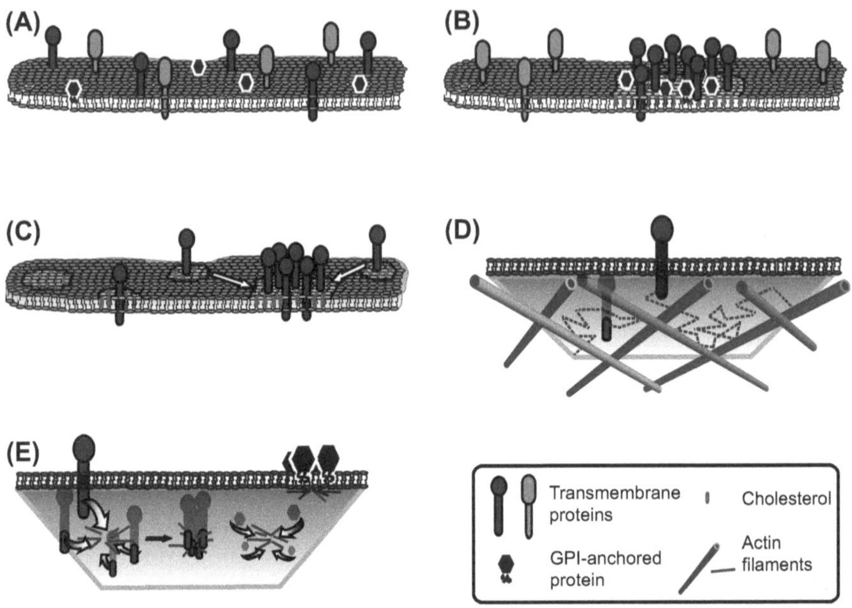

Figure 1 Models of cell membrane organization discussed in Section 1.1. (A) "Fluid mosaic" model. Proteins are distributed randomly through a homogenous phospholipid bilayer. (B) "Lipid raft" model. Sphingolipid and cholesterol patches are populated with proteins which have an affinity for these patches. Protein species can be raft associated or nonraft associated. (C) "Lipid shell" model. Some proteins will be targeted to self-assembled cholesterol and sphingolipid complexes which form a "lipid shell" around the protein. These "lipid shells" have an affinity for, and can coalesce with, larger lipid domains. (D) "Picket Fence" model. Transmembrane proteins are restricted in their diffusion by actin filaments (the "fence") which appose and run parallel to the cytoplasmic leaflet of the membrane, and by other transmembrane proteins bound to these filaments ("pickets", not shown). (E) "Active composite" model. Short actin filaments adjacent to the cytoplasmic membrane leaflet are arranged in "asters". Transmembrane proteins and GPI-anchored proteins are advected by actin and myosin to the centers of these asters, resulting in protein nanoclustering. See Section 1.1 for more detail. Readers please note that depictions of cell membranes here do not show as much protein (relative to lipid) as would be found in actual cell membranes. (See color plate)

membrane proteins would combine, and these mixtures could theoretically be transported to the apical membrane, mediating the sorting of sphingolipids and proteins in polarized epithelial cells (Simons & van Meer, 1988; Simons & Wandinger-Ness, 1990). Biochemical analyses appeared to support this model—the association of (glycophosphatidyl inositol) GPI-anchored proteins along with glycosphingolipids in cell lysate insoluble detergent fractions was taken as evidence of lipid–protein complexes in native membranes (Brown & Rose, 1992). In 1997 came a popular stating of one theory of lipid–protein complexes in the cell plasma membrane: the lipid raft model (Simons & Ikonen, 1997).

1.1.1.2 The lipid raft model

This model postulates that particular subsets of lipids can self-organize, forming discrete patches within the plasma membrane (see Figure 1(B)), believed to be enriched in cholesterol, sphingolipids, and GPI-anchored proteins. The affinity of particular species of membrane proteins for these self-organizing lipids would determine their inclusion into these patches, and in doing so, determine the spatial patterning of proteins in the cell plasma membrane. Sphingolipid self-association would occur through weak interactions between their head groups. Furthermore, cholesterol helps fill gaps between lipid molecules to reduce water permeability (Finkelstein & Cass, 1967). The result (within the context of the model) is phase-separated liquid ordered (Lo) domains (i.e., "lipid rafts") enriched in sphingolipids, cholesterol, and GPI-anchored proteins, within a surrounding liquid disordered (Ld) fluid phase enriched in unsaturated lipids and other proteins (Simons & Ikonen, 1997). The revised raft model posits that nanoscale assemblies of sphingolipids, cholesterol, and proteins are fluctuating, but can be stabilized into larger platforms important for signaling, viral infection, and membrane trafficking (Simons & Gerl, 2010). While these assemblies are now theorized as being dynamic and of variable size, the basic concept remains that preferential interactions between cholesterol, sphingolipids, and specific proteins are the central mechanism driving local heterogeneity of the cell plasma membrane (Simons & Gerl, 2010).

The lipid raft model is closely linked to the idea that membrane lipids can self-segregate into domains. In particular, separation of bilayers into Lo and Ld phases was proposed to explain membrane lateral heterogeneity in biological membranes (Brown & London, 1997, 1998a, 1998b; Simons & Ikonen, 1997). This idea is based on observations that in the absence of membrane proteins, and depending on lipid composition, temperature, and buffer conditions, artificially created lipid bilayers can separate into coexisting Lo, Ld, and/or solid (gel) phases (Honerkamp-Smith, Veatch, & Keller, 2009; Tanford, 1978; Veatch et al., 2008; Veatch, Gawrisch, & Keller, 2006; Veatch & Keller, 2003a, 2003b, 2005; Veatch, Polozov, Gawrisch, & Keller, 2004). Giant unilamellar vesicles (GUVs) (Angelova & Dimitrov, 1986), which are composed of a single bilayer, have been used extensively as a lipid model of liquid-ordered and liquid-disordered domains (Almeida, Pokorny, & Hinderliter, 2005; Bagatolli & Gratton,

1999, 2000; Bagatolli, Gratton, Khan, & Chong, 2000; Baumgart, Hammond, et al., 2007; Bagatolli, Sanchez, Hazlett, & Gratton, 2003; Baumgart, Hess, & Webb, 2003; Baumgart, Hunt, Farkas, Webb, & Feigenson, 2007; Dietrich et al., 2001; Korlach, Schwille, Webb, & Feigenson, 1999; Veatch 2007; Veatch et al., 2008; Veatch et al., 2006; Veatch & Keller, 2005). GUVs made from a mixture of 1:1:1 (unsaturated lipid:saturated lipid:cholesterol), for example, can be made to phase separate at or close to physiological temperature. In general, due to differences in fatty acid chain order and/or length, saturated and unsaturated lipids will under many conditions separate into liquid and gel phases unless sufficient sterol (cholesterol) is present to prevent demixing (Stottrup, Veatch, & Keller, 2004; Veatch & Keller, 2002, 2003a, 2003b). Intriguingly, disruption of interactions between the cell plasma membrane and the cortical actin cytoskeleton in a process called "blebbing" (Keller, Rentsch, & Hagmann, 2002; Tank, Wu, & Webb, 1982) has been shown to lead to lipid phase separation in bilayers attached to or derived from cells (Baumgart, Hammond, et al., 2007; Veatch et al., 2008). Thus, cell membranes can be caused to phase separate, but whether phase separation occurs in unperturbed cell membranes, and if so, under what conditions and on what length scales, is still unclear.

1.1.1.3 The lipid shell model

Anderson and Jacobson (Anderson & Jacobson, 2002) theorized that there must be some targeting motif encoded in proteins which determines inclusion into lipid domains. The lipid shell model stresses the self assembly of cholesterol–phospholipid complexes, and the differential tendency of proteins to associate with these complexes, either through direct interactions mediated by specific amino acid motifs in transmembrane domain sequences or electrostatic interactions between the head groups of phospholipids and oppositely charged amino acids in the protein. Proteins act as individual units each wrapped in a sphingolipid–cholesterol lipid shell (see Figure 1(C)); these lipid shells do not form a distinct lipid phase, but are thought to be mobile units in the membrane. However, larger scale lipid domains (such as caveolae and "lipid rafts") are proposed to have a characteristic Lo phase, and proteins in lipid shells would have an affinity for these domains due to the compatibility of lipids in the shell and the Lo phase of the lipid domain. Interactions between shelled proteins and those already in lipid domains may influence their time of residence in the domain, as may ligand binding.

1.1.2 *Proteins as organizers*

The importance of the cytoskeleton in erythrocyte membrane organization has been reported since the early 1980s (Koppel, 1981; Sheetz, 1983; Tsuji & Ohnishi, 1986) and specific models explaining the dynamics of proteins in living native cell membranes, and the potential for actin to create regions in membranes within which proteins and/or lipids are confined, concentrated, or excluded ("membrane domains"), have gained wider recognition.

1.1.2.1 The picket fence model

The basic underlying principle of this theory is that actin filaments, in a very close spatial association with the cytoplasmic leaflet of the plasma membrane, provide physical barriers which effectively compartmentalize the plasma membrane (Kusumi & Sako, 1996; Sako & Kusumi, 1994). The actin-based "membrane skeleton" (MSK), or "fence", creates compartments with diameters of 40–300 nm (Kusumi et al., 2012), and various transmembrane proteins bound to these actin filaments act as "pickets", which span the bilayer. Short-term diffusion of membrane lipids and non-MSK-bound membrane proteins can be confined within these compartments (see Figure 1(D)), with longer term trajectories punctuated by "hop" movements between adjacent compartments (termed "hop diffusion") (Fujiwara, Ritchie, Murakoshi, Jacobson, & Kusumi, 2002; Kusumi, et al., 2012). More recently, the model has been amended to include lipid rafts and protein oligomers, such that entire assemblies of proteins and lipids together may move within a compartment, but still be confined by the picket fence (Kusumi, et al., 2012).

1.1.2.2 Active composite model

Originally used to describe the distribution of GPI-anchored proteins in native cell membranes, this theory was developed to explain indirect observations which suggested these proteins existed either as monomers or in small, immobile, dense clusters, with an approximate average of four molecules per cluster (Sharma et al., 2004). These nanoclusters were reported to be transient and continually remodeling. However, they do not join or coalesce into larger domains, and the ratio of monomers (80%) to clusters (20%) is independent of total expression levels (Goswami et al., 2008). The authors propose the cytoplasmic side of the plasma membrane is abutted by a system of "short" (although estimated as ~250 nm in length) actin filaments arranged in "asters", (see Figure 1(E)) which are formed by actin

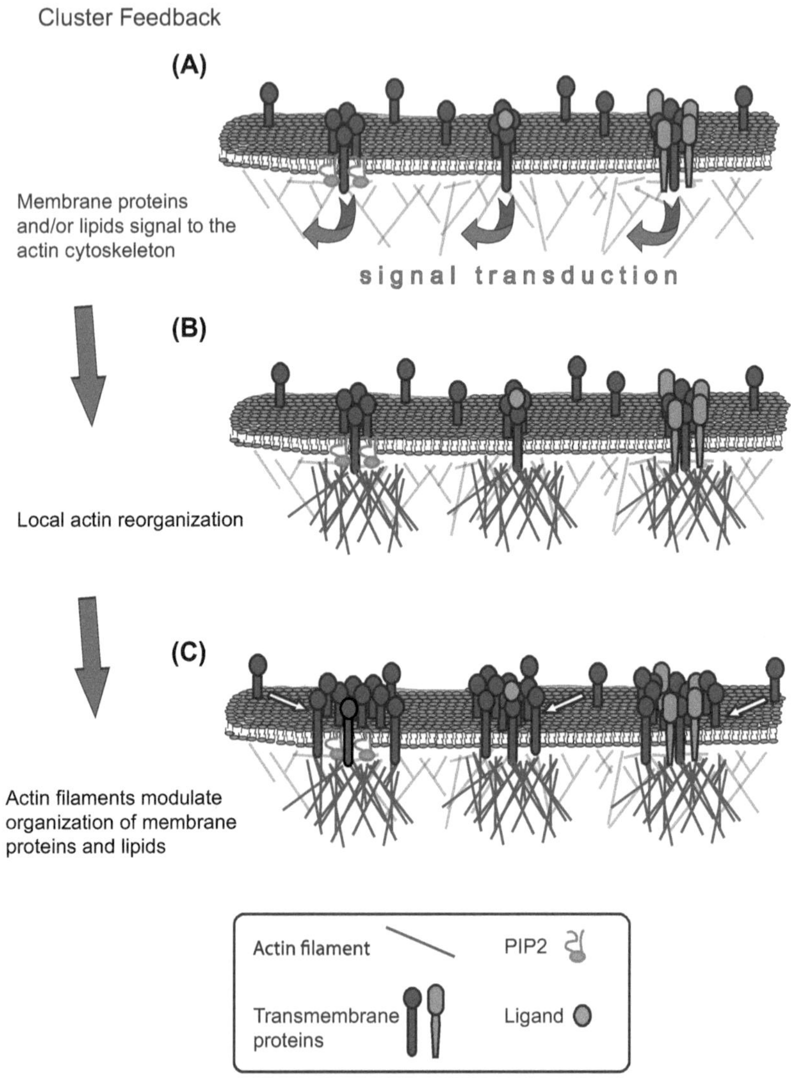

Figure 2 "Cluster Feedback" model of membrane organization. (A) Proteins cluster at the nanoscale in the plasma membrane. These clusters are protein–lipid (left, transmembrane protein and PIP2), mediated by ligand binding (middle), or protein–protein (right; depicted as a heterocluster; a homocluster oligomer is not shown). Of course, in clusters of any protein there exists the potential for local lipid clustering also. These protein–protein or protein–lipid clusters are collectively referred to hereafter as "membrane clusters." In each case, signaling to the actin cytoskeleton initiated by proteins and/or lipids in membrane clusters elicits the local remodeling of actin (B), either through the recruitment and binding of actin filaments to the membrane lipids or proteins, the de novo nucleation of new filaments or branches thereof, or both. ABPs which mediate these interactions are not shown, but see Section 3.1 for more detail. The

treadmilling and myosin contractility. Myosin contractility results in the barbed (or plus) ends of actin filaments pointing toward aster centers. In regions rich with myosin-bound actin filaments, peripheral actin filaments can be aligned with these other filaments, and pulled toward the most actin dense regions (the centers of asters). The authors posit that these forces are transmitted to (for example) GPI-anchored (exoplasmic leaflet) proteins, and proteins which bind F-actin, which are advected along with these filaments toward the centers of asters, forming nanoclusters at aster cores. Thus actin treadmilling and actin—myosin contraction would drive dynamic nanoscale clustering of proteins at the cell surface (Gowrishankar, Ghosh, Saha, Mayor, & Rao, 2012). The theoretical basis for this model is further discussed below.

1.1.2.3 Cluster feedback model

We propose the "cluster feedback" model (see Figure 2) whereby the reciprocal influence between two assemblies: (1) proteins and lipids in the membrane, and (2) the underlying actin cytoskeleton and associated actin binding proteins (ABPs), results in the formation of plasma membrane domains. We theorize that it is membrane-associated molecules which signal to induce local actin organization, and that these new actin platforms immediately adjacent to the membrane modulate the clustering of membrane proteins and/or lipids (Gudheti et al., 2013). This model does not discriminate between different hypotheses explaining the initial formation of membrane clusters (depicted in Figure 2(A)). We propose that when proteins are recruited into expanding membrane clusters they do so because of protein—protein interactions (either through other cluster proteins or by binding to actin filaments), or because actin filaments affect their mobility through steric interactions. The recruitment of new molecules into the cluster subsequent to local actin remodeling must, in some cases at least, contain a method by which this recruitment is differential, thereby sometimes

increase in actin density immediately adjacent to the membrane cluster then acts as a recruitment platform for other proteins (or lipids, not shown) diffusing in the membrane to join existing membrane clusters (C). This results in changes in the size, density, perimeter, and/or number of molecules within membrane clusters. Readers should note, the numbers of proteins depicted above are only for ease of communication—we do not hypothesize about the specific sizes (or numbers of molecules within) any of the membrane clusters here. Clusters may exist on the nano-, meso-, or micro scale; the Cluster Feedback model only predicts changes (here shown as increases) in cluster sizes from (A) to (C). See Sections 1.1 and 3.1 for more detail. (See color plate) favoring recruitment of "like" proteins into preexisting membrane clusters. This model is further discussed in Section 3.1.

1.1.3 Clarifying the theories of membrane organization: the need for direct evidence

1.1.3.1 The need for improved spatial and temporal resolution

The basic goal of science is to gain knowledge through the testing of falsifiable hypotheses. In many ways and for many years, membrane research has been prevented from achieving this goal. The lack of adequate methodology has resulted in only indirect measurements being used, which has often provided insufficient evidence to discriminate between competing models. As membrane domains are widely theorized as being submicron and dynamic, their visualization with light or electron microscopy (EM) has remained indefinite. In the absence of direct images of membrane domains, the above models of membrane organization have been incompletely tested. Many experimental approaches are controversial in their ability to distinguish membrane domains. For example, cold low density detergent-insoluble fractions of lysed cell membranes have been extensively used as a determinant of lipid rafts (from (Brown & Rose, 1992) to (Recktenwald et al., 2015)), however their representation of domains in native cell membranes is tenuous and yet to be verified (Edidin, 2003; Kraft, 2013; Lichtenberg, Goni, & Heerklotz, 2005).

Relying on cholesterol dependence to define raft-associated processes is also problematic. The cholesterol dependence of membrane domains is often tested by perturbing cholesterol, which is known to alter phosphatidylinositol (4,5)-bisphosphate (PIP2) availability, and in turn the organization of the actin cytoskeleton (Kwik et al., 2003), as well as other fundamental cellular processes such as endocytosis (Borroni et al., 2007). Additionally, perturbing cholesterol with cyclodextrins can result in phospholipid redistribution, and these drugs can interact with membrane proteins (Zidovetzki & Levitan, 2007). Additional hypotheses of cholesterol-mediated scaffolding of signaling proteins can also explain cholesterol dependence in lipid raft-*free* models of protein organization (see review in (Kraft, 2013)). Furthermore, cholesterol was found *not* to be enriched in membrane clusters of sphingolipids in fibroblast cells (Frisz, Klitzing, et al., 2013), and even in GUVs made from ternary lipid mixtures, the concentration of cholesterol was only modestly enriched in the Lo phase compared to the Ld phase (Veatch et al., 2006). Thus, despite its involvement in an abundance of biological functions and diseases (Maxfield & Menon, 2006;

Maxfield & Tabas, 2005; Michel & Bakovic, 2007; Pichler & Riezman, 2004; Rojo, Sjoberg, Hernandez, Zambrano, & Maccioni, 2006; Saini, Arneja, & Dhalla, 2004; Song, Kenworthy, & Sanders, 2014; Tabas, 2002), the roles of cholesterol in membrane lateral organization require further clarification. While the importance of cholesterol and other lipids has been heavily investigated in the field of membrane organization models, clearly membrane and cytoskeletal proteins also play a crucial role in cell membrane biology.

Many unanswered questions also remain about other models of membrane organization. Some aspects of these models are expressed as being mutually compatible. For example, a recent revision of the fluid mosaic membrane model envisages that actin fences, lipid rafts, and lipid shells (among other factors) can be congruous and all contribute to the segregation of integral membrane proteins (Nicolson, 2014). However, some aspects of these models are mutually exclusive: actin filaments in the "active composite" model are proposed as being ~250 nm long (Gowrishankar, et al., 2012), which would presumably result in asters with diameters ~500nm, too large to be bounded by actin membrane compartments of ≤300 nm diameter proposed by the "picket fence" model (Kusumi et al., 2012). Moreover, some tenets of these models are directly competing. For example, the "lipid shell" model hypothesizes that GPI-anchored protein clustering results from Lo phase coalescence between lipid shells and lipid domains (Anderson & Jacobson, 2002). Conversely, the "active composite" model hypothesizes that GPI-anchored protein clustering results from these proteins advecting with the movement of actin filaments toward the centers of actin asters (Gowrishankar et al., 2012). These are two very different mechanisms, each proposed to explain the same phenomenon, namely the clustering of GPI-anchored proteins.

Furthermore, the theoretical basis for some models also begs further elucidation. Actin monomers in treadmilling filaments move *from* the barbed *to* the pointed ends of filaments (Kirschner, 1980; Wegner, 1976). The "active composite" model (Gowrishankar et al., 2012) proposes a system of actin asters (see Figure 1(E)), formed by actin treadmilling and myosin contractility, with filament barbed ends pointing toward the centers of asters. If transmembrane proteins bound to F-actin are advected along with treadmilling actin filaments, and these filaments have their barbed ends pointing toward the center of an aster, then the motion of the transmembrane proteins will be toward the periphery of the aster, rather than toward the center (i.e., the treadmilling of actin will push the actin-bound transmembrane protein toward the actin filament pointed end, which is at the aster periphery). Presumably such movement would disperse transmembrane F-actin bound proteins, rather than clustering them.

The alignment of actin filaments through myosin contractility raises its own set of questions. Nonmuscle myosin II assembles into bipolar filaments that engage actin filaments of opposing polarity. These myosin II filaments can translocate oppositely oriented actin filaments, in an antiparallel manner, toward each other (Vicente-Manzanares, Ma, Adelstein, & Horwitz, 2009). In the case of the "active composite" model (see Section 1.1), myosin II is proposed to cross-link parallel actin filaments, and to translocate these actin filaments by walking along a pair of parallel actin filaments (Gowrishankar et al., 2012). Testing these and other hypotheses predicted by these models has been hampered by the inability of techniques used to directly image the structures in unperturbed living cell membranes.

1.1.3.2 Testing hypotheses about very small and very fast processes

Mounting evidence suggests that interactions between cell membranes and the actin cytoskeleton play a crucial role in controlling membrane organization. Studies of protein dynamics are highly informative, but considering the importance of the cytoskeleton, methods which can image multiple species (e.g., cytoskeletal and integral membrane proteins) simultaneously are highly desirable. Single particle tracking (SPT) can fulfill this role when combined with another microscopy method, or when adapted to allow multiple species to be tracked simultaneously (Weigel, Simon, Tamkun, & Krapf, 2011). This approach has already provided valuable information on the apparent restriction of FcεRI receptors by the actin cytoskeleton (Andrews et al., 2008), and this technique is a great tool for gathering precise information on dynamics of very small numbers of molecules. However, SPT tends to be used to quantify motions of individual particles, rather than specifically for imaging. The use of homo-Förster resonance energy transfer (homo-FRET) in describing cell surface distributions of proteins is also a powerful and dynamic method (Gowrishankar et al., 2012; Sharma et al., 2004), yet is limited in length scales to less than roughly twice the Förster radius (i.e., <10 nm) and to diffraction-limited scales (i.e., >200 nm), blocking direct observation of the size and shape of membrane domains.

Thus, methods of imaging membrane organization in living cells on length scales between 10 nm and the diffraction limited resolution of ~ 200 nm are highly desirable. The desire to study the interactions between membrane components or between membrane and cytoskeletal components has further motivated the development of super-resolution imaging methods.

2. METHODS

2.1 Principles of super-resolution microscopy

2.1.1 Diffraction limit

An optical imaging system, such as a microscope, is fundamentally limited in resolution due to the diffraction of light. The resolving power of a microscope is often given by the Rayleigh criterion,

$$r_0 = (0.61\ \lambda)/NA$$

where r_0 is the minimum distance needed between two objects to be independently identifiable, λ is the wavelength of light emitted from a sample, and NA is the numerical aperture of the objective lens of the microscope. For practical purposes, the numerical aperture of a lens is limited to ~1.45–1.65 and using the mean of the visible wavelengths of light ($\lambda = 550$ nm), the value of r_0 is ~200–230 nm. Although small compared to the size of a cell, this limit in resolution is substantial when compared to the sizes of proteins, viruses, and various intracellular processes. In order to resolve features below the diffraction limit, several super-resolution microscopy techniques have been developed which utilize fluorescent proteins and dyes.

2.1.2 Super-resolution imaging techniques

We here provide a brief overview of only some of a wide range of super-resolution techniques. For more in depth descriptions and comparisons of super-resolution methods, see (Coltharp & Xiao, 2012; Gould, Hess, & Bewersdorf, 2012; Leung & Chou, 2011; Rossy, Pageon, Davis, & Gaus, 2013; Schermelleh, Heintzmann, & Leonhardt, 2010).

2.1.2.1 Near-field scanning optical microscopy

Near-field scanning optical microscopy (NSOM), first developed in the 1980s (Lewis, Isaacson, Harootunian, & Muray, 1984; Pohl, Denk, & Lanz, 1984), uses a very small (subwavelength sized) aperture, placed very close to a fluorescently labeled sample to create a spot of illumination smaller than r_0 (Betzig & Trautman, 1992; Betzig, Trautman, Harris, Weiner, & Kostelak, 1991). This spot is then scanned across the sample and fluorescence intensity is recorded as a function of position to produce an image. Since the spot of illumination is very small, the emitted signal correspondingly represents a very small area of the sample and can be resolved much better than the diffraction limit would otherwise allow.

Lateral spatial resolutions of 20 nm and axial resolutions of 2–5 nm have been observed using NSOM (Oshikane et al., 2007). However, in order for the illumination spot to remain smaller than r_0, the aperture must be placed, and remain, much closer to the sample than the wavelength of light used. The resolving power of NSOM is quickly reduced as the aperture is moved away from the sample which primarily restricts NSOM to the imaging of membranes, sectioned samples, or other flat surfaces.

2.1.2.2 Stimulated emission depletion microscopy

Stimulated emission depletion (STED) microscopy (Hell, 2007; Hell & Wichmann, 1994; Klar, Jakobs, Dyba, Egner, & Hell, 2000) takes advantage of a process known as stimulated emission to suppress the usual fluorescence of fluorophores in an illumination region, except for within a very small spot. Stimulated emission refers to the phenomenon whereby a molecule in an excited state can interact with an incident photon and emit a second in-phase photon of the same wavelength, deexciting the molecule in the process. In STED, the fluorescence of molecules is interrupted (outcompeted) by stimulated emission so that the emission of fluorescent molecules not at the precise center of the focal volume can be quenched and the effective focal volume size reduced to a few tens of nanometers, or even smaller, in either two or three dimensions (Hein, Willig, & Hell, 2008; Meyer et al., 2008). STED has been used to image living systems (Westphal et al., 2008) and multiple fluorescent species (Meyer et al., 2008).

Typically, two lasers, often pulsed to achieve high intensity, are used in STED microscopes. An excitation laser, approximately Gaussian (actually an Airy pattern) in profile, serves to excite the fluorescent molecules of a sample. The second, a deexcitation laser, is shaped so that it reaches the sample with a toroidal profile and is timed to arrive within a few picoseconds of the excitation pulse. The intensity of the STED beam is increased to saturate the STED process until only a small spot in the sample is capable of fluorescing normally. This spot is then scanned across the sample and the emission recorded to generate an image with greatly increased resolution (Hell & Wichmann, 1994). STED microscopy has achieved better than 20 nm resolution (Gottfert et al., 2013).

The relatively high laser intensities used for STED may lead to issues with phototoxicity and excessive photobleaching (Eggeling, Widengren,

Rigler, & Seidel, 1998). However, the use of photostable dyes or proteins along with optimization of the STED lasers (intensity, wavelength, and scan time) can mitigate the risk of phototoxic effects (Hotta et al., 2010).

2.1.2.3 Structured illumination microscopy

Structured Illumination Microscopy (SIM) (Gustafsson, 2000) uses a periodic pattern of bands to illuminate a fluorescently labeled sample. This style of illumination generates interference between the fluorescent probes in the sample and the illumination itself to generate moiré fringes. These interference fringes are of lower spatial frequency than the illumination bands and can be more easily resolved. The pattern is rotated and phase shifted to obtain additional information from the sample. From this additional information, and using knowledge of the original illumination profile, a high resolution image can be reconstructed (Gustafsson, 2000). SIM is capable of imaging living samples at high frame rates (Kner, Chhun, Griffis, Winoto, & Gustafsson, 2009).

Conventional SIM is capable of resolutions around 100 nm, but with the introduction of nonlinear components to the system, such as saturation of the fluorophores or use of photoswitchable probes, nonlinear SIM is capable of achieving resolutions <50 nm (Gustafsson, 2005). SIM may be sensitive to rapid changes in structure during imaging and sample-induced aberrations (Gustafsson, 2005). Since SIM requires several rotations and phase shifts of the illumination pattern for optimal resolution, it is best used in conjunction with photostable probes so that photobleaching can be minimized.

2.1.2.4 Localization microscopy: FPALM, PALM, and STORM

Fluorescence photoactivation localization microscopy (FPALM) (Hess, Girirajan, & Mason, 2006), photoactivated localization microscopy (PALM) (Betzig et al., 2006), and stochastic optical reconstruction microscopy (STORM) (Rust, Bates, & Zhuang, 2006) are imaging techniques that take advantage of the properties of photoactivatable or photoswitchable fluorescent probes to improve resolution. These probes, often fluorescent proteins or dyes, are characterized by their ability to exhibit changes to their fluorescent properties, such as transitions from an inactive dark state to an active fluorescent state, when illuminated with a particular wavelength of light (often in the blue or ultraviolet wavelength range). Two lasers are typically used in combination, an activation laser to convert inactive (nonfluorescent) molecules to an activated state and a readout laser to drive the activated molecules to emit light. Since the activation of the fluorescent probes is a stochastic process, with probability proportional to the activation laser intensity, using low activation laser intensity can cause a small subset of the total population of fluorescent molecules in the sample to be activated at any given time. This subset of activated molecules will fluoresce when excited with a readout laser and then photobleach or return to a dark state. With low density of active molecules, there is a high probability that each fluorescent molecule will be resolvable from the others. When this occurs, each molecule can be localized by finding the centroid of the image or by fitting the image with an appropriate function, such as a two-dimensional Gaussian. The process of activation, readout, and photobleaching of the fluorescent probes can be repeated until there are no more fluorescent molecules in the sample or until a desired number of molecules has been recorded (Betzig et al., 2006; Hess et al., 2006; Rust et al., 2006).

FPALM was the first of these techniques used to image living cells (Hess et al., 2007); both PALM and STORM share this capability (Jones, Shim, He, & Zhuang, 2011; Shroff, Galbraith, Galbraith, & Betzig, 2008). Multiple fluorescent species can also be imaged (Bates, Huang, Dempsey, & Zhuang, 2007; Bossi et al., 2008; Curthoys, Mlodzianoski, Kim, & Hess, 2013; Gunewardene et al., 2011; Shroff et al., 2007). Furthermore, these types of localization microscopy can be used to produce images in three dimensions by introducing z-dependent elements to alter the detection point spread function (PSF). This z dependency can be added in a number of ways, such as by using a cylindrical lens to induce astigmatism (Huang, Wang, Bates, & Zhuang, 2008), by simultaneous bi-plane detection (Juette et al., 2008), or by introducing strong axial dependence of the PSF using a spatial light modulator (Jia, Vaughan, & Zhuang, 2014; Pavani et al., 2009) or 4Pi detection geometry (Shtengel et al., 2009).

The resolution of FPALM, PALM, and STORM depends largely on the number of photons detected per molecule, the background noise per pixel, the camera pixel size, and the total number of molecules localized. Using bright fluorescent probes, localization precisions <15 nm are possible in all three spatial dimensions (Jia et al., 2014).

Although these localization methods all rely on the same basic principle to achieve high resolutions, there are some subtle differences between methods. FPALM and PALM initially used fluorescent proteins (Betzig et al., 2006; Hess et al., 2006) while STORM initially used antibodies conjugated to organic dyes to label structures of interest (Rust et al., 2006). The intracellular introduction of antibodies generally requires cell membranes to be permeabilized in some manner, a consideration when

STORM is used for live cell studies. However, organic dyes have the advantage of being relatively bright and photostable, especially in the presence of an oxygen scavenging system. This allows for resampling of the same fluorophore several times before photobleaching. While this increases sampling density, and potentially resolution, this resampling can result in overcounting of the number of localized molecules in the sample. Fluorescent proteins, by comparison, sometimes emit fewer photons before photobleaching. Such characteristics reduce the rates of overcounting, but fewer detected photons per molecule can also limit localization precision and, in the presence of background, the number of molecules identified. However, fluorescent proteins are expressed by the cell, eliminating the need to introduce large antibodies through the cell membrane, and they also do not require use of reducing agents in the imaging buffer. Since these localization methods rely on constructing an image from many localized point sources, labeling density and specific probe photophysics tend to be the most limiting factors.

2.1.2.5 Super-resolution optical fluctuation imaging

Super-resolution optical fluctuation imaging (SOFI) is a software-based postprocessing technique that exploits random fluctuations of fluorescence intensities of fluorophores in order to enhance resolution and reduce background in experimental image data (Dertinger, Colyer, Iyer, Weiss, & Enderlein, 2009). SOFI can be applied to many fluorescence imaging modalities provided that the fluorescent label used has two or more distinguishable states (for example, a bright and dark state), and that each molecule switches stochastically and independently of the others. To achieve enhanced resolution, a movie of a fluorescently labeled sample is acquired, and an n-th order autocumulant is calculated, effectively filtering the acquisition such that only highly correlated fluctuations remain. This results in an image with greatly reduced background and a resolution that is the square root of 2 better than the diffraction limited case. Alternatively, cross-cumulants may be used instead for even larger increases in resolution (Dertinger, Colyer, Vogel, Enderlein, & Weiss, 2010). SOFI can be used in 3D imaging applications (Dertinger, Xu, Naini, Vogel, & Weiss, 2012), with living samples (Geissbuehler et al., 2014), and has been demonstrated with organic dyes (Dertinger, Heilemann, Vogel, Sauer, & Weiss, 2010).

Sophisticated postprocessing is required to provide resolution enhancement and improvement is limited without use of high-order cumulants, which require long acquisition times. However, SOFI can be performed with relatively short acquisition times and moderate laser intensities to obtain modest resolution improvement and has the advantage of being compatible with conventional fluorophores and microscopes (Dertinger, Colyer, et al., 2012).

2.1.2.6 Universal point accumulation for imaging in nanoscale topography

Universal point accumulation for imaging in nanoscale topography (uPAINT), a generalization of PAINT (Sharonov & Hochstrasser, 2006), makes use of continuous labeling while imaging to localize and track single molecules in living samples (Giannone et al., 2010). In uPAINT, highly specific fluorescent ligands are added to the sample via solution just before imaging. The ligands bind to target proteins of interest and, with an appropriately chosen concentration, are sparse enough to be localized. Excited fluorophores will eventually photobleach but are stochastically replenished by ligands from solution. An oblique laser illumination profile is used to excite the fluorophores in order to limit the focal volume, reduce background, and prevent bleaching of out of focus molecules. uPAINT is well suited to tracking single molecules for many time steps, and multiple dye molecules can be bound to a single ligand to further increase tracking duration (up to tens of seconds). uPAINT has been used to examine up to two species simultaneously (Winckler et al., 2013). Typical localization precisions are within the range of 40–50 nm (Giannone et al., 2010). Since the fluorescent labels are added exogenously to the sample, careful optimization of ligand concentration, laser intensity, and imaging rate is essential.

2.1.2.7 Super-resolution: worth the Nobel Prize

The development of these methods has fundamentally changed the capabilities of scientific researchers, and is already leading to crucial insights in a number of fields including biomedical research, and specifically cell membrane biology. Stefan W. Hell, Eric Betzig, and W.E. Moerner were awarded the Nobel Prize in Chemistry in 2014 for their contributions toward the development of super-resolution microscopy.

2.2 Tools for quantification of membrane organization and dynamics

To be able to distinguish and refine the existing models of membrane organization, quantification of the fundamentals of membrane biology are necessary. These fundamentals relate directly to the interactions between membrane components, such as proteins and lipids. While studies of lipid phase separation can provide some insight into the behavior of lipids in the absence of proteins, methods which quantify protein–protein, lipid–lipid, and lipid–protein interactions in the (preferably living) cellular environment are essential to advancing the understanding of membrane biology (Bar-On et al., 2012; Gudheti et al., 2013; Sengupta et al., 2011). Several methods (see Section 2.1) can image the spatial distributions and time-dependence of (labeled) proteins and lipids in intact cell membranes. However, imaging is really just the first step toward understanding the processes which have led to these distributions. Quantification of the distributions of membrane components is often the next step.

2.2.1 Ripley's K and the pair correlation function

Several statistical techniques are commonly used to identify and quantify clustering and codistribution of labeled proteins and lipids. Ripley's K-function and the pair correlation function (PCF) test spatial heterogeneity and clustering of objects (Gudheti et al., 2013; Hess et al., 2005; Philimonenko, Janacek, & Hozak, 2000; Plowman, Muncke, Parton, & Hancock, 2005; Ripley, 1977; Yeomans, 2002). The K-function quantifies the number of objects (e.g., imaged proteins or lipids) of a given type found within a circle of specified radius from each other object of the same type (or a different type for co-distribution measurements) and normalizes the amplitude by the overall density such that an amplitude of zero corresponds to a random distribution (Ripley, 1977). The PCF quantifies the number of probes at a given distance from each other molecule, normalizes by the area of each ring and the average molecular density, and averages over all molecules, such that an amplitude of 1 corresponds to random. Typically, these tests are performed over a range of radii in order to determine whether there is any clustering of the probes and if so, to elucidate properties such as mean cluster radii or mean cluster density (Gudheti et al., 2013; Hess et al., 2005; Philimonenko et al., 2000; Plowman et al., 2005). There are other tests to measure spatial heterogeneity such as the Getis G statistic (Getis & Ord, 1992), which includes different normalizations and weighing parameters dependent on molecular spatial position. These differences result in increased sensitivities to certain cluster parameters (Itano et al., 2014).

The Ripley's cross K-function (cross-K) and pair cross-correlation function (PCCF) quantify the spatial relationship between two differently labeled species in a sample. In principle, the algorithms work similarly to the K-function and PCF, by counting the measured numbers of probes of one species at a distance r from each molecule of the other species, then averaging over all molecules. These tests quantify the degree to which and the range of distances over which two labeled species colocalize (Gudheti et al., 2013; Philimonenko et al., 2000).

2.2.2 Nearest neighbor-based cluster analysis

Clustering can also be defined by nearest neighbor distance. Single-linkage cluster analysis (Sneath, 1957) defines all molecules within a user-defined distance of any other molecule in the set to be within the same cluster. This type of algorithm is useful because it precisely defines the set of molecules within each cluster, and thus the localized coordinates of those molecules can be quantified directly (Bar-On et al., 2012; Gudheti et al., 2013). The nearest neighbor distance distribution can also be used to determine the local density of the clustered probes. This information allows for comparison of cells that have been treated to disrupt or enhance clustering (Hess et al., 2005) or for comparison of nonclustered probes in different regions of the plasma membrane (Stryer, 1978; Zacharias, Violin, Newton, & Tsien, 2002). Additionally, an angular histogram (the angle between a molecule and its two nearest neighbors) reveals any preferential orientation of probes within the cluster. Orientational order can result from hexatic phases or interactions with other cellular components which may induce ordering of the proteins, such as cytoskeletal components (Hess et al., 2005).

2.2.3 Quantifying dynamics

Fluorescence microscopy techniques are capable of quantifying the dynamics of fluorescently labeled species in live cells. Tracking of single molecule trajectories over two or more consecutive frames enables calculation of molecular mobility (Manley et al., 2008). These high-density maps of short molecular trajectories provide both spatial and dynamic information, and in many ways are a high-density limit of SPT, where low densities of labeled molecules are typically used. Likewise, as single-molecule localization methods have advanced, computational methods have developed which enable larger numbers of SPT molecules in close proximity to be concurrently analyzed (Jaqaman et al., 2008). Longer molecular trajectories obtained by localization microscopy or SPT allow distinction between confined diffusion, free diffusion, active transport, and other types of motion.

Particle image correlation spectroscopy is an excellent way to quantify large numbers of (even relatively short) single molecule trajectories to test the type(s) of motion occurring in a system (Semrau & Schmidt, 2007). In addition, histograms of the turn angle (the vertex between the positions of a single molecule through three time points) of moving single molecules can reveal whether molecules have a preferred turning angle. Molecules undergoing confined diffusion (or other situations which lead to reversal of motion) are more likely to have turn angles near 180°, while freely diffusing molecules have no preferred turn angle (Gudheti et al., 2013; Hess et al., 2005).

3. BIOLOGICAL APPLICATIONS

3.1 Membrane–actin interactions

The actin cytoskeleton is now widely recognized as being key in regulating the spatial distribution of some membrane proteins (Chichili & Rodgers, 2009; Plowman et al., 2005) and membrane lipids (Frisz, Lou, et al., 2013; Head, Patel, & Insel, 2014; Liu & Fletcher, 2006). In return, membrane proteins and lipids can locally remodel (or create new) actin filaments (see review by (Saarikangas, Zhao, & Lappalainen, 2010)). While the precise mechanisms are still under debate, instrumental in many actin-membrane interactions are actin binding proteins (ABPs), which can variously connect actin filaments with membrane proteins or the lipid membrane itself, and modulate the local nucleation, branching, cross-linking, and dissociation of actin filaments. Due to constraints of space we here discuss a small subset of these ABPs.

3.1.1 Molecular participants

3.1.1.1 ABPs: the membrane—actin connection

Long since theorized as contributing to lateral heterogeneity in plasma membranes (Edwards & Crumpton, 1991), the annexins can bind negatively charged phospholipids (Edwards & Crumpton, 1991; Jackle et al., 1994), many membrane and signaling proteins (Cornely, Rentero, Enrich, Grewal, & Gaus, 2011), and through an F-actin binding domain (Hosoya, Kobayashi, Tsukita, & Matsumura, 1992) may act as a nexus between the actin cytoskeleton and the plasma membrane (Hayes, Shao, Grieve, Levine, Bailly, & Moss, 2009). Proteins of the ezrin/radixin/moesin (ERM) family (also including ABPs with homology in their 4.1 ezrin/radixin/moesin (FERM) domains, such as talin (see review in (Critchley, 2005)) can bind F-actin, phospholipids, and a range of transmembrane proteins (including those believed to corral membrane proteins into domains such as

CD317/tetherin (Rollason, Korolchuk, Hamilton, Jepson, & Banting, 2009)), and membrane-associated cytosolic proteins (see review in (Fehon, McClatchey, & Bretscher, 2010)). Some ERM proteins can also signal to other ABPs from the plasma membrane, e.g., ezrin to myosin (see review in (Manes & Viola, 2006)) to effect structural change.

3.1.1.2 Ezrin tethers, filamin traps

Viola and Gupta (2007) theorized that resting immune cell membranes contain small membrane protein clusters linked to cortical actin by ezrin. Antigen stimulation disconnects ezrin from these clusters and from actin (Gupta et al., 2006), and the theory holds that small protein clusters uncoupled from the actin cytoskeleton coalesce into large membrane domains which are then "tethered and trapped," via rebinding to the actin cytoskeleton through the now-activated ABP filamin (Viola & Gupta, 2007). Indeed, the filamin isoform FLNa is required for actin-mediated stabilization of membrane domains at the immunological synapse (Tavano et al., 2006), and for the internalization of caveolae (Muriel et al., 2011; Sverdlov, Shinin, Place, Castellon, & Minshall, 2009).

3.1.1.3 Spectrin, α-actinin, myosins (and a supervillin)

Also implicated in actin rearrangement at caveolae is the actin cross-linker α-actinin (Singleton, Dudek, Chiang, & Garcia, 2005), which is commonly isolated in detergent-resistant membranes (see review in (Chichili & Rodgers, 2009)). Among the first ABPs recognized in regulating membrane organization, spectrin family members (including α-actinin, and nonerythroid spectrin or fodrin) are able to bind F-actin and phospholipids (Davis & Bennett, 1994; Hartwig & DeSisto, 1991) and can signal to reorganize actin through various pathways (Machnicka, Czogalla, Hryniewicz-Jankowska, Boguslawska, Grochowalska, Heger, & Sikorski, 2014).

The membrane proximal actin cortex is commonly understood as being populated with myosin motors (Salbreux, Charras, & Paluch, 2012), and different myosin isoforms have been identified in DRMs (Nebl et al., 2002). Myosin II contractility may be needed for the coalescence of lipid domains along F-actin bridges (Jordan & Rodgers, 2003; Rodgers, Farris, & Mishra, 2005) and Myo1c is needed for the delivery and recycling of membrane domain-associated lipids (Brandstaetter, Kendrick-Jones, & Buss, 2012). Further connection between myosins and membrane domains may occur through binding of membrane-associated ABPs such as supervillin (Chen et al., 2003).

3.1.1.4 Arp2/3: pushing membranes around

One class of ABPs alters actin organization by initiating the formation of new actin filaments. These "nucleators" include actin-related protein 2/3 (Arp(2/3)) protein complexes, which create branched actin filament networks, and can be ultimately regulated by cooperative mechanisms including interactions with phospholipids (Derivery & Gautreau, 2010; Lebensohn & Kirschner, 2009). The proximity of these branched networks to cell membranes is central to their involvement in contexts of membrane movement—both at the whole cell level (underpinning the lamellipodial morphology (Stradal et al., 2004)) and at the level of the vesicle (aiding in endocytosis (Kaksonen, Toret, & Drubin, 2006)).

3.1.1.5 PIP2: the powerhouse

Actin filaments can be bound to membrane proteins and lipids through ABP conduits, but they can also influence the diffusion and distribution of membrane molecules through steric interactions (reviewed in (Alenghat & Golan, 2013)). Moreover, the nexus between membrane proteins and ABPs need not be direct for each to alter the spatial localization of the other. Beyond the scope of this chapter is a plethora of signaling pathways which regulate the activation of ABPs and their sorting to membrane domains, such as the Rho family GTPases (e.g., see reviews by (Bisi et al., 2013; de Curtis & Meldolesi, 2012)) and the family of Bin–amphipysin–Rvs167 (BAR) domain containing proteins (e.g., see reviews by (Aspenstrom, 2014; Frost, Unger, & De Camilli, 2009)).

We instead briefly discuss some regulatory influences of the anionic membrane phospholipid, phosphatidylinositol 4,5-bisphosphate (PIP2). By binding to PIP2, ABPs can be targeted to membranes (e.g., spectrin; see review in (Boguslawska, Machnicka, Hryniewicz-Jankowska, & Czogalla, 2014), ezrin/radixin/moesin proteins and talin (Barret, Roy, Montcourrier, Mangeat, & Niggli, 2000; Hao et al., 2009; Jayasundar et al., 2012; Saltel et al., 2009) and annexins (Hayes et al., 2009; Martin-Belmonte & Mostov, 2007)). PIP2 binding is also considered a necessary step in the activation of many ABPs (Wu et al., 2014), such as ezrin/radixin/moesin proteins (Yonemura, Matsui, & Tsukita, 2002), and Arp2/3 (via WASP) (Higgs & Pollard, 2000). However, PIP2 binding may also inhibit the cross-linking of actin by filamin (Furuhashi, Inagaki, Hatano, Fukami, & Takenawa, 1992), the bundling of actin by α-actinin (Fraley et al., 2003) and the severing of actin by cofilin (van Rheenen et al., 2007). Additionally, PIP2 may be in return regulated by ABPs (e.g., talin (Di Paolo et al., 2002)).

PIP2 also regulates signaling cascades upstream of ABPs and actin remodeling, and helps coordinate actin organization in a variety of contexts (see reviews by (Rocha-Perugini, Gordon-Alonso, & Sanchez-Madrid, 2014; Saarikangas et al., 2010; Sun, Thapa, Hedman, & Anderson, 2013)). See Section 3.4.2 for further discussion of PIP2 imaging.

We have, here, only described a small fraction of the myriad ways in which actin is bound to, and can be regulated by, membrane components. In all likelihood, there are a gross number of complicated yet precise methods by which cells can regulate how membrane proteins and lipids move and are distributed. However, some mechanisms appear crucial across multiple systems. By understanding the spatial and functional relationships between ABPs, actin, and the membrane, we can better understand exactly how cell systems are able to coordinate a staggering array of varied and distinct molecular processes at the membrane. Before assessing the future opportunities in this field, we first summarize the techniques which have led us to this point.

3.1.2 Early findings

Aside from the identification of ABPs in fractions of lysed cells (such as DRMs) to indicate their involvement in membrane domains (Jordan & Rodgers, 2003; Nebl et al., 2002; Singleton et al., 2005), classic techniques including coimmunoprecipitation and yeast two hybrid screens are used to gauge the interactions between ABPs and membrane proteins (e.g., (Sverdlov et al., 2009; Tavano et al., 2006)). Biochemical techniques such as these are useful for exploring binding relationships, but do not describe distributions of proteins (or lipids).

Diffraction-limited microscopy has been used to image membrane domains such as the immunological synapse (Tavano et al., 2006) and invadopodia (Antelmi et al., 2013), and specific inhibition of ABPs followed by widefield imaging has elucidated the importance of, for example, myosin motors in the movement of large (micrometers in length) domains into larger immunological synapses (Jordan & Rodgers, 2003).

3.1.2.1 FRET, FRAP, and FCS

Techniques such as FRAP have been extensively used to investigate the dynamics of the actin cortex (e.g., (Sund & Axelrod, 2000)), ABPs (e.g., ezrin (Coscoy et al., 2002)) or proteins moving through the plasma membrane (e.g., (Golan & Veatch, 1980)). Kenworthy et al. used FRAP very effectively to distinguish between membrane models with and without cytoskeletal interactions (Kenworthy et al., 2004), and discovered that the slow diffusion of the cholera toxin subunit B (CTxB) is due to confinement by the actin cytoskeleton (Day & Kenworthy, 2012). FRAP and SPT have been used together to identify anomalous diffusion of clustered membrane receptors, suggesting interactions with the cytoskeleton (Feder, Brust-Mascher, Slattery, Baird, & Webb, 1996). FRET has been used to investigate the coclustering of putative raft and nonraft markers (Chichili & Rodgers, 2007) and Goswami et al. were able to quantify an association between GPI-anchored protein clustering and cortical actin remodeling using homo-FRET (Goswami et al., 2008). However, neither FRET nor FRAP visualize (subdiffraction sized) cluster geometry directly, so results must be interpreted in terms of some kind of model, and FRET microscopy does not directly access clustering on length scales between ~10 and 250 nm.

Fluorescence correlation spectroscopy (FCS) (Magde, Elson, & Webb, 1972) is a powerful method able to provide information on microsecond and millisecond molecular dynamics within a (typically) diffraction-limited observation volume of ~0.3−1.0 μm in size (Hess & Webb, 2002). FCS uses the fluctuations in a fluorescence signal to measure molecular concentration, diffusion coefficient or transport rate, and molecular transitions between bright and dark states (Hess, Huang, Heikal, & Webb, 2002). FCS of lipid probes in cell membranes demonstrated anomalous diffusion (Schwille, Korlach, & Webb, 1999) consistent with cytoskeletal interactions or nanodomains. FCS measurements of diffusion as a function of observation volume size (Wawrezinieck, Rigneault, Marguet, & Lenne, 2005) have been very effectively used to quantify GPI-anchored protein clustering, and results also suggest that cytoskeletal interactions mediate membrane lateral organization (Lenne et al., 2006). Lasserre et al. used similar FCS measurements to quantify protein nanodomains related to activation of signaling pathways related to Akt and phosphoinositide-3 kinase, which can regulate the actin cytoskeleton (Lasserre et al., 2008). A super-resolution version of FCS which uses STED to confine the observation volume (Eggeling et al., 2009) has demonstrated that sphingolipids and GPI-anchored proteins are transiently trapped in small cholesterol-dependent complexes of <20 nm in size, while phosphoglycerolipids are not.

Conventional versions of FCS, FRAP, and FRET are, however, limited in their ability to resolve lateral heterogeneity on length scales <200 nm: FCS does not form an image at all, FRET cannot access length scales between ~10 and 200 nm, and conventional FRAP does not image structures below the diffraction limit.

Widefield TIRF microscopy has been used to study the colocalization of various proteins with markers of caveolae (e.g., tagged filamin and caveolin-1 (Muriel et al., 2011)), and ABPs have been tracked with respect to endosomal markers and other ABPs (e.g., tagged supervillin and tagged myosin (Fang et al., 2010)). While useful in their indication of gross colocalization, direct visualization of the spatial and dynamic distributions at the nanoscale would aid in understanding how distributions of membrane-associated ABPs are spatially and functionally related to the dynamics and clustering of membrane proteins and lipids.

3.1.3 Findings using super-resolution methods

3.1.3.1 Perturbing ABP expression and watching what happens

Many groups use super-resolution microscopy to report the spatial distributions of membrane proteins and how these change with actin disruption or cholesterol depletion. Recently, membrane protein distributions have also been imaged as a function of ABP disruption, indicating the reorganization of actin in NK cells lacking the actin depolymerizing protein coronin (Mace & Orange, 2014), and alterations in the distributions of putative "raft" and "nonraft" markers in a cell-dependent manner with varying annexin (anxA6) expression levels (Alvarez-Guaita et al., 2014). Moreover, ABPs can affect motility of membrane proteins; STORM imaging has shown that the mobility of single BCRs and entire assemblies is decreased in cells lacking ezrin (Pore et al., 2013).

Combined super-resolution imaging of the ABPs themselves together with conventional transmitted light (or differential interference contrast) imaging can show the nanoscale localization of these proteins relative to diffraction limited images of membranes and other cellular structures. For example, this has been explored for spectrin (Blunk et al., 2014; Zhong et al., 2014) and α-actinin (Hou et al., 2014), just as single-species imaging of actin can inform as to nanoscale actin distributions with respect to a larger structure or a whole cell (Izeddin et al., 2011; Xu, Babcock, & Zhuang, 2012).

3.1.3.2 Super-resolution multicolor imaging: picturing combinations of molecules

With the advent of multicolor localization microscopy, studies directly and simultaneously imaging nanoscale actin distributions and membrane proteins, or membrane proteins and membrane-associated ABP distributions, in fixed cells, are starting to emerge. As examples, actin and lytic granules (see also Section 3.4) have been imaged together with dual-color STED in cells lacking an isoform of the ABP coronin, fuelling a novel model of granule secretion (Mace & Orange, 2014). Dual color FPALM has shown that clusters of influenza hemagglutinin (HA) can either be colocalized with, or completely exclude, the ABP cofilin (Gudheti et al., 2013). Sequential dual color PALM/dSTORM imaging has revealed a tight spatial association between the post synaptic density and Arp2/3, which is functionally reliant on the Rho GTPase Rac (Chazeau et al., 2014). Dual-color PALM has shown that actin and GPI-anchored proteins (which have extensively been used to demonstrate aspects of various models of membrane organization (see Section 1.1)), cluster on the nanoscale *without* appreciable colocalization, until antibody cross-linking stimulates formation of GPI-anchored protein clusters many microns in size which *do* associate with actin (Sengupta et al., 2011). Dual-color STED imaging of PC12 membranes has shown that mesoscale membrane clusters contain many species of protein which are further spatially organized *within* a single cluster, and these clusters exclude, and are strikingly bordered by, actin and spectrin (Saka et al., 2014).

3.1.3.3 Super-resolution multicolor live cell imaging: watching the dance

Live cell multispecies imaging is also now allowing extraordinary views of the dynamic interplay between membrane components and the actin cytoskeleton. As examples, FPALM has already elucidated that clusters of the influenza protein HA are associated spatially and dynamically with clusters of actin (i.e., not contained in an actin poor compartment and bounded by a fence, but rather shuffling on top of a thick actin hedge) (Gudheti et al., 2013). Exciting avenues for testing between models of membrane organization are now being realized with the ability of super-resolution microscopy to visualize the dynamics of three protein species simultaneously, in living cells, at the nanoscale. Three-color live cell FPALM experiments have already shown that HA and transferrin receptors (previously considered putative "raft" and "nonraft" markers, respectively) each form spatially segregated nano- and microscale clusters, yet both species spatially and dynamically correlate with actin clustering (Gunewardene et al., 2011).

3.1.4 Proposed model: Cluster feedback

3.1.4.1 The cluster feedback model: two-way communication between actin and the membrane

The insights into membrane organization afforded by super-resolution microscopy, along with observations from decades of membrane research provide the basis for our cluster feedback model (see Section 1.1), whereby the relationship between (1) membrane proteins and lipids (including PIP2), and (2) actin and ABPs is bidirectional. The differential regulation of ABPs by membrane-associated molecules can induce the local formation and reorganization of actin filaments. These actin filaments can then work to locally alter the distribution and diffusion of membrane molecules, creating clusters of proteins and lipids in membrane domains. This model builds on ideas published in a range of membrane protein clustering contexts (Chichili & Rodgers, 2009; Gowrishankar et al., 2012; Gudheti et al., 2013; Jaumouille et al., 2014; Viola & Gupta, 2007; van Zanten et al., 2009). Experimental evidence for this model has come from observations that (1) increasing actin stability can increase coclustering of membrane proteins (e.g., (Chichili & Rodgers, 2007; Gudheti et al., 2013)), (2) ABPs are differentially localized with respect to clusters of membrane proteins (Chazeau et al., 2014; Gudheti et al., 2013; Saka et al., 2014), and (3) altering ABP expression can be sufficient to modulate membrane protein mobility (Pore et al., 2013) and clustering (e.g., (Alvarez-Guaita et al., 2014; Chazeau et al., 2014). Feedback has been demonstrated by, for example, work showing that Fcγ receptor activation can influence actin polymerization, which in turn affects Fcγ receptor mobility (Jaumouille et al., 2014), and in the T-cell activated remodeling of F-actin, which in turn mediates the organization of plasma-membrane bound signaling proteins (see review in (Kumari, Curado, Mayya, & Dustin, 2014)). Upstream regulators of ABPs have also been measured with respect to membrane protein organization, such as the elegant dual color NSOM study demonstrating that ligand binding of integrins induces clustering of, and colocalization with, GPI-anchored proteins (van Zanten et al., 2009). We hope to see more studies using super-resolution microscopy to directly image actin, membrane proteins, and the ABPs themselves, so that we may further test hypotheses of membrane protein—actin/ABP feedback predicted by the cluster feedback model.

3.1.4.2 Outlook

While various studies highlight the importance of ABPs in regulating protein distributions at the cell membrane, many ABPs can elicit change through multiple pathways. Similarly, actin can contribute to many processes which alter spatial organization of membrane proteins (e.g., endo- and exocytosis). Single species imaging of membrane proteins with or without ABPS (or actin disrupting drugs) *can* test for effects on membrane protein distribution, but cannot easily determine *how*. Better understanding of mechanisms can come from simultaneous imaging of the membrane protein and the ABP (and/or actin) being investigated. Research into other membrane platforms, such as focal adhesions, has been aided by super-resolution microscopy elucidating (for example) movement of single talin molecules with respect to β-integrins (Rossier et al., 2012) and the organization in three dimensions of multiple ABPs with respect to the membrane (Kanchanawong et al., 2010).

Despite the benefits that super-resolution microscopy has afforded membrane research, many competing models of how ABPs and actin influence distributions of membrane proteins and lipids are yet to be tested. The development of photoactivatable lipid tags (e.g. Mizuno et al., 2011) has already allowed exciting super-resolution imaging of membrane lipids (Abe et al., 2012; Honigmann et al., 2013), yet many opportunities for multichannel lipid/actin/membrane imaging remain relatively unexplored. Studies which visualize and analyze the three-dimensional organization of the membrane and actin cytoskeleton will continue to be extremely fruitful. We are now in a golden age of microscopy, where dreams of visualizing the precise dynamic steps of cellular function, including the previously enigmatic process of organizing membrane domains and the associated actin cytoskeleton, are now becoming reality. Studies combining super-resolution techniques will allow us to finally see, in three dimensions and real time, how many different players dance together to organize cellular membranes.

3.2 Virus infection

Viruses are responsible worldwide for significant illness across many species, and are able to induce changes in the organization of the plasma membrane to facilitate infection, replication, budding, release, and evasion of the host immune system (Manes, del Real, & Martinez, 2003). Exceptionally high resolution views of virus infection have been obtained with EM. For example, images of whole influenza virus (Ruigrok, Krijgsman, de Ronde-Verloop, & de Jong, 1985) and in particular the influenza fusion protein hemagglutinin (HA) (Hess et al., 2005) have been obtained and quantified by EM. While not yet rivaling the resolution obtained by EM, super-resolution microscopy is far better suited for imaging living, dynamic systems. Coupled with advances in temporal resolution (Huang et al., 2013; Nelson, Gunewardene, & Hess, 2014), super-resolution imaging gives investigators the tools capable of answering more questions about pathogen interactions with host cell membranes. Because this is a highly researched field, were here limit discussion to a small subset of studies only.

3.2.1 Super-resolution microscopy: viruses meet their match

Even though direct optical imaging of membrane domains with diffraction limited techniques is not always possible, many properties of viral assembly and function can be uncovered through indirect imaging methods. FRET microscopy has reported the association between viral proteins and putative lipid raft markers (Engel et al., 2010), and FRET has been quite useful in quantifying clustering of many membrane proteins on length scales <10 nm (Karpova et al., 2003; Kenworthy, 2001, 2005; Kenworthy & Edidin, 1998; Kenworthy, Petranova, & Edidin, 2000). FRAP (Elson, Schlessinger, Koppel, Axelrod, & Webb, 1976) has yielded insight into mechanisms of diffusion of viral membrane proteins at the cell membrane (Kenworthy et al., 2004), and nuclear magnetic resonance (NMR) has revealed how lipid phase changes as a function of temperature help protect viral stability (Polozov, Bezrukov, Gawrisch, & Zimmerberg, 2008). However, FRET is insensitive to length scales from ~10 to 200 nm, NMR does not provide an image, and FRAP does not give direct information about spatial organization below the diffraction limit. Using super-resolution imaging, investigators have been able to directly quantify shapes, sizes, and densities of membrane protein distributions, as well as the degree of spatial overlap between different species of proteins at the nanoscale (Gould et al., 2008; Gudheti et al., 2013; Gunewardene et al., 2011; Hess et al., 2007; Shroff et al., 2007). These capabilities have allowed researchers to answer previously inaccessible questions about viral assembly, viral protein trafficking, and viral interactions with host cell components, which help build our understanding of the infection process, and have the potential to reveal new anti-viral drug targets.

3.2.2 Influenza virus hemagglutinin: the versatile membrane protein hijacking your cells

The influenza virus is responsible for tens of thousands of deaths annually. Influenza can use host cell proteins to aid in infection, and mass spectrometry has shown that a number of host cell proteins are also preferentially incorporated into influenza virus released from infected cells (Shaw, Stone, Colangelo, Gulcicek, & Palese, 2008), leading to the question of how these associations occur. Some answers may be found with the influenza membrane protein hemagglutinin (HA), which is crucial in many steps of viral infection. HA binds sialic-acid containing cell surface receptors (Skehel & Wiley, 2000); HA catalyzes membrane fusion necessary for viral entry (Biswas, Yin, Blank, & Zimmerberg, 2008; Chernomordik, Frolov, Leikina,

Bronk, & Zimmerberg, 1998; Skehel & Wiley, 2000, 2002; Wiley & Skehel, 1987); and clustering of HA in the viral membrane is crucial for fusion to be accomplished (Bentz, 2000; Chernomordik et al., 1998; Chernomordik, Leikina, Frolov, Bronk, & Zimmerberg, 1997; Kozlov & Chernomordik, 1998; Kumar, Kenworthy, Roth, & Zimmerberg, 2001; Takeda, Leser, Russell, & Lamb, 2003). HA assembles with other viral components before budding (Nayak, Hui, & Barman, 2004), and in the late 1990s, biochemical experiments were able to show that the influenza virus buds from areas of the host membrane where viral components including HA and certain cell lipids are concentrated (Scheiffele, Rietveld, Wilk, & Simons, 1999; Scheiffele, Roth, & Simons, 1997). HA dynamics have been investigated with SPT, which helped elucidate the HA-dependent mechanism by which viral RNA traverses the nuclear envelope (Babcock, Chen, & Zhuang, 2004; Lakadamyali, Rust, Babcock, & Zhuang, 2003), and FRAP experiments measured the diffusion coefficient of HA $\sim 0.1\ \mu m^2/s$, and suggested an immobile fraction ($\sim 25\%$) of HA (Kenworthy et al., 2004). While these experiments have greatly helped shape our understanding of influenza infection, they have not fully clarified the nanoscale organization of viral and host cell components during infection.

Recently, super-resolution experiments have made substantial advances in the understanding of several aspects of the influenza virus life cycle. Super-resolution microscopy has also been used to determine the spatial distribution of the host cell protein CD81 (tetraspanin), which is recruited to assembling influenza viruses, and is concentrated at the growing tip and budding neck of progeny viruses (He et al., 2013). CD81 can control the progression of membrane protein distributions in, for example, immunological synapse formation (Rocha-Perugini et al., 2013), and form complexes with a number of signaling proteins and other master regulators such as the integrins (Berditchevski, 2001). Tetraspanin redistribution by influenza may be one method by which the virus is able to reorganize other host cell membrane proteins on the surface of budding virions.

3.2.3 *Influenza hemagglutinin and host cell actin: an unhealthy relationship?*

One other host cell protein which is greatly exploited by influenza, and viruses in general, is actin (Radtke, Dohner, & Sodeik, 2006). While it was shown that HA clusters (which are necessary precursors of viral budding (Scheiffele et al., 1997)) can persist over timescales of at least tens of seconds (Hess et al., 2007), high-speed FPALM showed that fluctuations in area, perimeter, and shape of these clusters can occur on timescales as short as 0.1 s (Nelson et al., 2014) leading to the question of how clusters are able to persist. Multicolor FPALM imaging in live cells has shown that HA mobility decreases with increased cortical actin density. Along with the discovery of two distinct populations of HA, one with low (non-zero) mobility and confined motion on 100–200 nm length scales (Gudheti et al., 2013), these findings suggest that local actin is influencing the dynamics of this membrane protein. Colocalization of HA clusters with actin clusters and the increase in HA cluster size upon treatment with actin-stabilizing jasplakinolide treatment do not seem consistent with a picket-fence description (Gudheti et al., 2013). Moreover, the intriguing nanoscale differential anti- and colocalization of HA and ABPs, including cofilin (which is strikingly excluded from some HA clusters, yet strongly colocalizes with others), suggests the relationship is more involved than HA molecules simply being confined between actin fences (Gudheti et al., 2013). Rather, these insights made possible by super-resolution microscopy suggest a dynamic "cluster feedback" between membrane HA and the underlying actin cytoskeleton (see Section 1.1 and Figure 2). Understanding these HA organizing mechanisms could be vital in identifying novel antiviral drug treatments, and understanding the HA/actin/ABP interplay may illuminate cellular processes which are used to organize the distributions of many other membrane proteins.

3.2.4 Role of Gag in HIV life cycle

The U.S. Centers for Disease Control (CDC) estimates that the Human Immunodeficiency Virus (HIV) infects tens of thousands of people in the United States each year. Understanding of the molecular dynamics involved in HIV infection and replication is critical for developing future medical treatments for infected individuals, as well as for preventing infection. Formation of the HIV immature capsid (and in turn budding, release, and maturation of the virus) depends on formation of a polyprotein assembly of the HIV protein group-specific antigen (Gag) (Lingappa, Reed, Tanaka, Chutiraka, & Robinson, 2014). Live cell sptPALM was used to compare the dynamic behaviors of Gag and the vesicular stomatitis virus (VSV) G protein (Manley et al., 2008). This allowed researchers to build dense "trajectory maps" of many different proteins to help understand the dynamic behavior of individual proteins, and entire populations and assemblies, in the plasma membrane. While the distributions and mobility of Gag and VSV-G proteins differed greatly, they were both found to be consistent with results obtained through diffraction-limited techniques (Jouvenet et al., 2006;

Kenworthy et al., 2004). Using STORM, researchers revealed that Gag recruits and corrals the HIV viral envelope protein (Env) into large immobile clusters on the plasma membrane, in a process dependent on the Env cytoplasmic tail (Roy, Chan, Lambele, & Thali, 2013). STED has also elucidated this relationship, revealing that clustering of Env on viral envelopes changed as a function of viral maturity, and that this clustering required the Gag-interacting Env tail (Chojnacki et al., 2012). Using sptPALM and PALM images, micro-RNA overexpression was shown to reduce Gag mobility, and also reduce Gag cluster size and density (Chen et al., 2014), which could in turn affect the clustering of HIV-1 Env and overall infectivity of the virus.

Multicolor super-resolution studies indicate that Gag colocalizes with a variety of host cell transmembrane proteins by interacting with basic motifs within their cytoplasmic tails (Grover, Veatch, & Ono, 2015). Correlative iPALM/EM images and multicolor 3D super-resolution imaging have beautifully shown host cell endosomal sorting complexes required for transport (ESCRT) machinery bundled up inside the Gag lattice of budding virus particles (Van Engelenburg et al., 2014), indicating that after hijacking ESCRT to help bud the host cell membrane, HIV virions swallow ESCRT whole.

3.2.5 Outlook

Ground-breaking progress in our understanding of virus infection has already emerged from the use of super-resolution microscopy. In particular, the manipulation of host membrane protein organization by viruses has been shown, at the nanoscale, to be imperative for some steps in viral infection. These microscopy methods could also be employed to help understand currently unknown mechanisms in past and rising health threats such as pox viruses (e.g., smallpox and vaccinia), coronaviruses (e.g., SARS), and filoviruses (e.g., Ebola and Marburg viruses). There are at least six membrane proteins known to be associated with the currently unclear process of virion formation in the vaccinia virus (Liu, Cooper, Howley, & Hayball, 2014). In filoviruses, the proteins responsible for viral assembly and virion production have been identified, but the mechanisms of interaction with the host cell remain unclear (Stahelin, 2014). Understanding the interplay between host and viral membrane protein organization at the nanoscale with super-resolution microscopy will undoubtedly continue to rapidly improve our understanding of virus infection, and so aid in the development of targeted and efficacious antiviral therapies. Super-resolution microscopy has also been used to study the defenses mounted by cells upon viral infection (see Section 3.4.1 for more detail on immune cells). FPALM imaging of zebrafish cells showed that snakehead rhabdovirus (SHRV) infection resulted in downregulated caveolin expression, which in turn dispersed the clustering of a zebrafish type I interferon receptor (IFN-R) homolog, the clustering of which was crucial for the innate immune response (Gabor et al., 2013). Exciting recent developments now show that FPALM can be used in vivo to image membrane structures within living zebrafish (Gabor, Kim, Kim, & Hess, 2015), suggesting that changes in membrane organization during viral infection, and many other possibilities, can now be investigated in vivo.

Many super-resolution viral studies have focused on assembly and related processes near the membrane. While these experiments provide invaluable information about the viral life cycle, they explore only a part of the full story. More research into the organization and dynamics of the virion envelope during binding, entry, and uncoating, could provide additional insights and help identify new antiviral drug targets. As many entire virions can be smaller than the diffraction limit, questions relating to the organization of host cell and viral proteins within the virion itself require the ability to image nanoscale structure. Super-resolution can see at the nano- and virus-scale, and we can now resolve host cell proteins (and lipids) in living cells as they are commandeered by the virus. These capabilities are well suited for understanding infection, so we can better develop methods to combat it. We look forward to seeing in real time at the nanoscale exactly how viruses use, abuse, and steal our membranes and associated proteins for their own infective purposes.

3.3 Applications to mitochondria

Mitochondrial dysfunction has been causally linked to a number of degenerative diseases including Huntington's disease, Alzheimer's disease, Parkinson's disease, and Amyotrophic Lateral Sclerosis (ALS), among others (Burte, Carelli, Chinnery, & Yu-Wai-Man; Palomo & Manfredi, 2014). Though dysfunction appears in a number of ways, it is often associated with changes in mitochondrial morphology and may be accompanied by altered regulation of mitochondrial fission and fusion proteins (Johri & Beal, 2012).

3.3.1 Mitochondrial fission: membrane contacts and protein helices

Mitochondrial fission in mammals is supported by the mitochondrial outer membrane (MOM) protein mitochondrial fission factor (Mff), which localizes to sites of fission (Otera et al., 2010). Mff is the MOM receptor essential for recruiting dynamin-related protein 1 (Drp1, or the yeast homolog

Dnm1) to sites of fission (Otera & Mihara, 2011), where it has long been believed that self-assembling Drp1 oligomers form constrictive helices which sever mitochondria (Labrousse, Zappaterra, Rube, & van der Bliek, 1999).

Conventional microscopy images of dividing mitochondria indicated that these Drp1 sites were smaller than the diffraction limit (e.g., (Labrousse et al., 1999)), while EM studies showed Drp1 (or Dnm1) formed tubular membrane clusters (Yoon, Pitts, & McNiven, 2001), and also helices ~100–130 nm in diameter at sites of extreme mitochondrial constriction and division (Ingerman et al., 2005). Recent multicolor PALM imaging has illustrated that Drp1 helix size reduces significantly as mitochondria are constricted and fission progresses (Rosenbloom et al., 2014). These helices represent functional membrane-associated Drp1 clusters. However, fission is also spatially correlated with sites of interaction between mitochondria and endoplasmic reticulum (ER).

EM and conventional fluorescence microscopy have shown that mitochondria maintain numerous contacts with the ER which can mediate mitochondrial constriction prior to Drp1 recruitment and mitochondrial division (Friedman et al., 2011). Super-resolution imaging (using a modified SIM geometry) in living astrocytes visualized these ER-mitochondrial contacts, which comprised fine tubules of ER (~100 nm in width) wound around mitochondria (Brunstein, Wicker, Herault, Heintzmann, & Oheim, 2013), in agreement with multicolor STORM imaging in living kidney epithelial cells which showed ER tubules (average widths 80–130 nm) were often found at sites of mitochondrial division (Shim et al., 2012).

3.3.2 Mitochondrial membranes within membranes: organization on many levels

Mitochondria themselves contain multiple membrane structures. Mitochondrial DNA is packaged into nucleoids, and the relative alignment of nucleoids to cristae (convolutions of the inner mitochondrial membrane (IMM)) is thought to have functional impacts, since mtDNA encodes for oxidative phosphorylation system proteins which require coordinated membrane insertion (van den Heuvel & Smeitink, 2001). STED imaging indicated nucleoids were of a uniform size and shape and were themselves clustered (Kukat et al., 2011); however, PALM imaging indicated a great variability in nucleoid sizes and shapes, and suggested physical interactions between nucleoids and the IMM (Brown, Tkachuk, et al., 2011). Subsequent correlation of EM and iPALM images has illustrated the three-dimensional interplay of nucleoids and the IMM: nucleoids can intertwine with cristae, or seemingly directly contact cristae tips or sides (Kopek, Shtengel, Xu, Clayton, & Hess, 2012).

The maintenance of the highly convoluted and distinctive IMM requires protein subunits of the mitochondrial inner membrane organizing system (MINOS) (Zerbes et al., 2012). Very little was known about the distribution of these proteins within mammalian mitochondria, although EM indicated their enrichment at cristae junctions of budding yeast mitochondria (Harner et al., 2011; Rabl et al., 2009).

Multicolor STED imaging of mitochondrial membranes and MINOS units in human cell lines revealed a highly ordered distribution of the subunit mitofilin, which consistently formed clusters at cristae junctions. Mitofilin was highly colocalized with MINOS subunits MINOS1 and CHCHD3, in a distinct, regular ("rail-like") pattern of clusters within or adjacent to the IMM at the rim of mitochondria, indicating a coordinated targeting and organization of MINOS with respect to the mitochondrial membrane (Jans et al., 2013).

This super-resolution work has illustrated the compartmentalization of mitochondrial membranes. FPALM tracking of individual mitochondrial membrane proteins (Tom20 and ATP synthase) in living cells further illustrated their altered mobility within (and transitions between) microcompartments, and these domains were present on both the IMM and MOM (Appelhans et al., 2012).

3.3.3 Outlook

Super-resolution imaging of mitochondria has already allowed the quantification of membrane—membrane interfaces in two and three dimensions, and the distributions of membrane bound and associated proteins. This heterogeneity appears to be imperative to specific organelle processes (such as mitochondrial fission and oxidative phosphorylation). Furthermore, it has revealed that mitochondrial membranes are partitioned into distinct microcompartments, within which the movement of membrane proteins is restricted. Further investigation into the distributions of proteins at the nanoscale will inform as to the processes which allow the highly ordered physical arrangements of convoluted mitochondrial membranes in the meso- and microscale. Applicable to a range of diseases in which mitochondria are involved, studies such as these also enlighten us as to the functional organization of these fundamental miniature energy conversion factories.

3.4 Signaling

3.4.1 Immune system function

Immune cell biology is at the cutting edge of emerging technologies. Super-resolution microscopy has afforded some unprecedented advances in cellular immunology, and its use has resulted in true paradigm shifts in the field.

B-Cells, T-Cells, natural killer (NK) cells, and mast cells have all been used as systems to investigate the recruitment and reorganization of membrane proteins in response to activating molecules. The nanometer localization precision provided by super-resolution techniques has enabled new and previously impossible observations of fundamental processes, such as the organization of receptors in the immunological synapse (IS), and how actin is organized with respect to the IS and lytic granules. Because this field is vast and diverse, we have limited discussion to a small number of super-resolution studies.

3.4.1.1 Mast cell signaling: interplay of molecular dynamics, clustering, and signaling

Mast cells are populated with FcεRI, a high affinity membrane receptor for antigen-specific IgE. When multivalent antigen cross-links these IgE–FcεRI complexes, mast cells are stimulated to release preformed granules containing immune system mediators (see review in Siraganian, 2003). EM imaging has demonstrated that antigen cross-linking induces FcεRI–IgE complex aggregation in fixed cells (Stump, Pfeiffer, Seagrave, & Oliver, 1988; Veatch, Chiang, Sengupta, Holowka, & Baird, 2012). In live cells, antigen cross-linking induces altered dynamics as demonstrated with SPT (Spendier, Lidke, Lidke, & Thomas, 2012), and with nonimaging methods (e.g., FCS and FRAP with stationary spot illumination) (Larson, Gosse, Holowka, Baird, & Webb, 2005; Menon, Holowka, Webb, & Baird, 1986) but super-resolution imaging provides a means of directly imaging the protein dynamics *and* clustering, and their context as a function of time in living cells.

The kinetics following antigen cross-linking of IgE–FcεRI complexes was measured in living rat basophilic leukemia (RBL)-2H3 mast cells at the nanoscale with STORM, first reported in 2013 (Shelby, Holowka, Baird, & Veatch, 2013). Importantly, this allowed the imaging of a distinct time series in receptor organization. Prior to antigen addition, IgE-bound FcεRI receptors are uniformly distributed throughout the membrane. Within 2 minutes of antigen addition, the mobility of receptors slows dramatically and the receptors themselves are laterally confined. It is not until 5 minutes post antigen addition that receptors organize into ~70 nm clusters, which are too small to resolve using conventional microscopy. Furthermore, the mobility of single receptors after antigen addition was highly dependent on the surrounding receptor density, even before strong clustering occurred. Single receptors reversibly associated with small and slowly moving receptor clusters soon after antigen addition, and before calcium mobilization, suggesting initial signaling promoted these molecular associations. Antigen-mediated clustering could be reversed; addition of an antigen-competitive monovalent ligand restored the uniform distribution of receptors in nonantigen-treated cells. This elegant and robust study is a fine example of the profound biological insight which can be gained from carefully quantified super-resolution data.

3.4.1.2 NK cells: pass a lytic granule cannonball through the eye of a needle

One long-standing question has faced immunity research for years: how can NK cells pass a cannonball through the eye of a needle? NK cells (and cytotoxic T cells) can directly eliminate virus-infected or tumor cells through specialized secretory lysosomes known as lytic granules (Bryceson et al., 2011; Lieberman, 2003), with diameters around 250 nm (Brown, Oddos, et al., 2011). Confocal studies suggested granules were secreted through a large (many microns in diameter) central clearance in an otherwise hyper-dense actin barrier (Orange et al., 2003). Recent super-resolution imaging of NK cells showed actin filament accumulation at the IS upon NK cell stimulation, which is not centrally cleared as previously thought ((SIM; Brown, Oddos, et al., 2011; STED; Rak et al., 2011)). Rather, the holes in the actin mesh (only present upon cell activation) are only large enough to accommodate a single granule (Brown, Dobbie, Alakoskela, Davis, & Davis, 2012; Brown, Oddos, et al., 2011; Rak et al., 2011). Two-color STED showed the close spatial association of actin and granules adjacent to the IS (Mace & Orange, 2012a), and that the actin ABP coronin aided actin hole formation, facilitating granule secretion through the otherwise dense actin cortex (Mace & Orange, 2014). So, the lytic granule cannonball can pass through, not because of a large barrier clearance, but because the eye of the dense actin mesh needle locally expands to accommodate it. This fundamental change in ideas is due to the insight made possible by imaging advances: "The use of super-resolution imaging was critical to the identification of the granule-sized clearances formed in response to activating signal" (Mace & Orange, 2012b).

3.4.1.3 T-cell signaling: are clusters pre-formed in the cell?

We here present only a subset of super-resolution studies investigating T-cell membrane organization; for further discussion, see recent reviews (Garcia-Parajo, Cambi, Torreno-Pina, Thompson, & Jacobson, 2014; Rossy et al., 2013; Sherman, Barr, & Samelson, 2013).

When T-Cells engage peptides on the surfaces of antigen-presenting cells, subunits of plasma membrane-bound T-Cell receptors (TCR) become phosphorylated by the Src family kinase Lck or Fyn, and in turn recruit the Syk family kinase, ζ-associated protein-70 (ZAP-70). ZAP-70 and Lck mediate the phosphorylation of the scaffolding proteins linker for activation of T-cells (LAT) and SH2 domain-containing leukocyte protein-76 (SLP76) (for review see (Le Floc'h & Huse, 2015)). These multimolecular signaling complexes (along with other associated effectors and adaptors) can further group into microclusters, and these complexes and microclusters drive signal amplification and underpin T-cell activation (for review see (Sherman et al., 2013)). As such, the spatial distributions and dynamics of these proteins, and TCR-dependent signaling microclusters, are of tremendous interest. Because many aspects of their organization span length scales smaller than the diffraction-limited resolution, immense advances in T-cell membrane organization research have been made with super-resolution microscopy.

In activated T-cells, imaging of phosphorylated LAT microclusters with single color dSTORM (Owen et al., 2010) and of ZAP-70 with single-color PALM (Hsu & Baumgart, 2011) showed microclusters of each were in fact comprised of many smaller clusters, unresolvable in conventional TIRF images (Hsu & Baumgart, 2011; Owen et al., 2010). In resting T-cells, plasma membrane-bound TCR and LAT were shown with single color PALM to precluster into independent "protein islands." Following T-cell activation, these islands appeared to concatenate, forming TCR and LAT microclusters, respectively (Lillemeier et al., 2010). Further use of dual color PALM also indicated that TCR and ZAP-70 resided in separate clusters which mixed upon T-cell activation (Sherman et al., 2011). Surprisingly, the same study indicated that clusters of LAT and TCR *decreased* in their spatial overlap upon T-cell activation (Sherman et al., 2011). Similarly, dual-color PALM imaging of ZAP-70 and another of its substrates, SLP-76, indicated that ZAP-70 and SLP-76 clusters *segregated* as a function of time after activation (Hsu & Baumgart, 2011). Debate ensued. How were T-cells regulating the precise clustering, mixing, and segregation of signaling proteins in response to activation?

Through the use of super-resolution microscopy, a story has begun to emerge. Single-color PALM revealed that following activation, T-cell plasma membranes contained dramatic increases in the numbers of clusters of LAT (Hsu & Baumgart, 2011; Williamson et al., 2011), SLP-76, and ZAP-70 (Hsu & Baumgart, 2011). Dual-color dSTORM and PALM imaging suggested that T-cell activation involved docking of subsynaptic vesicles and subsequent localized delivery of LAT molecules to plasma membrane activation sites, forming new, immobile LAT clusters (Williamson et al., 2011). Dual-color PALM showed SLP-76 accumulation at the rims of LAT clusters in activated T-cells (Sherman et al., 2011), and dual color dSTORM-TIRF imaging showed that upon T-cell activation, phosphorylated (p)LAT formed vesicular fusion-dependent "signaling nanoterritories," elongated pLAT nanoclusters adjacent to (within 20 nm), but spatially segregated from clusters of phosphorylated SLP-76 (Soares et al., 2013). Conversely, pLAT that was preclustered in the plasma membranes of resting T-cells (in morphologically distinct circular nanoclusters) rarely associated with pSLP76 (Soares et al., 2013). These results suggest that two different populations of LAT exist; with (1) activation induced fusion of vesicular LAT to the plasma membrane predominantly contributing to signaling complexes, and (2) preexisting plasma membrane-bound LAT clusters remaining relatively inert with respect to signal transduction.

The roles of the actin cytoskeleton in microcluster arrangement have also been studied using super-resolution techniques. PALM imaging has shown the distributions of SLP76 proximal to, but at the peripheries of LAT clusters, were abolished with actin disruption (Sherman et al., 2011), and actin depolymerization severely impaired TCR and LAT microcluster structure, resulting in deformed clusters with reduced circularity (Hsu & Baumgart, 2011). In resting T-cells, single color PALM imaging has indicated that actin may work to physically segregate "protein islands," as depolymerization induced the formation of large TCR and large LAT aggregations (Lillemeier et al., 2010).

Super-resolution microscopy continues to allow visualization of previously unresolvable molecular organizations, providing data which is invaluable in advancing the understanding of dynamic nanoscale clustering crucial to T-cell function. Moreover, results from these studies have far-reaching implications for understanding general molecular processes which guide membrane protein clustering and the potential influences of various pools of protein on the organization at the plasma membrane.

3.4.1.4 B-cell signaling: nanoscale changes can direct global signaling

B-cell plasma membranes undergo considerable reorganization during antigen recognition, and the formation of signaling-competent microclusters of B-cell receptors (BCRs) is mediated by dynamic modulation of the actin cytoskeleton by the ezrin/radixin/moesin family of ABPs (Harwood & Batista, 2010; Treanor, Depoil, Bruckbauer, & Batista, 2011). Imaging by tracking single molecules with STORM has shown that individual BCRs, and BCR microclusters, each exhibit reduced mobility in ezrin deficient cells (Pore et al., 2013). Imaging with dSTORM showed for the first time that BCR isotypes IgM and IgG, and the co-factor CD19, are differentially assembled into preformed clusters on the naive B-cell plasma membrane (Mattila et al., 2013). Antigen activation induces a local coalescence of nanoclusters of BCR and the cofactor CD19, with the membrane spanning molecular organizer CD81 (tetraspanin) determining reorganization. Mattila et al. propose that CD19 is sequestered from BCR nanoclusters by CD81, and B-cell activation induces actin remodeling which allows CD19 to converge with BCRs and elicit cell signaling (Mattila et al., 2013). The alterations in the clustering of these proteins occur at the nanoscale, without alterations to their global distributions. These studies highlight the insights allowed by super-resolution in understanding immune processes which are otherwise invisible (see review by (Grove, 2014)).

3.4.1.5 Outlook

The studies presented here have demonstrated dramatic advances made in understanding functional membrane domains at the nanoscale, and interactions between these domains and cellular components such as the actin cytoskeleton. The super-resolution tools available are not just allowing visualization of structures with better detail; they have enabled numerous discoveries which have led to fundamental changes in the understanding of these systems.

3.4.2 Phosphoinositides and associated pathways

3.4.2.1 Phosphoinositides: in many places and doing many things

Phosphoinositides are phosphorylated phosphatidylinositols commonly found in the cytoplasmic (inner) leaflet of the cell membrane (Balla, 2013). Mammals produce seven different types of phosphoinositide, each of which is found in a different domain of the membrane, and each of which is differentiated by the location of the phosphorylation of the inositol ring (Di Paolo & De Camilli,

2006). Phosphoinositides are involved in, and regulate, an immense number of cellular events, including cell signaling, membrane transport and regulation, cytokinesis, and organelle distinction (Di Paolo & De Camilli, 2006). They can do this by directly binding to proteins, by activating proteins upstream of structural change, and by mediating a number of signaling pathways. One particularly important phosphoinositide is phosphatidylinositol (4,5)-biphosphate (PIP2), which is also one of the more prevalent subspecies (Liu & Bankaitis, 2010). Of particular interest in membrane organization research, PIP2 has the potential to be a master regulator, as it can function to remodel the actin cytoskeleton in many different ways (see Section 3.1 for more detail), and may be a functional conduit between membrane proteins, lipids, and membrane-associated actin. Though these lipids are essential for cellular functionality, data suggest that PIP2-mediated processes are altered by nanoscale spatial distributions not resolvable by conventional microscopy. Super-resolution microscopy has allowed nanoscale investigation to begin.

3.4.2.2 Imaging PIP2 with super-resolution microscopy

STED imaging of a PIP2-binding pleckstrin-homology (PH) domain construct has indicated that PIP2 forms circular and linear clusters of approximately 73 nm in size in PC12 plasma membrane sheets (van den Bogaart et al., 2011). This is in agreement with cluster sizes of PIP2 imaged with dSTORM of antibody-labeled PIP2 (Wang & Richards, 2012). Simultaneous dual-color STED revealed, stunningly, that PIP2 was almost never found in the same clusters as phosphatidylinositol (3,4,5)-trisphosphate (PIP3), and that PIP3 clusters (~100 nm size) were significantly larger than PIP2 clusters (Wang & Richards, 2012).

PIP2 has also been shown to affect the nanoscale organization of membrane components. For example, PIP2 depletion from PC12 cells was shown to reduce the clustering of syntaxin-1A, a protein which normally clusters at sites of synaptic vesicle exocytosis (van den Bogaart et al., 2011), and STED-FCS has indicated that the PIP2 phosphatase synaptotagmin binds selectively to mobile syntaxin membrane clusters, estimated to contain PIP2 (Honigmann et al., 2013). Multispecies PALM/dSTORM imaging in fixed cells has shown that clusters of PIP2 (a cytoplasmic leaflet lipid) tightly colocalize with, and are dependent on, sphingomyelin (an exoplasmic leaflet lipid). This may be a result of localized PIP2 production, as further dual-color imaging showed that the PIP2-producing kinase, phosphatidylinositol 4-phosphate 5-kinase (PIP5Kβ, located at the cytoplasmic leaflet), also colocalizes with sphingomyelin clusters (Abe et al., 2012).

These collective images of PIP2 in membrane sheets and whole cell plasma membranes have yielded data on PIP2 distributions, in fixed cells, with resolutions of ~30–70 nm. The direct imaging of PIP2 and other lipid species at the nanoscale in living cells is a tantalizing prospect. Due to the compatibility of super-resolution microscopy with live cell imaging, much can be learned about biological membranes; for example, how are PIP2 and PIP3 molecules spatially segregated? Are molecular motions confined within clusters, or do certain molecules just spend longer dwell times within? Localization microscopy can provide single-molecule information such as high-density molecular trajectories (Gudheti et al., 2013; Manley et al., 2008) within the context of other membrane components, information which can be invaluable for quantification of membrane organization and dynamics to address a wide range of interesting questions.

Interactions between PIP2 and other lipids, PIP2 and membrane proteins, and PIP2 and the actin cytoskeleton remain greatly underexplored using super-resolution imaging methods. The exciting developments we have described are just part of the story, and these studies demonstrate that a multitude of fundamental processes can (and hopefully will) be better understood by using super-resolution imaging of phosphoinositides.

3.5 Future directions

3.5.1 Technical considerations

For localization microscopy itself, improving time resolution is still an important goal, especially for live cell imaging. Biological processes occur not just on the minute, second, and millisecond timescale, but movements of lipids and proteins can be exceptionally fast, and methods which can image on submillisecond timescales will yield valuable and previously invisible insights into how cells are really organized in time.

As is frequently the case in fluorescence methods, the properties of fluorescent probes themselves can be limiting, and can restrict the speed at which fast images can be acquired. Ideally, probes would emit very large numbers of photons (i.e., $>10^5$) before switching to a dark state or bleaching, they would have minimal blinking on millisecond and microsecond timescales (as this can lead to erroneously increased estimates of molecule number), and probes would not disrupt biological function. Of course, increased numbers of detected photons can in principle improve localization precision, although many other factors are involved at length scales below 20 nm, such as localization errors due to emission dipole orientation effects (Enderlein, Toprak, & Selvin, 2006), camera pixel heterogeneities (Pertsinidis, Zhang,

& Chu, 2010), and sample/stage drift (Mlodzianoski et al., 2011). However, to gain substantial increases in imaging speeds, using probes with high peak emission rates (i.e., 10^6 photons detected per molecule per second), may be the most important (and perhaps overlooked) property. While organic dyes can emit very large numbers of photons, they are still limited in emission rate by blinking, flicker, and saturation. Furthermore, very little data on maximum emission rate per molecule is available—most studies focus on number of detected photons, measured extinction coefficient and quantum yield, which do not address emission rate at saturation. To optimize use and choice of probes, there is considerable need for systematic photophysical studies of probes used for localization microscopy, such as those being carried out by Markus Sauer (van de Linde & Sauer, 2014), Carlos Bustamante (Lee, Shin, Lee, & Bustamante, 2012), Dominique Bourgeois (Avilov et al., 2014), and others.

Improvements in imaging speed have been enabled by multiemitter fitting algorithms (Cox & Jones, 2013; Cox et al., 2012; Holden, Uphoff, & Kapanidis, 2011; Huang et al., 2013), although analysis times may increase significantly when such algorithms are used. Development of increasingly efficient algorithms for larger numbers of spatially overlapping PSFs is of great interest, and can enable live cell imaging with speeds fivefold faster or more, especially when coupled to use of labels with high emission rates (Huang et al., 2013). In addition, high-speed cameras capable of imaging the action are also crucial, and recent developments in sCMOS and EMCCD sensor technology and electronics for readout have helped the field tremendously. We look forward to the ongoing technological advances in sensor development which allow for faster acquisition, and so facilitate greater temporal resolution.

The issue of photodamage needs to be explored further. As all interactions between light and a biological sample can potentially cause disruption of biological function, super-resolution techniques (including both STED and localization-based methods) seem to be particularly susceptible; the high intensities often used are expected to cause more damage, and the short length scales typically of interest are also more likely to manifest changes (however subtle). Notwithstanding, there is very little consensus in the field related to the intensity or dose thresholds for photodamage. Currently it seems that each group has their own method for testing for possible photodamage, and there seems not to be any standardization, making generalizations extremely difficult. Thus there is considerable need for systematic exploration of this crucial issue. Once better understanding of the most common damage mechanisms has been established, methods can be developed to mitigate photodamage while still allowing nanoscale structures to be imaged.

The use of super-resolution localization microscopy to count numbers of molecules is currently possible, but difficult, largely due to the need to control for several complicating factors. These factors include fluorophore blinking (Lee et al., 2012), inefficient fluorescent protein folding/activation (Durisic, Laparra-Cuervo, Sandoval-Alvarez, Borbely, & Lakadamyali, 2014), molecules lost to thresholds and excitation/activation polarization effects (whereby certain populations are excited or activated more efficiently because their transition dipoles are oriented along the direction of the incident laser electric field), and potential complications when both exogenous and endogenous molecular populations are present. Current work in the field is helping to overcome these challenges, so it is likely that in the near future, localization microscopy will become used more widely for quantification of molecular numbers, information which is not directly accessible by a number of other techniques. Moreover, discovery and engineering of new methods to tag cellular molecules (e.g., reductions in the size of tags, whether fluorescent proteins or antibodies) will bring us closer to the prospect of imaging with minimal disruption to cellular function.

Additionally, as a larger variety of photoactivatable probes becomes available, the prospect of simultaneously imaging large numbers of different species comes into reach. Simultaneous multicolor imaging can be limited by the spectra of probes used to tag them. New tags (e.g., new species of photoactivatable proteins, or newly engineered organic dyes) are emerging which are less overlapping in their emission spectra. Also emerging are new ways to reliably identify tagged species when there may be spectral overlap (including ratiometric analyses, but also the development of color recognition in sensor systems). Together these advances bring us toward simultaneous super-resolution imaging of five or more species in living cells. Some day soon, we really will be able to see, in real time, *all* of the dancers about which we hypothesize.

3.5.2 Biological considerations

The field of cell biology now has some truly excellent tools to understand membrane organization. We are fortunate indeed to be able to directly image nanoscale membrane clusters in living cells! As has been forecasted by the Nobel Committee in 2014, the power of super-resolution microscopy is only going to grow as time progresses.

Perhaps one of the highest priorities will be to put the current models to the test, with the aim of generating some consensus in the field. Of course, this requires cooperation between those posing models and those designing experiments; generation of testable, falsifiable hypotheses must remain of utmost importance, even as the complexity of models increases. While many models propose a starting and ending point in a dynamic process, the steps and dynamics which occur between those points can now be visualized directly. More attention should be spent on these "intermediate points," which can strongly govern *how* a biological system approaches a given outcome, even when multiple models predict the same final outcome. For example, very little has been done in the way of super-resolution timelapse imaging of the a priori formation of membrane protein clusters, or their subsequent dispersal.

In the direction of refining membrane models to improve our understanding of cell membranes, it will almost certainly be useful to further investigate the coupling between signaling and membrane organization, especially considering the potential for two-way interactions, such as the way membrane proteins can "talk to" actin through ABPs or through certain lipids, and so cause changes in actin organization, which can in turn cause changes in the membrane protein clusters themselves.

Some models of membrane organization consider the bilayer as a two-dimensional plane, and interactions with other components such as the actin cytoskeleton as perturbations to the two-dimensional energy landscape. While this can be useful, there is much more to be learned from studying the three-dimensional structure of the membrane and its interactions with the actin cytoskeleton. Not only can this strategy be helpful in discerning differences between models of how membrane proteins are regulated in their spatial distributions, but also it can allow a fuller understanding of the richness of cellular functions and interactions which are possible in such systems. For example, considering the actin cytoskeleton as its native three-dimensional self allows for ABP coupling between actin, membrane proteins, lipids, lateral organization, and three-dimensional membrane topology. That is to say, *both* proteins and lipids are clearly important. Likewise, as many cellular organelles have three-dimensional structures and complex membrane topologies, it is clear that we must approach such systems three-dimensionally, while also further investigating how lateral organization (i.e., domains) can occur within the membrane. Thus, there are a number of challenges to solve as we endeavor to understand membranes. Just a decade ago, many of these challenges seemed insurmountable. Yet, with the capabilities of super-resolution microscopy, a fundamentally better understanding of biological membranes is within reach.

AUTHORS' NOTE

Due to space constraints, we have not included a section on imaging of membranes in neurons. However, we believe the field is deserving of a dedicated chapter! While some super-resolution studies of neurons are mentioned here, we direct readers to recent reviews (MacGillavry & Blanpied, 2013; Maglione & Sigrist, 2013; Okabe, 2012; Sigrist & Sabatini, 2012).

ACKNOWLEDGMENTS

The authors thank Julie Gosse, Greg Innes, Mary Kraft, Joshua Zimmerberg, and Markus Sauer for crucial discussions, and Patricia Byard for administrative assistance. This work was funded by NIH R15 GM094713, NIH R15 ES024593, NIH R01 AR054170, Maine Technology Institute MTAF 1106 and 2061, the University of Maine Office of the Vice President for Research, and the Maine Economic Improvement Fund.

REFERENCES

Abe, M., Makino, A., Hullin-Matsuda, F., Kamijo, K., Ohno-Iwashita, Y., Hanada, K., et al. (2012). A role for sphingomyelin-rich lipid domains in the accumulation of phosphatidylinositol-4,5-bisphosphate to the cleavage furrow during cytokinesis. *Molecular and Cellular Biology, 32*(8), 1396–1407.

Alberts, B. (2002). *Molecular biology of the cell* (4th ed.). New York: Garland Science.

Alenghat, F. J., & Golan, D. E. (2013). Membrane protein dynamics and functional implications in mammalian cells. *Current Topics in Membranes, 72*, 89–120.

Almeida, P. F. F., Pokorny, A., & Hinderliter, A. (2005). Thermodynamics of membrane domains (Review) *Biochimica et Biophysica Acta—Biomembranes, 1720*(1–2), 1–13.

Alvarez-Guaita, A., Vila de Muga, S., Owen, D. M., Williamson, D., Magenau, A., Garcia-Melero, A., et al. (2014). Evidence for Annexin A6-dependent plasma membrane remodelling of lipid domains. *British Journal Pharmacology, 19*(10), 13022.

Anderson, R. G. W., & Jacobson, K. (2002). A role for lipid shells in targeting proteins to caveolae, rafts, and other lipid domains. *Science, 296*(5574), 1821–1825.

Andrews, N. L., Lidke, K. A., Pfeiffer, J. R., Burns, A. R., Wilson, B. S., Oliver, J. M., et al. (2008). Actin restricts FcepsilonRI diffusion and facilitates antigen-induced receptor immobilization. *Nature Cell Biology, 10*(8), 955–963.

Angelova, M. I., & Dimitrov, D. S. (1986). Liposome electroformation. *Faraday Discuss, 81*, 303–311.

Antelmi, E., Cardone, R. A., Greco, M. R., Rubino, R., Di Sole, F., Martino, N. A., et al. (2013). Beta1 integrin binding phosphorylates ezrin at T567 to activate a lipid raft signalsome driving invadopodia activity and invasion. *PLoS One, 8*(9).

Appelhans, T., Richter, C. P., Wilkens, V., Hess, S. T., Piehler, J., & Busch, K. B. (2012). Nanoscale organization of mitochondrial microcompartments revealed by combining tracking and localization microscopy. *Nano Letters, 12*(2), 610–616.

Aspenstrom, P. (2014). BAR domain proteins regulate Rho GTPase signaling. *Small GTPases, 5*(2), 1–8.

Avilov, S., Berardozzi, R., Gunewardene, M. S., Adam, V., Hess, S. T., & Bourgeois, D. (2014). In cellulo evaluation of phototransformation quantum yields in fluorescent proteins used as markers for single-molecule localization microscopy. *PLoS One, 9*(6), e98362.

Babcock, H. P., Chen, C., & Zhuang, X. (2004). Using single-particle tracking to study nuclear trafficking of viral genes. *Biophysical Journal, 87*(4), 2749–2758.

Bagatolli, L. A., & Gratton, E. (1999). Two-photon fluorescence microscopy observation of shape changes at the phase transition in phospholipid giant unilamellar vesicles. *Biophysical Journal, 77*(4), 2090–2101.

Bagatolli, L. A., & Gratton, E. (2000). A correlation between lipid domain shape and binary phospholipid mixture composition in free standing bilayers: a two-photon fluorescence microscopy study. *Biophysical Journal, 79*(1), 434–447.

Bagatolli, L., Gratton, E., Khan, T. K., & Chong, P. L. G. (2000). Two-photon fluorescence microscopy studies of bipolar tetraether giant liposomes from thermoacidophilic archaebacteria Sulfolobus acidocaldarius. *Biophysical Journal, 79*(1), 416–425.

Bagatolli, L. A., Sanchez, S. A., Hazlett, T., & Gratton, E. (2003). Giant vesicles, Laurdan, and two-photon fluorescence microscopy: evidence of lipid lateral separation in bilayers. *Biophotonics, Pt A, 360*, 481–500. San Diego: Academic Press Inc.

Balla, T. (2013). Phosphoinositides: tiny lipids with giant impact on cell regulation. *Physiological Reviews, 93*(3), 1019–1137.

Bar-On, D., Wolter, S., van de Linde, S., Heilemann, M., Nudelman, G., Nachliel, E., et al. (2012). Super-resolution imaging reveals the internal architecture of nano-sized syntaxin clusters. *The Journal of Biological Chemistry, 287*(32), 27158–27167.

Barret, C., Roy, C., Montcourrier, P., Mangeat, P., & Niggli, V. (2000). Mutagenesis of the phosphatidylinositol 4,5-bisphosphate (PIP(2)) binding site in the NH(2)-terminal domain of ezrin correlates with its altered cellular distribution. *The Journal of Cell Biology, 151*(5), 1067–1080.

Bates, M., Huang, B., Dempsey, G. T., & Zhuang, X. (2007). Multicolor super-resolution imaging with photo-switchable fluorescent probes. *Science, 317*(5845), 1749–1753.

Baumgart, T., Hammond, A. T., Sengupta, P., Hess, S. T., Holowka, D. A., Baird, B. A., et al. (2007). Large-scale fluid/fluid phase separation of proteins and lipids in giant plasma membrane vesicles. *Proceedings of the National Academy of Sciences USA, 104*(9), 3165–3170.

Baumgart, T., Hess, S. T., & Webb, W. W. (2003). Imaging coexisting fluid domains in biomembrane models coupling curvature and line tension. *Nature, 425*(6960), 821–824.

Baumgart, T., Hunt, G., Farkas, E. R., Webb, W. W., & Feigenson, G. W. (2007). Fluorescence probe partitioning between L-o/L-d phases in lipid membranes. *Biochimica et Biophysica Acta—Biomembranes, 1768*(9), 2182–2194.

Bentz, J. (2000). Minimal aggregate size and minimal fusion unit for the first fusion pore of influenza hemagglutinin-mediated membrane fusion. *Biophysical Journal, 78*(1), 227–245.

Berditchevski, F. (2001). Complexes of tetraspanins with integrins: more than meets the eye. *Journal of Cell Science, 114*(Pt 23), 4143–4151.

Betzig, E., Patterson, G. H., Sougrat, R., Lindwasser, O. W., Olenych, S., Bonifacino, J. S., et al. (2006). Imaging intracellular fluorescent proteins at nanometer resolution. *Science, 313*(5793), 1642–1645.

Betzig, E., & Trautman, J. K. (1992). Near-field optics: microscopy, spectroscopy, and surface modification beyond the diffraction limit. *Science, 257*(5067), 189–195.

Betzig, E., Trautman, J. K., Harris, T. D., Weiner, J. S., & Kostelak, R. L. (1991). Breaking the diffraction barrier – optical microscopy on a nanometric scale. *Science, 251*(5000), 1468–1470.

Bisi, S., Disanza, A., Malinverno, C., Frittoli, E., Palamidessi, A., & Scita, G. (2013). Membrane and actin dynamics interplay at lamellipodia leading edge. *Current Opinion in Cell Biology, 25*(5), 565–573.

Biswas, S., Yin, S. R., Blank, P. S., & Zimmerberg, J. (2008). Cholesterol promotes hemifusion and pore widening in membrane fusion induced by influenza hemagglutinin. *The Journal of General Physiology, 131*(5), 503–513.

Blunk, A. D., Akbergenova, Y., Cho, R. W., Lee, J., Walldorf, U., Xu, K., et al. (2014). Postsynaptic actin regulates active zone spacing and glutamate receptor apposition at the Drosophila neuromuscular junction. *Molecular and Cellular Neuroscience, 61*, 241–254.

van den Bogaart, G., Meyenberg, K., Risselada, H. J., Amin, H., Willig, K. I., Hubrich, B. E., et al. (2011). Membrane protein sequestering by ionic protein-lipid interactions. *Nature, 479*(7374), 552–555.

Boguslawska, D. M., Machnicka, B., Hryniewicz-Jankowska, A., & Czogalla, A. (2014). Spectrin and phospholipids – the current picture of their fascinating interplay. *Cellular and Molecular Biology Letters, 19*(1), 158–179.

Borroni, V., Baier, C. J., Lang, T., Bonini, I., White, M. M., Garbus, I., et al. (2007). Cholesterol depletion activates rapid internalization of submicron-sized acetylcholine receptor domains at the cell membrane. *Molecular Membrane Biology, 24*(1), 1–15.

Bossi, M., Folling, J., Belov, V. N., Boyarskiy, V. P., Medda, R., Egner, A., et al. (2008). Multicolor far-field fluorescence nanoscopy through isolated detection of distinct molecular species. *Nano Letters, 8*(8), 2463–2468.

Brandstaetter, H., Kendrick-Jones, J., & Buss, F. (2012). Myo1c regulates lipid raft recycling to control cell spreading, migration and Salmonella invasion. *Journal of Cell Science, 125*(Pt 8), 1991–2003.

Brown, A. C., Dobbie, I. M., Alakoskela, J. M., Davis, I., & Davis, D. M. (2012). Super-resolution imaging of remodeled synaptic actin reveals different synergies between NK cell receptors and integrins. *Blood, 120*(18), 3729–3740.

Brown, D. A., & London, E. (1997). Structure of detergent-resistant membrane domains: does phase separation occur in biological membranes? *Biochemical and Biophysical Research Communications, 240*(1), 1–7.

Brown, D. A., & London, E. (1998a). Functions of lipid rafts in biological membranes (Review) *Annual Review of Cell and Developmental Biology, 14*, 111–136.

Brown, D. A., & London, E. (1998b). Structure and origin of ordered lipid domains in biological membranes (Review) *The Journal of Membrane Biology, 164*(2), 103–114.

Brown, A. C., Oddos, S., Dobbie, I. M., Alakoskela, J. M., Parton, R. M., Eissmann, P., et al. (2011). Remodelling of cortical actin where lytic granules dock at natural killer cell immune synapses revealed by super-resolution microscopy. *PLoS Biology, 9*(9), 13.

Brown, D. A., & Rose, J. K. (1992). Sorting of GPI-anchored proteins to glycolipid-enriched membrane subdomains during transport to the apical cell surface. *Cell, 68*(3), 533–544.

Brown, T. A., Tkachuk, A. N., Shtengel, G., Kopek, B. G., Bogenhagen, D. F., Hess, H. F., et al. (2011). Superresolution fluorescence imaging of mitochondrial nucleoids reveals their spatial range, limits, and membrane interaction. *Molecular and Cellular Biology, 31*(24), 4994–5010.

Brunstein, M., Wicker, K., Herault, K., Heintzmann, R., & Oheim, M. (2013). Full-field dual-color 100-nm super resolution imaging reveals organization and dynamics of mitochondrial and ER networks. *Optics Express, 21*(22), 26162–26173.

Bryceson, Y. T., Chiang, S. C., Darmanin, S., Fauriat, C., Schlums, H., Theorell, J., et al. (2011). Molecular mechanisms of natural killer cell activation. *Journal of Innate Immunity, 3*(3), 216–226.

Burte, F., Carelli, V., Chinnery, P. F., & Yu-Wai-Man, P. (2014). Disturbed mitochondrial dynamics and neurodegenerative disorders. *Nature Reviews Neurology, 11*, 11–24.

Chazeau, A., Mehidi, A., Nair, D., Gautier, J. J., Leduc, C., Chamma, I., et al. (2014). Nanoscale segregation of actin nucleation and elongation factors determines dendritic spine protrusion. *EMBO Journal, 33*(23), 2745–2764.

Chen, A. K., Sengupta, P., Waki, K., Van Engelenburg, S. B., Ochiya, T., Ablan, S. D., et al. (2014). MicroRNA binding to the HIV-1 Gag protein inhibits Gag assembly and virus production. *Proceedings of the National Academy of Sciences USA, 111*(26), E2676–E2683.

Chen, Y., Takizawa, N., Crowley, J. L., Oh, S. W., Gatto, C. L., Kambara, T., et al. (2003). F-actin and myosin II binding domains in supervillin. *The Journal of Biological Chemistry, 278*(46), 46094–46106.

Chernomordik, L. V., Frolov, V. A., Leikina, E., Bronk, P., & Zimmerberg, J. (1998). The pathway of membrane fusion catalyzed by influenza hemagglutinin: restriction of lipids, hemifusion, and lipidic fusion pore formation. *The Journal of Cell Biology, 140*(6), 1369–1382.

Chernomordik, L. V., Leikina, E., Frolov, V., Bronk, P., & Zimmerberg, J. (1997). An early stage of membrane fusion mediated by the low pH conformation of influenza hemagglutinin depends upon membrane lipids. *The Journal of Cell Biology, 136*(1), 81–93.

Chichili, G. R., & Rodgers, W. (2007). Clustering of membrane raft proteins by the actin cytoskeleton. *The Journal of Biological Chemistry, 282*(50), 36682–36691.

Chichili, G., & Rodgers, W. (2009). Cytoskeleton–membrane interactions in membrane raft structure. *Cellular and Molecular Life Sciences, 66*(14), 2319–2328.

Chojnacki, J., Staudt, T., Glass, B., Bingen, P., Engelhardt, J., Anders, M., et al. (2012). Maturation-dependent HIV-1 surface protein redistribution revealed by fluorescence nanoscopy. *Science, 338*(6106), 524–528.

Coltharp, C., & Xiao, J. (2012). Superresolution microscopy for microbiology. *Cellular Microbiology, 14*(12), 1808–1818.

Cornely, R., Rentero, C., Enrich, C., Grewal, T., & Gaus, K. (2011). Annexin A6 is an organizer of membrane microdomains to regulate receptor localization and signalling. *IUBMB Life, 63*(11), 1009–1017.

Coscoy, S., Waharte, F., Gautreau, A., Martin, M., Louvard, D., Mangeat, P., et al. (2002). Molecular analysis of microscopic ezrin dynamics by two-photon FRAP. *Proceedings of the National Academy of Sciences USA, 99*(20), 12813–12818.

Cox, S., & Jones, G. E. (2013). Imaging cells at the nanoscale. *The International Journal of Biochemistry and Cell Biology, 45*(8), 1669–1678.

Cox, S., Rosten, E., Monypenny, J., Jovanovic-Talisman, T., Burnette, D. T., Lippincott-Schwartz, J., et al. (2012). Bayesian localization microscopy reveals nanoscale podosome dynamics. *Nature Methods, 9*(2), 195–200.

Critchley, D. R. (2005). Genetic, biochemical and structural approaches to talin function. *Biochemical Society Transactions, 33*(Pt 6), 1308–1312.

Curthoys, N. M., Mlodzianoski, M. J., Kim, D., & Hess, S. T. (2013). Simultaneous multicolor imaging of biological structures with fluorescence photoactivation localization microscopy. *Journal of Visualized Experiments, 9*(82), 50680.

de Curtis, I., & Meldolesi, J. (2012). Cell surface dynamics – how Rho GTPases orchestrate the interplay between the plasma membrane and the cortical cytoskeleton. *Journal of Cell Science, 125*(Pt 19), 4435–4444.

Davis, L. H., & Bennett, V. (1994). Identification of two regions of beta G spectrin that bind to distinct sites in brain membranes. *The Journal of Biological Chemistry, 269*(6), 4409–4416.

Day, C. A., & Kenworthy, A. K. (2012). Mechanisms underlying the confined diffusion of cholera toxin B-subunit in intact cell membranes. *PLoS One*, 7(4), e34923.

Derivery, E., & Gautreau, A. (2010). Generation of branched actin networks: assembly and regulation of the N-WASP and WAVE molecular machines. *Bioessays, 32*(2), 119–131.

Dertinger, T., Colyer, R., Iyer, G., Weiss, S., & Enderlein, J. (2009). Fast, background-free, 3D super-resolution optical fluctuation imaging (SOFI). *Proceedings of the National Academy of Sciences USA, 106*(52), 22287–22292.

Dertinger, T., Colyer, R., Vogel, R., Enderlein, J., & Weiss, S. (2010). Achieving increased resolution and more pixels with Superresolution Optical Fluctuation Imaging (SOFI). *Optics Express, 18*(18), 18875–18885.

Dertinger, T., Colyer, R., Vogel, R., Heilemann, M., Sauer, M., Enderlein, J., et al. (2012). Superresolution optical fluctuation imaging (SOFI). *Advances in Experimental Medicine and Biology, 733*, 17–21.

Dertinger, T., Heilemann, M., Vogel, R., Sauer, M., & Weiss, S. (2010). Superresolution optical fluctuation imaging with organic dyes. *Angewandte Chemie International Edition in English, 49*(49), 9441–9443.

Dertinger, T., Xu, J., Naini, O. F., Vogel, R., & Weiss, S. (2012). SOFI-based 3D superresolution sectioning with a widefield microscope. *Optical Nanoscopy, 1*(2), 2.

Di Paolo, G., & De Camilli, P. (2006). Phosphoinositides in cell regulation and membrane dynamics. *Nature, 443*(7112), 651–657.

Di Paolo, G., Pellegrini, L., Letinic, K., Cestra, G., Zoncu, R., Voronov, S., et al. (2002). Recruitment and regulation of phosphatidylinositol phosphate kinase type 1 gamma by the FERM domain of talin. *Nature, 420*(6911), 85–89.

Dietrich, C., Bagatolli, L. A., Volovyk, Z. N., Thompson, N. L., Levi, M., Jacobson, K., et al. (2001). Lipid rafts reconstituted in model membranes. *Biophysical Journal, 80*(3), 1417–1428.

Durisic, N., Laparra-Cuervo, L., Sandoval-Alvarez, A., Borbely, J. S., & Lakadamyali, M. (2014). Single-molecule evaluation of fluorescent protein photoactivation efficiency using an in vivo nanotemplate. *Nature Methods, 11*(2), 156–162.

Edidin, M. (2003). The state of lipid rafts: from model membranes to cells. *Annual Review of Biophysics and Biomolecular Structure, 32*, 257–283.

Edwards, H. C., & Crumpton, M. J. (1991). Ca(2+)-dependent phospholipid and arachidonic acid binding by the placental annexins VI and IV. *European Journal of Biochemistry, 198*(1), 121–129.

Eggeling, C., Ringemann, C., Medda, R., Schwarzmann, G., Sandhoff, K., Polyakova, S., et al. (2009). Direct observation of the nanoscale dynamics of membrane lipids in a living cell. *Nature, 457*(7233), 1159–1162.

Eggeling, C., Widengren, J., Rigler, R., & Seidel, C. A. (1998). Photobleaching of fluorescent dyes under conditions used for single-molecule detection: evidence of two-step photolysis. *Analytical Chemistry, 70*(13), 2651–2659.

Elson, E. L., Schlessinger, J., Koppel, D. E., Axelrod, D., & Webb, W. W. (1976). Measurement of lateral transport on cell surfaces. *Progress in Clinical and Biological Research, 9*, 137–147.

Enderlein, J., Toprak, E., & Selvin, P. R. (2006). Polarization effect on position accuracy of fluorophore localization. *Optics Express, 14*(18), 8111–8120.

Engel, S., Scolari, S., Thaa, B., Krebs, N., Korte, T., Herrmann, A., et al. (2010). FLIM-FRET and FRAP reveal association of influenza virus haemagglutinin with membrane rafts. *Biochemical Journal, 425*(3), 567–573.

Fang, Z., Takizawa, N., Wilson, K. A., Smith, T. C., Delprato, A., Davidson, M. W., et al. (2010). The membrane-associated protein, supervillin, accelerates F-actin-dependent rapid integrin recycling and cell motility. *Traffic, 11*(6), 782–799.

Feder, T. J., Brust-Mascher, I., Slattery, J. P., Baird, B., & Webb, W. W. (1996). Constrained diffusion or immobile fraction on cell surfaces: a new interpretation. *Biophysical Journal, 70*(6), 2767–2773.

Fehon, R. G., McClatchey, A. I., & Bretscher, A. (2010). Organizing the cell cortex: the role of ERM proteins. *Nature Reviews Molecular Cell Biology, 11*(4), 276–287.

Finkelstein, A., & Cass, A. (1967). Effect of cholesterol on the water permeability of thin lipid membranes. *Nature, 216*(5116), 717–718.

Fraley, T. S., Tran, T. C., Corgan, A. M., Nash, C. A., Hao, J., Critchley, D. R., et al. (2003). Phosphoinositide binding inhibits alpha-actinin bundling activity. *The Journal of Biological Chemistry, 278*(26), 24039–24045.

Friedman, J. R., Lackner, L. L., West, M., DiBenedetto, J. R., Nunnari, J., & Voeltz, G. K. (2011). ER tubules mark sites of mitochondrial division. *Science, 334*(6054), 358–362.

Frisz, J. F., Klitzing, H. A., Lou, K., Hutcheon, I. D., Weber, P. K., Zimmerberg, J., et al. (2013). Sphingolipid domains in the plasma membranes of fibroblasts are not enriched with cholesterol. *The Journal of Biological Chemistry, 288*(23), 16855–16861.

Frisz, J. F., Lou, K., Klitzing, H. A., Hanafin, W. P., Lizunov, V., Wilson, R. L., et al. (2013). Direct chemical evidence for sphingolipid domains in the plasma membranes of fibroblasts. *Proceedings of the National Academy of Sciences USA, 110*(8), E613–E622.

Frost, A., Unger, V. M., & De Camilli, P. (2009). The BAR domain superfamily: membrane-molding macromolecules. *Cell, 137*(2), 191–196.

Fujiwara, T., Ritchie, K., Murakoshi, H., Jacobson, K., & Kusumi, A. (2002). Phospholipids undergo hop diffusion in compartmentalized cell membrane. *The Journal of Cell Biology, 157*(6), 1071–1081.

Furuhashi, K., Inagaki, M., Hatano, S., Fukami, K., & Takenawa, T. (1992). Inositol phospholipid-induced suppression of F-actin-gelating activity of smooth muscle filamin. *Biochemical and Biophysical Research Communications, 184*(3), 1261–1265.

Gabor, K. A., Kim, D., Kim, C. H., & Hess, S. T. (2015). Nanoscale imaging of caveolin-1 membrane domains in vivo. *PLoS One, 10*(2), e0117225.

Gabor, K. A., Stevens, C. R., Pietraszewski, M. J., Gould, T. J., Shim, J., Yoder, J. A., et al. (2013). Super resolution microscopy reveals that caveolin-1 is required for spatial organization of CRFB1 and subsequent antiviral signaling in zebrafish. *PLoS One, 8*(7), e68759.

Garcia-Parajo, M. F., Cambi, A., Torreno-Pina, J. A., Thompson, N., & Jacobson, K. (2014). Nanoclustering as a dominant feature of plasma membrane organization. *Journal of Cell Science, 127*(23), 4995–5005.

Geissbuehler, S., Sharipov, A., Godinat, A., Bocchio, N. L., Sandoz, P. A., Huss, A., et al. (2014). Live-cell multiplane three-dimensional super-resolution optical fluctuation imaging. *Nature Communications, 5*, 5830.

Getis, A., & Ord, J. K. (1992). The analysis of spatial association by use of distance statistics. *Geographical Analysis, 24*(3), 189–206.

Giannone, G., Hosy, E., Levet, F., Constals, A., Schulze, K., Sobolevsky, A. I., et al. (2010). Dynamic superresolution imaging of endogenous proteins on living cells at ultra-high density. *Biophysical Journal, 99*(4), 1303–1310.

Golan, D. E., & Veatch, W. (1980). Lateral mobility of band 3 in the human erythrocyte membrane studied by fluorescence photobleaching recovery: evidence for control by cytoskeletal interactions. *Proceedings of the National Academy of Sciences USA,* 77(5), 2537–2541.

Goswami, D., Gowrishankar, K., Bilgrami, S., Ghosh, S., Raghupathy, R., Chadda, R., et al. (2008). Nanoclusters of GPI-anchored proteins are formed by cortical actin-driven activity. *Cell, 135*(6), 1085–1097.

Gottfert, F., Wurm, C. A., Mueller, V., Berning, S., Cordes, V. C., Honigmann, A., et al. (2013). Coaligned dual-channel STED nanoscopy and molecular diffusion analysis at 20 nm resolution. *Biophysical Journal, 105*(1), L01–L03.

Gould, T. J., Gunewardene, M. S., Gudheti, M. V., Verkhusha, V. V., Yin, S.-R., Gosse, J. A., et al. (2008). Nanoscale imaging of molecular positions and anisotropies. *Nature Methods, 5*(12), 1027–1030.

Gould, T. J., Hess, S. T., & Bewersdorf, J. (2012). Optical nanoscopy: from acquisition to analysis. *Annual Review of Biomedical Engineering, 14*, 231–254.

Gowrishankar, K., Ghosh, S., Saha, S. C. R., Mayor, S., & Rao, M. (2012). Active remodeling of cortical actin regulates spatiotemporal organization of cell surface molecules. *Cell, 149*(6), 1353–1367.

Grove, J. (2014). Super-resolution microscopy: a virus' eye view of the cell. *Viruses, 6*(3), 1365–1378.

Grover, J. R., Veatch, S. L., & Ono, A. (2015). Basic motifs target PSGL-1, CD43, and CD44 to plasma membrane sites where HIV-1 assembles. *Journal of Virology, 89*(1), 454–467.

Gudheti, M. V., Curthoys, N. M., Gould, T. J., Kim, D., Gunewardene, M. S., Gabor, K. A., et al. (2013). Actin mediates the nanoscale membrane organization of the clustered membrane protein influenza hemagglutinin. *Biophysical Journal, 104*(10), 2182–2192.

Gunewardene, M. S., Subach, F. V., Gould, T. J., Penoncello, G. P., Gudheti, M. V., Verkhusha, V. V., et al. (2011). Super-resolution imaging of multiple fluorescent proteins with highly overlapping emission spectra in living cells. *Biophysical Journal, 101*(6), 1522–1528.

Gupta, N., Wollscheid, B., Watts, J. D., Scheer, B., Aebersold, R., & DeFranco, A. L. (2006). Quantitative proteomic analysis of B cell lipid rafts reveals that ezrin regulates antigen receptor-mediated lipid raft dynamics. *Nature Immunology, 7*(6), 625–633.

Gustafsson, M. G. (2000). Surpassing the lateral resolution limit by factor of two using structured illumination microscopy. *Journal of Microscopy, 198*, 82–87.

Gustafsson, M. G. (2005). Nonlinear structured-illumination microscopy: wide-field fluorescence imaging with theoretically unlimited resolution. *Proceedings of the National Academy of Sciences USA, 102*(37), 13081–13086.

Hao, J. J., Liu, Y., Kruhlak, M., Debell, K. E., Rellahan, B. L., & Shaw, S. (2009). Phospholipase C-mediated hydrolysis of PIP2 releases ERM proteins from lymphocyte membrane. *The Journal of Cell Biology, 184*(3), 451–462.

Harner, M., Korner, C., Walther, D., Mokranjac, D., Kaesmacher, J., Welsch, U., et al. (2011). The mitochondrial contact site complex, a determinant of mitochondrial architecture. *EMBO Journal, 30*(21), 4356–4370.

Hartwig, J. H., & DeSisto, M. (1991). The cytoskeleton of the resting human blood platelet: structure of the membrane skeleton and its attachment to actin filaments. *The Journal of Cell Biology, 112*(3), 407–425.

Harwood, N. E., & Batista, F. D. (2010). Early events in B cell activation. *Annual Review of Immunology, 28*, 185–210.

Hayes, M. J., Shao, D. M., Grieve, A., Levine, T., Bailly, M., & Moss, S. E. (2009). Annexin A2 at the interface between F-actin and membranes enriched in phosphatidylinositol 4,5,-bisphosphate. *Biochimica et Biophysica Acta, 6*(95), 29.

He, J., Sun, E., Bujny, M. V., Kim, D., Davidson, M. W., & Zhuang, X. (2013). Dual function of CD81 in influenza virus uncoating and budding. *PLoS Pathogens, 9*(10), e1003701.

Head, B. P., Patel, H. H., & Insel, P. A. (2014). Interaction of membrane/lipid rafts with the cytoskeleton: impact on signaling and function: membrane/lipid rafts, mediators of cytoskeletal arrangement and cell signaling. *Biochimica et Biophysica Acta, 2*, 532–545.

Hein, B., Willig, K. I., & Hell, S. W. (2008). Stimulated emission depletion (STED) nanoscopy of a fluorescent protein-labeled organelle inside a living cell. *Proceedings of the National Academy of Sciences USA, 105*(38), 14271–14276.

Hell, S. W. (2007). Far-field optical nanoscopy (Review) *Science, 316*(5828), 1153–1158.

Hell, S. W., & Wichmann, J. (1994). Breaking the diffraction resolution limit by stimulated emission: stimulated-emission depletion fluorescence microscopy. *Optics Letters, 19*(11), 780–782.

Hess, S. T., Girirajan, T. P., & Mason, M. D. (2006). Ultra-high resolution imaging by fluorescence photoactivation localization microscopy. *Biophysical Journal, 91*(11), 4258–4272.

Hess, S. T., Gould, T. J., Gudheti, M. V., Maas, S. A., Mills, K. D., & Zimmerberg, J. (2007). Dynamic clustered distribution of hemagglutinin resolved at 40 nm in living cell membranes discriminates between raft theories. *Proceedings of the National Academy of Sciences USA, 104*(44), 17370–17375.

Hess, S. T., Huang, S., Heikal, A. A., & Webb, W. W. (2002). Biological and chemical applications of fluorescence correlation spectroscopy: a review. *Biochemistry, 41*(3), 697–705.

Hess, S. T., Kumar, M., Verma, A., Farrington, J., Kenworthy, A., & Zimmerberg, J. (2005). Quantitative electron microscopy and fluorescence spectroscopy of the membrane distribution of influenza hemagglutinin. *The Journal of Cell Biology, 169*(6), 965–976.

Hess, S. T., & Webb, W. W. (2002). Focal volume optics and experimental artifacts in confocal fluorescence correlation spectroscopy. *Biophysical Journal, 83*(4), 2300–2317.

van den Heuvel, L., & Smeitink, J. (2001). The oxidative phosphorylation (OXPHOS) system: nuclear genes and human genetic diseases. *Bioessays, 23*(6), 518–525.

Higgs, H. N., & Pollard, T. D. (2000). Activation by Cdc42 and PIP(2) of Wiskott-Aldrich syndrome protein (WASp) stimulates actin nucleation by Arp2/3 complex. *The Journal of Cell Biology, 150*(6), 1311–1320.

Holden, S. J., Uphoff, S., & Kapanidis, A. N. (2011). DAOSTORM: an algorithm for high-density super-resolution microscopy. *Nature Methods, 8*(4), 279–280.

Honerkamp-Smith, A. R., Veatch, S. L., & Keller, S. L. (2009). An introduction to critical points for biophysicists; observations of compositional heterogeneity in lipid membranes (Review) *Biochimica et Biophysica Acta—Biomembranes, 1788*(1), 53–63.

Honigmann, A., van den Bogaart, G., Iraheta, E., Risselada, H. J., Milovanovic, D., Mueller, V., et al. (2013). Phosphatidylinositol 4,5-bisphosphate clusters act as molecular beacons for vesicle recruitment. *Nature Structural and Molecular Biology, 20*(6), 679–686.

Hosoya, H., Kobayashi, R., Tsukita, S., & Matsumura, F. (1992). Ca(2+)-regulated actin and phospholipid binding protein (68 kD-protein) from bovine liver: identification as a homologue for annexin VI and intracellular localization. *Cell Motility Cytoskeleton, 22*(3), 200–210.

Hotta, J., Fron, E., Dedecker, P., Janssen, K. P., Li, C., Mullen, K., et al. (2010). Spectroscopic rationale for efficient stimulated-emission depletion microscopy fluorophores. *Journal of American Chemical Society, 132*(14), 5021–5023.

Hou, Y., Crossman, D. J., Rajagopal, V., Baddeley, D., Jayasinghe, I., & Soeller, C. (2014). Super-resolution fluorescence imaging to study cardiac biophysics: alpha-actinin distribution and Z-disk topologies in optically thick cardiac tissue slices. *Progress in Biophysics and Molecular Biology, 115*(2–3), 328–339.

Hsu, C. J., & Baumgart, T. (2011). Spatial association of signaling proteins and F-actin effects on cluster assembly analyzed via photoactivation localization microscopy in T cells. *PLoS One, 6*(8), 24.

Huang, F., Hartwich, T. M., Rivera-Molina, F. E., Lin, Y., Duim, W. C., Long, J. J., et al. (2013). Video-rate nanoscopy using sCMOS camera-specific single-molecule localization algorithms. *Nature Methods, 10*(7), 653–658.

Huang, B., Wang, W. Q., Bates, M., & Zhuang, X. W. (2008). Three-dimensional super-resolution imaging by stochastic optical reconstruction microscopy. *Science, 319*(5864), 810–813.

Ingerman, E., Perkins, E. M., Marino, M., Mears, J. A., McCaffery, J. M., Hinshaw, J. E., et al. (2005). Dnm1 forms spirals that are structurally tailored to fit mitochondria. *The Journal of Cell Biology, 170*(7), 1021–1027.

Itano, M. S., Graus, M. S., Pehlke, C., Wester, M. J., Liu, P., Lidke, K. A., et al. (2014). Super-resolution imaging of C-type lectin spatial rearrangement within the dendritic cell plasma membrane at fungal microbe contact sites. *Frontiers in Physics, 2*, 46.

Izeddin, I., Specht, C. G., Lelek, M., Darzacq, X., Triller, A., Zimmer, C., et al. (2011). Super-resolution dynamic imaging of dendritic spines using a low-affinity photoconvertible actin probe. *PLoS One, 6*(1), 0015611.

Jackle, S., Beisiegel, U., Rinninger, F., Buck, F., Grigoleit, A., Block, A., et al. (1994). Annexin VI, a marker protein of hepatocytic endosomes. *The Journal of Biological Chemistry, 269*(2), 1026–1032.

Jans, D. C., Wurm, C. A., Riedel, D., Wenzel, D., Stagge, F., Deckers, M., et al. (2013). STED super-resolution microscopy reveals an array of MINOS clusters along human mitochondria. *Proceedings of the National Academy of Sciences USA, 110*(22), 8936–8941.

Jaqaman, K., Loerke, D., Mettlen, M., Kuwata, H., Grinstein, S., Schmid, S. L., et al. (2008). Robust single-particle tracking in live-cell time-lapse sequences. *Nature Methods, 5*(8), 695–702.

Jaumouille, V., Farkash, Y., Jaqaman, K., Das, R., Lowell, C. A., & Grinstein, S. (2014). Actin cytoskeleton reorganization by Syk regulates Fcgamma receptor responsiveness by increasing its lateral mobility and clustering. *Developmental Cell, 29*(5), 534–546.

Jayasundar, J. J., Ju, J. H., He, L., Liu, D., Meilleur, F., Zhao, J., et al. (2012). Open conformation of ezrin bound to phosphatidylinositol 4,5-bisphosphate and to F-actin revealed by neutron scattering. *The Journal of Biological Chemistry, 287*(44), 37119–37133.

Jia, S., Vaughan, J. C., & Zhuang, X. (2014). Isotropic 3D super-resolution imaging with a self-bending point spread function. *Nature Photonics, 8*, 302–306.

Johri, A., & Beal, M. F. (2012). Mitochondrial dysfunction in neurodegenerative diseases. *Journal of Pharmacology and Experimental Therapeutics, 342*(3), 619–630.

Jones, S. A., Shim, S. H., He, J., & Zhuang, X. (2011). Fast, three-dimensional super-resolution imaging of live cells. *Nature Methods, 8*(6), 499–508.

Jordan, S., & Rodgers, W. (2003). T cell glycolipid-enriched membrane domains are constitutively assembled as membrane patches that translocate to immune synapses. *The Journal of Immunology, 171*(1), 78–87.

Jouvenet, N., Neil, S. J., Bess, C., Johnson, M. C., Virgen, C. A., Simon, S. M., et al. (2006). Plasma membrane is the site of productive HIV-1 particle assembly. *PLoS Biology, 4*(12), e435.

Juette, M. F., Gould, T. J., Lessard, M. D., Mlodzianoski, M. J., Nagpure, B. S., Bennett, B. T., et al. (2008). Three-dimensional sub-100 nm resolution fluorescence microscopy of thick samples. *Nature Methods, 5*(6), 527–529.

Kaksonen, M., Toret, C. P., & Drubin, D. G. (2006). Harnessing actin dynamics for clathrin-mediated endocytosis. *Nature Reviews Molecular Cell Biology, 7*(6), 404–414.

Kanchanawong, P., Shtengel, G., Pasapera, A. M., Ramko, E. B., Davidson, M. W., Hess, H. F., et al. (2010). Nanoscale architecture of integrin-based cell adhesions. *Nature, 468*(7323), 580–584.

Karnovsky, M. J., Kleinfeld, A. M., Hoover, R. L., & Klausner, R. D. (1982). The concept of lipid domains in membranes. *The Journal of Cell Biology, 94*(1), 1–6.

Karpova, T. S., Baumann, C. T., He, L., Wu, X., Grammer, A., Lipsky, P., et al. (2003). Fluorescence resonance energy transfer from cyan to yellow fluorescent protein detected by acceptor photobleaching using confocal microscopy and a single laser. *Journal of Microscopy–Oxford, 209*, 56–70.

Keller, H., Rentsch, P., & Hagmann, J. (2002). Differences in cortical actin structure and dynamics document that different types of blebs are formed by distinct mechanisms. *Experimental Cell Research, 277*(2), 161–172.

Kenworthy, A. K. (2001). Imaging protein-protein interactions using fluorescence resonance energy transfer microscopy. *Methods, 24*(3), 289–296.

Kenworthy, A. K. (2005). Fleeting glimpses of lipid rafts: how biophysics is being used to track them. *Journal of Investigative Medicine, 53*(6), 312–317.

Kenworthy, A. K., & Edidin, M. (1998). Distribution of a glycosylphosphatidylinositol-anchored protein at the apical surface of MDCK cells examined at a resolution of <100 angstrom using imaging fluorescence resonance energy transfer. *The Journal of Cell Biology, 142*(1), 69–84.

Kenworthy, A. K., Nichols, B. J., Remmert, C. L., Hendrix, G. M., Kumar, M., Zimmerberg, J., et al. (2004). Dynamics of putative raft-associated proteins at the cell surface. *The Journal of Cell Biology, 165*(5), 735–746.

Kenworthy, A. K., Petranova, N., & Edidin, M. (2000). High-resolution FRET microscopy of cholera toxin B-subunit and GPI-anchored proteins in cell plasma membranes. *Molecular Biology of the Cell, 11*(5), 1645–1655.

Kirschner, M. W. (1980). Implications of treadmilling for the stability and polarity of actin and tubulin polymers in vivo. *The Journal of Cell Biology, 86*(1), 330–334.

Klar, T. A., Jakobs, S., Dyba, M., Egner, A., & Hell, S. W. (2000). Fluorescence microscopy with diffraction resolution barrier broken by stimulated emission. *Proceedings of the National Academy of Sciences USA, 97*(15), 8206–8210.

Kner, P., Chhun, B. B., Griffis, E. R., Winoto, L., & Gustafsson, M. G. (2009). Super-resolution video microscopy of live cells by structured illumination. *Nature Methods, 6*(5), 339–342.

Kopek, B. G., Shtengel, G., Xu, C. S., Clayton, D. A., & Hess, H. F. (2012). Correlative 3D superresolution fluorescence and electron microscopy reveal the relationship of mitochondrial nucleoids to membranes. *Proceedings of the National Academy of Sciences USA, 109*(16), 6136–6141.

Koppel, D. E. (1981). Association dynamics and lateral transport in biological membranes. *Journal of Supramolecular Structure and Cellular Biochemistry, 17*(1), 61–67.

Korlach, J., Schwille, P., Webb, W. W., & Feigenson, G. W. (1999). Characterization of lipid bilayer phases by confocal microscopy and fluorescence correlation spectroscopy. *Proceedings of the National Academy of Sciences USA, 96*(15), 8461–8466.

Kozlov, M. M., & Chernomordik, L. V. (1998). A mechanism of protein-mediated fusion: coupling between refolding of the influenza hemagglutinin and lipid rearrangements. *Biophysical Journal, 75*(3), 1384–1396.

Kraft, M. L. (2013). Plasma membrane organization and function: moving past lipid rafts. *Molecular Biology of the Cell, 24*(18), 2765–2768.

Kukat, C., Wurm, C. A., Spahr, H., Falkenberg, M., Larsson, N. G., & Jakobs, S. (2011). Super-resolution microscopy reveals that mammalian mitochondrial nucleoids have a uniform size and frequently contain a single copy of mtDNA. *Proceedings of the National Academy of Sciences USA, 108*(33), 13534–13539.

Kumar, M., Kenworthy, A. K., Roth, M. G., & Zimmerberg, J. (2001). Raft association plays an important role in influenza hemagglutinin-mediated fusion (Meeting abstract) *Molecular Biology of the Cell, 12*, 74A.

Kumari, S., Curado, S., Mayya, V., & Dustin, M. L. (2014). T cell antigen receptor activation and actin cytoskeleton remodeling. *Biochimica et Biophysica Acta, 2*, 546–556.

Kusumi, A., Fujiwara, T. K., Chadda, R., Xie, M., Tsunoyama, T. A., Kalay, Z., et al. (2012). Dynamic organizing principles of the plasma membrane that regulate signal transduction: commemorating the fortieth anniversary of Singer and Nicolson's fluid-mosaic model. *Annual Review of Cell and Developmental Biology, 28*, 215–250.

Kusumi, A., & Sako, Y. (1996). Cell surface organization by the membrane skeleton. *Current Opinion in Cell Biology, 8*(4), 566–574.

Kwik, J., Boyle, S., Fooksman, D., Margolis, L., Sheetz, M. P., & Edidin, M. (2003). Membrane cholesterol, lateral mobility, and the phosphatidylinositol 4,5-bisphosphate-dependent organization of cell actin. *Proceedings of the National Academy of Sciences USA, 100*(24), 13964–13969.

Labrousse, A. M., Zappaterra, M. D., Rube, D. A., & van der Bliek, A. M. (1999). *C. elegans* dynamin-related protein DRP-1 controls severing of the mitochondrial outer membrane. *Molecular Cell, 4*(5), 815–826.

Lakadamyali, M., Rust, M. J., Babcock, H. P., & Zhuang, X. W. (2003). Visualizing infection of individual influenza viruses. *Proceedings of the National Academy of Sciences USA, 100*(16), 9280–9285.

Larson, D. R., Gosse, J. A., Holowka, D. A., Baird, B. A., & Webb, W. W. (2005). Temporally resolved interactions between antigen-stimulated IgE receptors and Lyn kinase on living cells. *The Journal of Cell Biology, 171*(3), 527–536.

Lasserre, R., Guo, X. J., Conchonaud, F., Hamon, Y., Hawchar, O., Bernard, A. M., et al. (2008). Raft nanodomains contribute to Akt/PKB plasma membrane recruitment and activation. *Nature Chemical Biology, 4*(9), 538–547.

Le Floc'h, A., & Huse, M. (2015). Molecular mechanisms and functional implications of polarized actin remodeling at the T cell immunological synapse. *Cellular and Molecular Life Sciences, 72*(3), 537–556.

Lebensohn, A. M., & Kirschner, M. W. (2009). Activation of the WAVE complex by coincident signals controls actin assembly. *Molecular Cell, 36*(3), 512–524.

Lee, S. H., Shin, J. Y., Lee, A., & Bustamante, C. (2012). Counting single photoactivatable fluorescent molecules by photoactivated localization microscopy (PALM). *Proceedings of the National Academy of Sciences USA, 109*(43), 17436–17441.

Lehninger, A. L., Nelson, D. L., & Cox, M. M. (2013). *Principles of biochemistry* (6th ed.). New York: W.H. Freeman.

Lenne, P. F., Wawrezinieck, L., Conchonaud, F., Wurtz, O., Boned, A., Guo, X. J., et al. (2006). Dynamic molecular confinement in the plasma membrane by microdomains and the cytoskeleton meshwork. *EMBO Journal, 25*(14), 3245–3256.

Leung, B. O., & Chou, K. C. (2011). Review of super-resolution fluorescence microscopy for biology. *Applied Spectroscopy, 65*(9), 967–980.

Lewis, A., Isaacson, M., Harootunian, A., & Muray, A. (1984). Development of a 500 Å spatial resolution light microscope: I. Light is efficiently transmitted through λ/16 diameter apertures. *Ultramicroscopy, 13*(3), 227–231.

Lichtenberg, D., Goni, F. M., & Heerklotz, H. (2005). Detergent-resistant membranes should not be identified with membrane rafts. *Trends in Biochemical Sciences, 30*(8), 430–436.

Lieberman, J. (2003). The ABCs of granule-mediated cytotoxicity: new weapons in the arsenal. *Nature Reviews Immunology, 3*(5), 361–370.

Lillemeier, B. F., Mortelmaier, M. A., Forstner, M. B., Huppa, J. B., Groves, J. T., & Davis, M. M. (2010). TCR and Lat are expressed on separate protein islands on T cell membranes and concatenate during activation. *Nature Immunology, 11*(1), 90–96.

van de Linde, S., & Sauer, M. (2014). How to switch a fluorophore: from undesired blinking to controlled photoswitching. *Chemical Society Reviews, 43*(4), 1076–1087.

Lingappa, J. R., Reed, J. C., Tanaka, M., Chutiraka, K., & Robinson, B. A. (2014). How HIV-1 Gag assembles in cells: putting together pieces of the puzzle. *Virus Research, 193*, 89–107.

Liu, Y., & Bankaitis, V. A. (2010). Phosphoinositide phosphatases in cell biology and disease. *Progress in Lipid Research, 49*(3), 201–217.

Liu, L., Cooper, T., Howley, P. M., & Hayball, J. D. (2014). From crescent to mature virion: vaccinia virus assembly and maturation. *Viruses, 6*(10), 3787–3808.

Liu, A. P., & Fletcher, D. A. (2006). Actin polymerization serves as a membrane domain switch in model lipid bilayers. *Biophysical Journal, 91*(11), 4064–4070.

Mace, E. M., & Orange, J. S. (2012a). Dual channel STED nanoscopy of lytic granules on actin filaments in natural killer cells. *Communicative and Integrative Biology, 5*(2), 184–186.

Mace, E. M., & Orange, J. S. (2012b). New views of the human NK cell immunological synapse: recent advances enabled by super- and high-resolution imaging techniques. *Frontiers in Immunology, 3*, 421.
Mace, E. M., & Orange, J. S. (2014). Lytic immune synapse function requires filamentous actin deconstruction by Coronin 1A. *Proceedings of the National Academy of Sciences USA, 111*(18), 6708–6713.
MacGillavry, H. D., & Blanpied, T. A. (2013). Single-molecule tracking photoactivated localization microscopy to map nano-scale structure and dynamics in living spines. *Current Protocols in Neuroscience, 2*(220), 1–2.
Machnicka, B., Czogalla, A., Hryniewicz-Jankowska, A., Boguslawska, D. M., Grochowalska, R., Heger, E., et al. (2014). Spectrins: a structural platform for stabilization and activation of membrane channels, receptors and transporters. *Biochimica et Biophysica Acta—Biomembranes, 1838*(2), 620–634.
Magde, D., Elson, E. L., & Webb, W. W. (1972). Thermodynamic fluctuations in a reacting system—measurement by fluorescence correlation spectroscopy. *Physical Review Letters, 29*, 705–708.
Maglione, M., & Sigrist, S. J. (2013). Seeing the forest tree by tree: super-resolution light microscopy meets the neurosciences. *Nature Neuroscience, 16*(7), 790–797.
Manes, S., del Real, G., & Martinez, C. (2003). Pathogens: raft hijackers. *Nature Reviews Immunology, 3*, 557–568.
Manes, S., & Viola, A. (2006). Lipid rafts in lymphocyte activation and migration. *Molecular Membrane Biology, 23*(1), 59–69.
Manley, S., Gillette, J. M., Patterson, G. H., Shroff, H., Hess, H. F., Betzig, E., et al. (2008). High-density mapping of single-molecule trajectories with photoactivated localization microscopy. *Nature Methods, 5*(2), 155–157.
Martin-Belmonte, F., & Mostov, K. (2007). Phosphoinositides control epithelial development. *Cell Cycle, 6*(16), 1957–1961.
Mattila, P. K., Feest, C., Depoil, D., Treanor, B., Montaner, B., Otipoby, K. L., et al. (2013). The actin and tetraspanin networks organize receptor nanoclusters to regulate B cell receptor-mediated signaling. *Immunity, 38*(3), 461–474.
Maxfield, F. R., & Menon, A. K. (2006). Intracellular sterol transport and distribution. *Current Opinion in Cell Biology, 18*(4), 379–385.
Maxfield, F. R., & Tabas, I. (2005). Role of cholesterol and lipid organization in disease (Review) *Nature, 438*(7068), 612–621.
Menon, A. K., Holowka, D., Webb, W. W., & Baird, B. (1986). Cross-linking of receptor-bound IgE to aggregates larger than dimers leads to rapid immobilization. *The Journal of Cell Biology, 102*(2), 541–550.
Meyer, L., Wildanger, D., Medda, R., Punge, A., Rizzoli, S. O., Donnert, G., et al. (2008). Dual-color STED microscopy at 30-nm focal-plane resolution. *Small, 4*(8), 1095–1100.
Michel, V., & Bakovic, M. (2007). Lipid rafts in health and disease (Review) *Biology of the Cell, 99*(3), 129–140.
Mizuno, H., Abe, M., Dedecker, P., Makino, A., Rocha, S., Ohno-Iwashita, Y., et al. (2011). Fluorescent probes for superresolution imaging of lipid domains on the plasma membrane. *Chemical Science, 2*(8), 1548–1553. http://dx.doi.org/10.1039/C1SC00169H.
Mlodzianoski, M. J., Schreiner, J. M., Callahan, S. P., Smolkova, K., Dlaskova, A., Santorova, J., et al. (2011). Sample drift correction in 3D fluorescence photoactivation localization microscopy. *Optics Express, 19*(16), 15009–15019.
Moore, B. M., Lentz, B. R., & Meissner, G. (1978). Effects of sarcoplasmic reticulum Ca^{2+}-ATPase on phospholipid bilayer fluidity: boundary lipid. *Biochemistry, 17*(24), 5248–5255.

Muriel, O., Echarri, A., Hellriegel, C., Pavon, D. M., Beccari, L., & Del Pozo, M. A. (2011). Phosphorylated filamin A regulates actin-linked caveolae dynamics. *Journal of Cell Science, 124*(Pt 16), 2763–2776.

Nayak, D. P., Hui, E. K., & Barman, S. (2004). Assembly and budding of influenza virus. *Virus Research, 106*(2), 147–165.

Nebl, T., Pestonjamasp, K. N., Leszyk, J. D., Crowley, J. L., Oh, S. W., & Luna, E. J. (2002). Proteomic analysis of a detergent-resistant membrane skeleton from neutrophil plasma membranes. *The Journal of Biological Chemistry, 277*(45), 43399–43409.

Nelson, A. J., Gunewardene, M. S., & Hess, S. T. (2014). High speed fluorescence photo-activation localization microscopy imaging. In *Proc. SPIE 9169, Nanoimaging and Nanospectroscopy II, 9169*, 91690p.

Nicolson, G. L. (2014). The Fluid–Mosaic model of membrane structure: still relevant to understanding the structure, function and dynamics of biological membranes after more than 40 years. *Biochimica et Biophysica Acta–Biomembranes, 1838*(6), 1451–1466.

Okabe, S. (2012). Molecular dynamics of the excitatory synapse. *Advances in Experimental Medicine and Biology, 970*, 131–152.

Orange, J. S., Harris, K. E., Andzelm, M. M., Valter, M. M., Geha, R. S., & Strominger, J. L. (2003). The mature activating natural killer cell immunologic synapse is formed in distinct stages. *Proceedings of the National Academy of Sciences USA, 100*(24), 14151–14156.

Oshikane, Y., Kataoka, T., Okuda, M., Hara, S., Inoue, H., & Nakano, M. (2007). Observation of nanostructure by scanning near-field optical microscope with small sphere probe. *Science and Technology of Advanced Materials, 8*, 181–185.

Otera, H., & Mihara, K. (2011). Discovery of the membrane receptor for mitochondrial fission GTPase Drp1. *Small GTPases, 2*(3), 167–172.

Otera, H., Wang, C., Cleland, M. M., Setoguchi, K., Yokota, S., Youle, R. J., et al. (2010). Mff is an essential factor for mitochondrial recruitment of Drp1 during mitochondrial fission in mammalian cells. *The Journal of Cell Biology, 191*(6), 1141–1158.

Owen, D. M., Rentero, C., Rossy, J., Magenau, A., Williamson, D., Rodriguez, M., et al. (2010). PALM imaging and cluster analysis of protein heterogeneity at the cell surface. *Journal of Biophotonics, 3*(7), 446–454.

Palomo, G. M., & Manfredi, G. (2014). Exploring new pathways of neurodegeneration in ALS: the role of mitochondria quality control. *Brain Research, 6*(11), 01330–01339.

Pavani, S. R. P., Thompson, M. A., Biteen, J. S., Lord, S. J., Liu, N., Twieg, R. J., et al. (2009). Three-dimensional, single-molecule fluorescence imaging beyond the diffraction limit by using a double-helix point spread function. *Proceedings of the National Academy of Sciences USA, 106*(9), 2995–2999.

Pertsinidis, A., Zhang, Y., & Chu, S. (2010). Subnanometre single-molecule localization, registration and distance measurements. *Nature, 466*(7306), 647–651.

Philimonenko, A. A., Janacek, J., & Hozak, P. (2000). Statistical evaluation of colocalization patterns in immunogold labeling experiments. *Journal of Structural Biology, 132*(3), 201–210.

Pichler, H., & Riezman, H. (2004). Where sterols are required for endocytosis (Review) *Biochimica et Biophysica Acta–Biomembranes, 1666*(1–2), 51–61.

Plowman, S. J., Muncke, C., Parton, R. G., & Hancock, J. F. (2005). H-ras, K-ras, and inner plasma membrane raft proteins operate in nanoclusters with differential dependence on the actin cytoskeleton. *Proceedings of the National Academy of Sciences USA, 102*(43), 15500–15505.

Pohl, D. W., Denk, W., & Lanz, M. (1984). Optical stethoscopy: image recording with resolution λ/20. *Applied Physics Letters, 44*(7), 651–653.

Polozov, I. V., Bezrukov, L., Gawrisch, K., & Zimmerberg, J. (2008). Progressive ordering with decreasing temperature of the phospholipids of influenza virus. *Nature Chemical Biology, 4*(4), 248–255.

Pore, D., Parameswaran, N., Matsui, K., Stone, M. B., Saotome, I., McClatchey, A. I., et al. (2013). Ezrin tunes the magnitude of humoral immunity. *The Journal of Immunology, 191*(8), 4048–4058.

Rabl, R., Soubannier, V., Scholz, R., Vogel, F., Mendl, N., Vasiljev-Neumeyer, A., et al. (2009). Formation of cristae and crista junctions in mitochondria depends on antagonism between Fcj1 and Su e/g. *The Journal of Cell Biology, 185*(6), 1047–1063.

Radtke, K., Dohner, K., & Sodeik, B. (2006). Viral interactions with the cytoskeleton: a hitchhiker's guide to the cell. *Cellular Microbiology, 8*(3), 387–400.

Rak, G. D., Mace, E. M., Banerjee, P. P., Svitkina, T., & Orange, J. S. (2011). Natural killer cell lytic granule secretion occurs through a pervasive actin network at the immune synapse. *PLoS Biology, 9*(9), 13.

Recktenwald, C. V., Lichtenfels, R., Wulfaenger, J., Müller, A., Dressler, S. P., & Seliger, B. (2015). Impact of the mitogen-activated protein kinase pathway on the subproteome of detergent-resistant microdomains of colon carcinoma cells. *Proteomics, 15*(1), 77–88.

van Rheenen, J., Song, X., van Roosmalen, W., Cammer, M., Chen, X., Desmarais, V., et al. (2007). EGF-induced PIP2 hydrolysis releases and activates cofilin locally in carcinoma cells. *The Journal of Cell Biology, 179*(6), 1247–1259.

Ripley, B. D. (1977). Modeling spatial patterns. *Journal of Royal Statistical Society: Series B—Methodological, 39*(2), 172–212.

Rocha-Perugini, V., Gordon-Alonso, M., & Sanchez-Madrid, F. (2014). PIP2: choreographer of actin-adaptor proteins in the HIV-1 dance. *Trends in Microbiology, 22*(7), 379–388.

Rocha-Perugini, V., Zamai, M., Gonzalez-Granado, J. M., Barreiro, O., Tejera, E., Yanez-Mo, M., et al. (2013). CD81 controls sustained T cell activation signaling and defines the maturation stages of cognate immunological synapses. *Molecular and Cellular Biology, 33*(18), 3644–3658.

Rodgers, W., Farris, D., & Mishra, S. (2005). Merging complexes: properties of membrane raft assembly during lymphocyte signaling. *Trends in Immunology, 26*(2), 97–103.

Rojo, L., Sjoberg, M. K., Hernandez, P., Zambrano, C., & Maccioni, R. B. (2006). Roles of cholesterol and lipids in the etiopathogenesis of Alzheimer's disease (Review) *Journal of Biomedicine and Biotechnology, 2006*(73976), 1–17.

Rollason, R., Korolchuk, V., Hamilton, C., Jepson, M., & Banting, G. (2009). A CD317/tetherin-RICH2 complex plays a critical role in the organization of the subapical actin cytoskeleton in polarized epithelial cells. *The Journal of Cell Biology, 184*(5), 721–736.

Rosenbloom, A. B., Lee, S. H., To, M., Lee, A., Shin, J. Y., & Bustamante, C. (2014). Optimized two-color super resolution imaging of Drp1 during mitochondrial fission with a slow-switching Dronpa variant. *Proceedings of the National Academy of Sciences USA, 111*(36), 13093–13098.

Rossier, O., Octeau, V., Sibarita, J. B., Leduc, C., Tessier, B., Nair, D., et al. (2012). Integrins beta1 and beta3 exhibit distinct dynamic nanoscale organizations inside focal adhesions. *Nature Cell Biology, 14*(10), 1057–1067.

Rossy, J., Pageon, S. V., Davis, D. M., & Gaus, K. (2013). Super-resolution microscopy of the immunological synapse. *Current Opinion in Immunology, 25*(3), 307–312.

Roy, N. H., Chan, J., Lambele, M., & Thali, M. (2013). Clustering and mobility of HIV-1 Env at viral assembly sites predict its propensity to induce cell-cell fusion. *Journal of Virology, 87*(13), 7516–7525.

Ruigrok, R. W., Krijgsman, P. C., de Ronde-Verloop, F. M., & de Jong, J. C. (1985). Natural heterogeneity of shape, infectivity and protein composition in an influenza A (H3N2) virus preparation. *Virus Research, 3*, 69–76.

Rust, M. J., Bates, M., & Zhuang, X. (2006). Sub-diffraction-limit imaging by stochastic optical reconstruction microscopy (STORM). *Nature Methods, 3*(10), 793–795.

Saarikangas, J., Zhao, H., & Lappalainen, P. (2010). Regulation of the actin cytoskeleton-plasma membrane interplay by phosphoinositides. *Physiological Reviews, 90*(1), 259–289.

Saini, H. K., Arneja, A. S., & Dhalla, N. S. (2004). Role of cholesterol in cardiovascular dysfunction (Review) *Canadian Journal of Cardiology, 20*(3), 333–346.
Saka, S. K., Honigmann, A., Eggeling, C., Hell, S. W., Lang, T., & Rizzoli, S. O. (2014). Multi-protein assemblies underlie the mesoscale organization of the plasma membrane. *Nature Communications, 5*, 4509.
Sako, Y., & Kusumi, A. (1994). Compartmentalized structure of the plasma membrane for receptor movements as revealed by a nanometer-level motion analysis. *The Journal of Cell Biology, 125*(6), 1251–1264.
Salbreux, G., Charras, G., & Paluch, E. (2012). Actin cortex mechanics and cellular morphogenesis. *Trends in Cell Biology, 22*(10), 536–545.
Saltel, F., Mortier, E., Hytonen, V. P., Jacquier, M. C., Zimmermann, P., Vogel, V., et al. (2009). New PI(4,5)P2- and membrane proximal integrin-binding motifs in the talin head control beta3-integrin clustering. *The Journal of Cell Biology, 187*(5), 715–731.
Scheiffele, P., Rietveld, A., Wilk, T., & Simons, K. (1999). Influenza viruses select ordered lipid domains during budding from the plasma membrane. *The Journal of Biological Chemistry, 274*(4), 2038–2044.
Scheiffele, P., Roth, M. G., & Simons, K. (1997). Interaction of influenza virus haemagglutinin with sphingolipid-cholesterol membrane domains via its transmembrane domain. *EMBO Journal, 16*(18), 5501–5508.
Schermelleh, L., Heintzmann, R., & Leonhardt, H. (2010). A guide to super-resolution fluorescence microscopy. *The Journal of Cell Biology, 190*(2), 165–175.
Schwille, P., Korlach, J., & Webb, W. W. (1999). Anomalous subdiffusion of proteins and lipids in membranes observed by fluorescence correlation spectroscopy. *Biophysical Journal, 76*(1), A391.
Semrau, S., & Schmidt, T. (2007). Particle image correlation spectroscopy (PICS): retrieving nanometer-scale correlations from high-density single-molecule position data. *Biophysical Journal, 92*(2), 613–621.
Sengupta, P., Jovanovic-Talisman, T., Skoko, D., Renz, M., Veatch, S. L., & Lippincott-Schwartz, J. (2011). Probing protein heterogeneity in the plasma membrane using PALM and pair correlation analysis. *Nature Methods, 8*(11), 969–975.
Sharma, P., Varma, R., Sarasij, R. C., Ira, Gousset, K., Krishnamoorthy, G., et al. (2004). Nanoscale organization of multiple GPI-anchored proteins in living cell membranes. *Cell, 116*(4), 577–589.
Sharonov, A., & Hochstrasser, R. M. (2006). Wide-field subdiffraction imaging by accumulated binding of diffusing probes. *Proceedings of the National Academy of Sciences USA, 103*(50), 18911–18916.
Shaw, M. L., Stone, K. L., Colangelo, C. M., Gulcicek, E. E., & Palese, P. (2008). Cellular proteins in influenza virus particles. *Plos Pathogens, 4*(6), e1000085.
Sheetz, M. P. (1983). Membrane skeletal dynamics: role in modulation of red cell deformability, mobility of transmembrane proteins, and shape. *Seminars in Hematology, 20*(3), 175–188.
Shelby, S. A., Holowka, D., Baird, B., & Veatch, S. L. (2013). Distinct stages of stimulated FcepsilonRI receptor clustering and immobilization are identified through superresolution imaging. *Biophysical Journal, 105*(10), 2343–2354.
Sherman, E., Barr, V., Manley, S., Patterson, G., Balagopalan, L., Akpan, I., et al. (2011). Functional nanoscale organization of signaling molecules downstream of the T cell antigen receptor. *Immunity, 35*(5), 705–720.
Sherman, E., Barr, V., & Samelson, L. E. (2013). Super-resolution characterization of TCR-dependent signaling clusters. *Immunological Reviews, 251*(1), 21–35.
Shim, S. H., Xia, C., Zhong, G., Babcock, H. P., Vaughan, J. C., Huang, B., et al. (2012). Super-resolution fluorescence imaging of organelles in live cells with photoswitchable membrane probes. *Proceedings of the National Academy of Sciences USA, 109*(35), 13978–13983.

Shroff, H., Galbraith, C. G., Galbraith, J. A., & Betzig, E. (2008). Live-cell photoactivated localization microscopy of nanoscale adhesion dynamics. *Nature Methods, 5*(5), 417–423.

Shroff, H., Galbraith, C. G., Galbraith, J. A., White, H., Gillette, J., Olenych, S., et al. (2007). Dual-color superresolution imaging of genetically expressed probes within individual adhesion complexes. *Proceedings of the National Academy of Sciences USA, 104*(51), 20308–20313.

Shtengel, G., Galbraith, J. A., Galbraith, C. G., Lippincott-Schwartz, J., Gillette, J. M., Manley, S., et al. (2009). Interferometric fluorescent super-resolution microscopy resolves 3D cellular ultrastructure. *Proceedings of the National Academy of Sciences USA, 106*(9), 3125–3130.

Sigrist, S. J., & Sabatini, B. L. (2012). Optical super-resolution microscopy in neurobiology. *Current Opinion in Neurobiology, 22*(1), 86–93.

Simons, K., & Gerl, M. J. (2010). Revitalizing membrane rafts: new tools and insights. *Nature Reviews Molecular Cell Biology, 11*(10), 688–699.

Simons, K., & Ikonen, E. (1997). Functional rafts in cell membranes. *Nature, 387*(6633), 569–572.

Simons, K., & van Meer, G. (1988). Lipid sorting in epithelial cells. *Biochemistry, 27*(17), 6197–6202.

Simons, K., & Wandinger-Ness, A. (1990). Polarized sorting in epithelia. *Cell, 62*(2), 207–210.

Singer, S. J. (1971). The molecuar organization of biological membranes. In L. I. Rothfield (Ed.), *Structure and function of biological membranes* (pp. 145–222). New York: Academic Press.

Singer, S. J., & Nicolson, G. L. (1972). The fluid mosaic model of the structure of cell membranes. *Science, 175*, 720–731.

Singleton, P. A., Dudek, S. M., Chiang, E. T., & Garcia, J. G. (2005). Regulation of sphingosine 1-phosphate-induced endothelial cytoskeletal rearrangement and barrier enhancement by S1P1 receptor, PI3 kinase, Tiam1/Rac1, and alpha-actinin. *FASEB Journal, 19*(12), 1646–1656.

Siraganian, R. P. (2003). Mast cell signal transduction from the high-affinity IgE receptor. *Current Opinion in Immunology, 15*(6), 639–646.

Skehel, J. J., & Wiley, D. C. (2000). Receptor binding and membrane fusion in virus entry: the influenza hemagglutinin (Review) *Annual Review of Biochemistry, 69*, 531–569.

Skehel, J. J., & Wiley, D. C. (2002). Influenza haemagglutinin. *Vaccine, 20*, S51–S54.

Sneath, P. H. (1957). The application of computers to taxonomy. *Journal of General Microbiology, 17*(1), 201–226.

Soares, H., Henriques, R., Sachse, M., Ventimiglia, L., Alonso, M. A., Zimmer, C., et al. (2013). Regulated vesicle fusion generates signaling nanoterritories that control T cell activation at the immunological synapse. *The Journal of Experimental Medicine, 210*(11), 2415–2433.

Song, Y., Kenworthy, A. K., & Sanders, C. R. (2014). Cholesterol as a co-solvent and a ligand for membrane proteins. *Protein Science, 23*(1), 1–22.

Spendier, K., Lidke, K. A., Lidke, D. S., & Thomas, J. L. (2012). Single-particle tracking of immunoglobulin E receptors (FcepsilonRI) in micron-sized clusters and receptor patches. *FEBS Letters, 586*(4), 416–421.

Stahelin, R. V. (2014). Could the Ebola virus matrix protein VP40 be a drug target? *Expert Opinion on Therapeutic Targets, 18*(2), 115–120.

Stottrup, B. L., Veatch, S. L., & Keller, S. L. (2004). Nonequilibrium behavior in supported lipid membranes containing cholesterol. *Biophysical Journal, 86*(5), 2942–2950.

Stradal, T. E., Rottner, K., Disanza, A., Confalonieri, S., Innocenti, M., & Scita, G. (2004). Regulation of actin dynamics by WASP and WAVE family proteins. *Trends in Cell Biology, 14*(6), 303–311.

Stryer, L. (1978). Fluorescence energy-transfer as a spectroscopic ruler (Review) *Annual Review of Biochemistry, 47*, 819–846.

Stump, R. F., Pfeiffer, J. R., Seagrave, J., & Oliver, J. M. (1988). Mapping gold-labeled IgE receptors on mast cells by scanning electron microscopy: receptor distributions revealed by silver enhancement, backscattered electron imaging, and digital image analysis. *Journal of Histochemistry and Cytochemistry, 36*(5), 493–502.

Sun, Y., Thapa, N., Hedman, A. C., & Anderson, R. A. (2013). Phosphatidylinositol 4,5-bisphosphate: targeted production and signaling. *Bioessays, 35*(6), 513–522.

Sund, S. E., & Axelrod, D. (2000). Actin dynamics at the living cell submembrane imaged by total internal reflection fluorescence photobleaching. *Biophysical Journal, 79*(3), 1655–1669.

Sverdlov, M., Shinin, V., Place, A. T., Castellon, M., & Minshall, R. D. (2009). Filamin A regulates caveolae internalization and trafficking in endothelial cells. *Molecular Biology of the Cell, 20*(21), 4531–4540.

Tabas, I. (2002). Cholesterol in health and disease. *The Journal of Clinical Investigation, 110*(5), 583–590.

Takeda, M., Leser, G. P., Russell, C. J., & Lamb, R. A. (2003). Influenza virus hemagglutinin concentrates in lipid raft microdomains for efficient viral fusion. *Proceedings of the National Academy of Sciences USA, 100*(25), 14610–14617.

Tanford, C. (1978). Hydrophobic effect and organization of living Matter. *Science, 200*(4345), 1012–1018.

Tanford, C. (1991). *The hydrophobic effect: Formation of micelles and biological membranes* (2nd ed.). Malabar, Fla: Krieger.

Tank, D. W., Wu, E. S., & Webb, W. W. (1982). Enhanced molecular diffusibility in muscle membrane blebs: release of lateral constraints. *The Journal of Cell Biology, 92*(1), 207–212.

Tavano, R., Contento, R. L., Baranda, S. J., Soligo, M., Tuosto, L., Manes, S., et al. (2006). CD28 interaction with filamin-A controls lipid raft accumulation at the T-cell immunological synapse. *Nature Cell Biology, 8*(11), 1270–1276.

Thompson, T. E., & Tillack, T. W. (1985). Organization of glycosphingolipids in bilayers and plasma membranes of mammalian cells. *Annual Review of Biophysics and Biophysical Chemistry, 14*, 361–386.

Treanor, B., Depoil, D., Bruckbauer, A., & Batista, F. D. (2011). Dynamic cortical actin remodeling by ERM proteins controls BCR microcluster organization and integrity. *The Journal of Experimental Medicine, 208*(5), 1055–1068.

Tsuji, A., & Ohnishi, S. (1986). Restriction of the lateral motion of band 3 in the erythrocyte membrane by the cytoskeletal network: dependence on spectrin association state. *Biochemistry, 25*(20), 6133–6139.

Van Engelenburg, S. B., Shtengel, G., Sengupta, P., Waki, K., Jarnik, M., Ablan, S. D., et al. (2014). Distribution of ESCRT machinery at HIV assembly sites reveals virus scaffolding of ESCRT subunits. *Science, 343*(6171), 653–656.

Veatch, S. L. (2007). From small fluctuations to large-scale phase separation: lateral organization in model membranes containing cholesterol (Review) *Seminars in Cell and Developmental Biology, 18*(5), 573–582.

Veatch, S. L., Chiang, E. N., Sengupta, P., Holowka, D. A., & Baird, B. A. (2012). Quantitative nanoscale analysis of IgE-FcepsilonRI clustering and coupling to early signaling proteins. *The Journal of Physical Chemistry B, 116*(23), 6923–6935.

Veatch, S. L., Cicuta, P., Sengupta, P., Honerkamp-Smith, A., Holowka, D., & Baird, B. (2008). Critical fluctuations in plasma membrane vesicles. *ACS Chemical Biology, 3*(5), 287–293.

Veatch, S. L., Gawrisch, K., & Keller, S. L. (2006). Closed-loop miscibility gap and quantitative tie-lines in ternary membranes containing diphytanoyl PC. *Biophysical Journal, 90*(12), 4428–4436.

Veatch, S. L., & Keller, S. L. (2002). Organization in lipid membranes containing cholesterol. *Physical Review Letters, 89*(26), 268101.

Veatch, S. L., & Keller, S. L. (2003a). A closer look at the canonical "raft mixture" in model membrane studies (Letter) *Biophysical Journal, 84*(1), 725–726.

Veatch, S. L., & Keller, S. L. (2003b). Separation of liquid phases in giant vesicles of ternary mixtures of phospholipids and cholesterol. *Biophysical Journal, 85*(5), 3074–3083.

Veatch, S. L., & Keller, S. L. (2005). Miscibility phase diagrams of giant vesicles containing sphingomyelin. *Physical Review Letters, 94*(14), 148101.

Veatch, S. L., Polozov, I. V., Gawrisch, K., & Keller, S. L. (2004). Liquid domains in vesicles investigated by NMR and fluorescence microscopy. *Biophysical Journal, 86*(5), 2910–2922.

Vicente-Manzanares, M., Ma, X., Adelstein, R. S., & Horwitz, A. R. (2009). Non-muscle myosin II takes centre stage in cell adhesion and migration. *Nature Reviews Molecular Cell Biology, 10*(11), 778–790.

Viola, A., & Gupta, N. (2007). Tether and trap: regulation of membrane-raft dynamics by actin-binding proteins (Review) *Nature Reviews Immunology,* 7(11), 889–896.

Wang, J., & Richards, D. A. (2012). Segregation of PIP2 and PIP3 into distinct nanoscale regions within the plasma membrane. *Biology Open, 1*(9), 857–862.

Wawrezinieck, L., Rigneault, H., Marguet, D., & Lenne, P. F. (2005). Fluorescence correlation spectroscopy diffusion laws to probe the submicron cell membrane organization. *Biophysical Journal, 89*(6), 4029–4042.

Wegner, A. (1976). Head to tail polymerization of actin. *Journal of Molecular Biology, 108*(1), 139–150.

Weigel, A. V., Simon, B., Tamkun, M. M., & Krapf, D. (2011). Ergodic and nonergodic processes coexist in the plasma membrane as observed by single-molecule tracking. *Proceedings of the National Academy of Sciences USA, 108*(16), 6438–6443.

Westphal, V., Rizzoli, S. O., Lauterbach, M. A., Kamin, D., Jahn, R., & Hell, S. W. (2008). Video-rate far-field optical nanoscopy dissects synaptic vesicle movement. *Science, 320*(5873), 246–249.

Wiley, D. C., & Skehel, J. J. (1987). The structure and function of the hemagglutinin membrane glycoprotein of influenza-virus (Review) *Annual Review of Biochemistry, 56*, 365–394.

Williamson, D. J., Owen, D. M., Rossy, J., Magenau, A., Wehrmann, M., Gooding, J. J., et al. (2011). Pre-existing clusters of the adaptor Lat do not participate in early T cell signaling events. *Nature Immunology, 12*(7), 655–662.

Winckler, P., Lartigue, L., Giannone, G., De Giorgi, F., Ichas, F., Sibarita, J. B., et al. (2013). Identification and super-resolution imaging of ligand-activated receptor dimers in live cells. *Scientific Reports, 3*, 2387.

Wu, C. Y., Lin, M. W., Wu, D. C., Huang, Y. B., Huang, H. T., & Chen, C. L. (2014). The role of phosphoinositide-regulated actin reorganization in chemotaxis and cell migration. *British Journal Pharmacology, 171*(24), 5541–5554.

Xu, K., Babcock, H. P., & Zhuang, X. (2012). Dual-objective STORM reveals three-dimensional filament organization in the actin cytoskeleton. *Nature Methods, 9*(2), 185–188.

Yeomans, J. M. (2002). *Statistical mechanics of phase transitions*. Oxford: Oxford University Press.

Yonemura, S., Matsui, T., & Tsukita, S. (2002). Rho-dependent and -independent activation mechanisms of ezrin/radixin/moesin proteins: an essential role for polyphosphoinositides in vivo. *Journal of Cell Science, 115*(Pt 12), 2569–2580.

Yoon, Y., Pitts, K. R., & McNiven, M. A. (2001). Mammalian dynamin-like protein DLP1 tubulates membranes. *Molecular Biology of the Cell, 12*(9), 2894–2905.

Zacharias, D. A., Violin, J. D., Newton, A. C., & Tsien, R. Y. (2002). Partitioning of lipid-modified monomeric GFPs into membrane microdomains of live cells. *Science, 296*(5569), 913–916.

van Zanten, T. S., Cambi, A., Koopman, M., Joosten, B., Figdor, C. G., & Garcia-Parajo, M. F. (2009). Hotspots of GPI-anchored proteins and integrin nanoclusters function as nucleation sites for cell adhesion. *Proceedings of the National Academy of Sciences USA, 106*(44), 18557–18562.

Zerbes, R. M., van der Klei, I. J., Veenhuis, M., Pfanner, N., van der Laan, M., & Bohnert, M. (2012). Mitofilin complexes: conserved organizers of mitochondrial membrane architecture. *Biological Chemistry, 393*(11), 1247–1261.

Zhong, G., He, J., Zhou, R., Lorenzo, D., Babcock, H. P., Bennett, V., et al. (2014). Developmental mechanism of the periodic membrane skeleton in axons. *eLife, 23*(3), 04581.

Zidovetzki, R., & Levitan, I. (2007). Use of cyclodextrins to manipulate plasma membrane cholesterol content: evidence, misconceptions and control strategies. *Biochimica et Biophysica Acta, 6*(24), 6.

CHAPTER FOUR

The Nanoscale Organization of Signaling Domains at the Plasma Membrane

Juliette Griffié, Garth Burn and Dylan M. Owen*
Department of Physics and Randall Division of Cell and Molecular Biophysics, King's College London, London, UK
*Corresponding author: E-mail: dylan.owen@kcl.ac.uk

Contents

Current Topics in Membranes, Volume 75
ISSN 1063-5823
http://dx.doi.org/10.1016/bs.ctm.2015.03.004

Abstract

In this chapter, we present an overview of the role of the nanoscale organization of signaling domains in regulating key cellular processes. In particular, we illustrate the importance of protein and lipid nanodomains as triggers and mediators of cell signaling. As particular examples, we summarize the state of the art of understanding the role of nanodomains in the mounting of an immune response, cellular adhesion, intercellular communication, and cell proliferation. Thus, this chapter underlines the essential role the nanoscale organization of key signaling proteins and lipid domains. We will also see how nanodomains play an important role in the lifecycle of many pathogens relevant to human disease and therefore illustrate how these structures may become future therapeutic targets.

Abbreviations

AIDS	Acquired immunodeficiency syndrome
APC	Antigen presenting cell
BAM	Brain activity mapping
BCR	B cell receptor
BDNF	Brain-derived neurotrophic factor
CAM	Cell adhesion molecule
DAF	Decay accelerating factor
EM	Electron microscopy
FRET	Förster resonance energy transfer
HIV	Human immunodeficiency virus
IFN	Interferon
IS	Immunological synapse
ITAM	Intracellular tyrosine-based activation motif
LAT	Linker for activation of T cells
NK	Natural killer
PALM	Photoactivated localization microscopy
PH	Pleckstrin homology
pMHC	Peptide-major histocompatibility complex
Ras	Rat sarcoma
SH	Src-homology
SIM	structured illumination microscopy
SMLM	Single molecule localization microscopy
SNAP	Soluble *N*-ethylmaleimide-sensitive factor attachment protein
SNARE	Soluble *N*-ethylmaleimide-sensitive factor attachment protein receptor
STED	stimulated emission depletion
(d)STORM	(direct) Stochastic optical reconstruction microscopy
SV40	Simian virus 40
SyV	Synaptic vesicles
TCR	T cell receptor
TLR	Toll-like receptors
ZAP70	Zeta-chain associated protein kinase of 70 kD

1. INTRODUCTION

The plasma membrane (Singer & Nicolson, 1972) is the interface between a cell and its environment and is therefore an essential cellular organelle. Although its most obvious function is to delineate and contain intracellular activity by separating the intracellular complexes from the extracellular environment, it has been shown that it also plays the role of a key communication platform (Kusumi et al., 2012). Understanding how cells signal via the plasma membrane has been, and remains still today, an open question. The heterogeneity of the plasma membrane, and in particular, its nanoscale organization has been shown to be an important regulatory component associated with signaling. Nanoscale domains are of three types: lipid order-specific nanodomains in the plasma membrane itself (lipid rafts) (Pike, 2006; Sengupta, Baird, & Holowka, 2007; Simons & Ikonen, 1997; Simons & Vaz, 2004), protein clusters associated with lipid domains, (lipid-protein nanodomains), and protein—protein complexes. Therefore, these nanodomains result from three main molecular interactions: lipid—lipid, lipid-protein, and protein—protein. In fact, around 30% of the proteins encoded in our genome encode membrane proteins. Thus, signaling nanoclusters are frequently associated with the plasma membrane.

Indeed, the spatiotemporal organization of the plasma membrane, especially molecular clustering in signaling nanodomains, has been shown to be a key component in various cellular activities such as cell activation, migration and survival, as well as intercellular signal transmission. These molecular processes also appear to have larger scale repercussions, for example, the triggering of an immune response or neuronal activity. Thus, disrupting the organization of these signaling domains can modify cellular activity leading to pathology; autoimmune disease, neurodegenerative disease, or cancer. Understanding and characterizing these domains has therefore become a major research topic, profiting from the development of laboratory techniques in biophysics, which provide new tools for the study and analysis of these processes. As the organization of these domains is typically on the nanoscale (below the diffraction limit of conventional microscopy), a detailed description of their behavior has been dependent on technological development (Jacobson & Dietrich, 1999; Jacobson, Mouritsen, & Anderson, 2007; Owen, Magenau, Williamson, & Gaus, 2012). In the past, their study has heavily relied on biochemistry (Babiychuk & Draeger,

2006; Garner, Smith, & Hooper, 2008; Lingwood & Simons, 2007), however, the development of new fluorescent microscopy techniques, in particular super resolution methods (Huang, 2010; Lippincott-Schwartz & Manley, 2009; Schermelleh, Heintzmann, & Leonhardt, 2010), has allowed scientists to visualize these signaling domains for the first time.

Although a better understanding of signaling domains is emerging from work based on super resolution, their characteristics and function remain mainly to be discovered. The case of lipid rafts are a vivid example of the issues that the field has been facing. Indeed the existence of rafts remains a source of controversy despite studies tending to suggest their existence (Eggeling et al., 2009; Gomez-Mouton et al., 2004; Lingwood & Simons, 2010; Owen, Magenau, et al., 2012; Pike, 2003, 2006). We will show that recent studies suggest that protein clustering as a result of raft association constitutes a core principle in signal regulation. Proteins can interact with the plasma membrane via post translation modifications that lead to the addition of myristoyl or palmitoyl fatty acid chains, for example, which has been shown to be an especially important regulatory element (Brown, 2006; Rocks et al., 2010; Smotrys & Linder, 2004).

In this chapter, we will present the current state of the art of understanding the nanoscale organization of signaling domains at the plasma membrane, focusing on the role of membrane lipid nanodomains. Here, we will use the following definition of these structures; that lipid nanodomains (or rafts) are highly dynamic, cholesterol- and sphingolipid-enriched nanoscale lipid domains within the plasma membrane (Lingwood & Simons, 2010; Owen, Magenau, et al., 2012; Pike, 2006). More generally, we will present the role of protein nanoclusters in cellular communication and signaling processes resulting from protein—protein interactions as a result of nanocluster formation. We will also discuss the dynamics of such nanoclusters and focus on the role of signaling nanoclusters in the activation of immune cells, the organization of focal adhesions, and in intercellular signal transmission and transduction (Cebecauer, Spitaler, Serge, & Magee, 2010; Garcia-Parajo, Cambi, Torreno-Pina, Thompson, & Jacobson, 2014). We will also discuss signaling nanodomains from a therapeutic perspective, summarizing their exploitation by certain pathogens, particularly viruses, which use these domains as an entry point into intracellular compartments (Jury, Flores-Borja, & Kabouridis, 2007; Maxfield & Tabas, 2005; Simons & Ehehalt, 2002).

2. SIGNALING COMPLEX FORMATION IN CELL ACTIVATION: THE CASE OF T CELLS

2.1 Introduction

Signaling pathways associated with cell activation as part of the immune system have been extensively studied. Their main components, such as the various cell receptors, kinases, adaptor proteins, and so on are today well identified (Dustin & Groves, 2012; Weiss & Littman, 1994). Nonetheless, the role of plasma membrane heterogeneity, in particular the existence of ordered-phase lipid nanodomains and signaling protein clusters for the most part remains to be characterized (Cheng, Dykstra, Mitchell, & Pierce, 1999; Dykstra, Cherukuri, Sohn, Tzeng, & Pierce, 2003). We will summarize here the state of art on this topic focusing in particular on T cells, as it is a powerful illustration of the key role of nanodomains in signal regulation. It has been shown that these signaling domains play a major role in the triggering of an immune response (Choudhuri et al., 2014; Dustin & Groves, 2012; Sherman et al., 2011). For example, when patrolling the body via the blood or the lymph system, immune cells such T and B lymphocytes or natural killer (NK) cells are constantly self-regulating their signaling machinery via nanoscale protein and lipid domains. However, it is the reorganization of signaling domains on the nanoscale, both spatially and structurally that leads to the activation of the cell (Davis & Dustin, 2004; Pizzo & Viola, 2004).

Indeed, the immune system is the body's main defense against pathogens and malignant cells. It is composed of both an innate and adaptive system. Unlike the innate system, which does not adapt to new threats, immune cells associated with the adaptive system (that is to say, with unique antigen specificity), are continuously patrolling the body in search of such threats, testing antigen presenting cells (APCs) for signs of infection. If an immune cell recognizes specific antigen displayed on the surface of an APC in the form of a peptide-major histocompatibility complex (pMHC), it will activate by the formation of an immune synapse and a downstream signaling cascade. Activation, in particular of a T Lymphocyte, results in proliferation and differentiation of T and B cells, which will either eliminate the threat and/or activate (via soluble factors) other cells of the immune system such as NK cells or phagocytes.

Impressively, this essential mechanism is initiated by a unique interaction between two nanoscale receptor complexes: the engagement of a

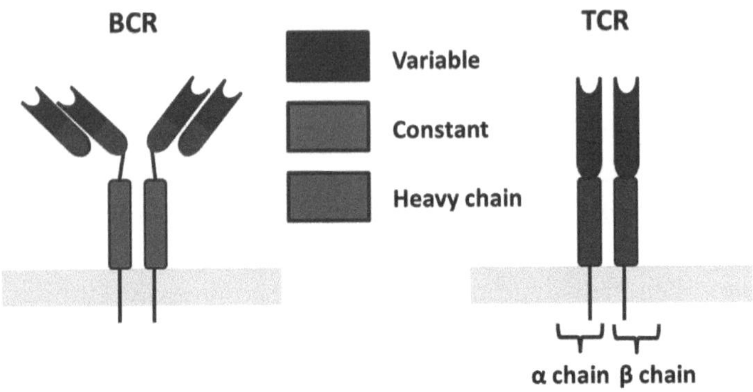

Figure 1 B cell receptor (BCR) and T cell receptor (TCR) main subunits, highlighting their variable components that enable the creation of a vast repertoire. (See color plate)

receptor complex on the immune cell membrane with the peptide presented on the MHC of an APC. In fact, it has been shown that each immune cell expresses a very specific kind of receptor complex on its surface; for instance T cell receptor (TCR) in the case of T cells or the B cell receptor (BCR) in the case of B cells (Figure 1). In contrast, the cells of the innate system (such as NK cells) display multiple types of receptors on their surface (for example, toll-like receptors (TLR)); a repertoire capable of recognizing limited threats as well as warnings from abnormal cell behavior and adaptive immune cell signaling. Therefore, these receptor complexes, composed of various subunits, play the role of signaling domains and are key to the immune response (Dustin & Groves, 2012; Grakoui et al., 1999).

Interestingly, the generation of a large-scale, stable signaling platform between these immune cells and their targets (i.e., an immunological synapse (IS)) is characterized by extensive protein and lipid reorganization. The formation of a mature immune synapse functioning as a stable signaling platform and leading eventually the initiation of the immune response is critically dependent on the nanoscale organization of receptors, kinases, phosphatases, adaptors, and their associated lipid-ordered nanodomains (Davis et al., 2003; Grakoui, et al., 1999; Mossman, Campi, Groves, & Dustin, 2005; Owen, Oddos, et al., 2010; Purbhoo et al., 2010; Rossy, Williamson, & Gaus, 2012; Xavier, Brennan, Li, McCormack, & Seed, 1998). Understanding the behavior of signaling domains at the IS, in particular TCR signaling complexes, therefore appear crucial in immunology. We will focus on this aspect in the following section.

2.2 Receptor complexes engagement and the formation of the immune synapse

The principal signaling nanodomains of cells of the adaptive immune system are their receptor complexes; TCR in the case of a T cell and BCR for B cells. The capacity of an immune system to recognize a pathogen essentially depends on the efficiency of these antigen-specific receptor complexes, in particular, the affinity of binding and the efficiency of a signal cascade initiation. Each of these receptors (TCR or BCR) is unique and therefore adapted to a very specific antigen (Goldrath & Bevan, 1999). There are of the order of 10^7 different TCRs and BCRs expressed in the human body. The production of such unique complexes relies on post-transcriptional modification of the gene sequence, a process called somatic recombination, allowing the generation of a potential repertoire of more than 10^{14} different TCR. Such a variety explains why each individual will react differently to an infection depending on the TCR repertoire available. The aim is to produce, stochastically, a vast variety of receptor shapes and affinities. This process allows the human body to face most existing pathogen threats, moreover providing a pool of receptors capable of detecting almost any unknown threat. TCR complexes can be found in pre-existing clusters which undergo spatial rearrangement following engagement (Crites et al., 2014; Dustin & Groves, 2012). These complexes are formed from the TCR heterodimer itself, 4 CD3 subunits, and 2 zeta chains. The CD3 subunits and zeta chains contain the intracellular tyrosine-based activation motifs (ITAMs) through which signaling occurs (Figure 2).

Following the key step of receptor complex recognition, the immune synapse plays a major role as a signaling interface. Early signaling occurs on, or very close to, this interface. The first step is the phosphorylation of the ITAMs on the TCR-CD3 complex. Indeed, the TCR subunits themselves have no signaling capabilities (Figure 2). If engagement is indeed necessary for the triggering of TCR signaling pathway, the signal itself is initiated via the CD3 and zeta chains via phosphorylation by the kinase Lck. Phosphorylated ITAMs then recruit the kinase ZAP70 which in turn recruits adaptor proteins such as linker for activation of T cells (LAT) and SLP76 forming larger protein complexes. Therefore, following TCR engagement, it is at the immune synapse that nanodomains amplify the initial signal associated to the receptor engagement by bringing together clusters of the main proteins required for the initialization of a signaling cascade.

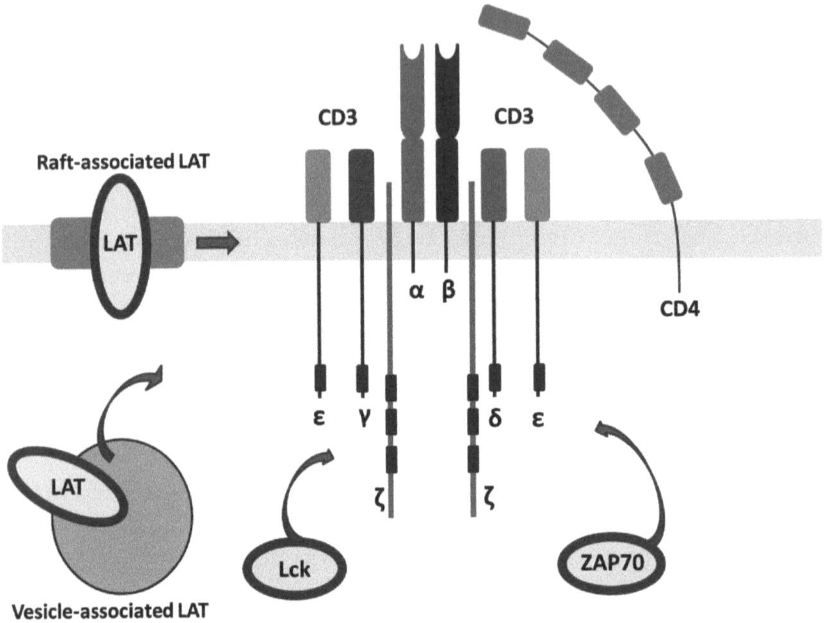

Figure 2 The T cell receptor early signaling pathway. Illustration of the recruitment of key signaling proteins in the form of nanoclusters either sustained by protein–protein interactions or protein-lipid interactions in the case of linker for activation of T cells (LAT). The 10 intracellular tyrosine-based activation motif domains have also been highlighted in red on the CD3 and zeta subunits. (See color plate)

Although we focus here on T cells, such receptor complexes play a key role in most immune cells, whether from the innate or adaptive system. Like TCR signaling pathways, BCR engagement is followed by the reorganization of the receptor on the B cell surface into clusters (Mattila et al., 2013). It has been shown that ligand mobility plays a key role in the regulation of signaling, suggesting that the spatial organization of this two subunit complex into clusters is necessary for activation (Ketchum, Miller, Song, & Upadhyaya, 2014; Treanor et al., 2010). It also appears that lipid raft plays an active role in early signaling through the engagement of IgE receptor in the case of Mast cells for instance (Davey, Walvick, Liu, Heikal, & Sheets, 2007).

2.3 T cell activation

Even though TCR engagement with its specific peptide–MHC signaling complex on an APC is the necessary trigger for an adaptive response, it has been shown that early signaling also relies on the trafficking of other key signaling proteins at the IS as well as their phosphorylation. Thus, following TCR engagement, Lck clusters and Lck/CD4 complexes are recruited to the vicinity of engaged TCR. Lck is a protein from the Src family, a group of tyrosine kinase proteins specialized in mediating membrane-associated receptor signaling and therefore essential for the activation of T and B cells. Lck in its active form can phosphorylate the 10 ITAM domains of the TCR–CD3 complex subunits. Interestingly, Lck has been shown to associate and cluster in membrane rafts due to its double palmitoylation sites at the N-terminus (Ventimiglia & Alonso, 2013). More generally, the presence of Src-homology (SH)2 and SH3 or pleckstrin homology (PH) domains or a combination of these facilitate their interaction and anchoring with ordered-phase lipid domains (Denny, Patai, & Straus, 2000) as well as protein–protein complex formation (Palacios & Weiss, 2004). Rossy, Owen, Williamson, Yang, and Gaus (2013) suggest that Lck nanoorganization and conformational changes regulates TCR signaling. They have based their study on photo-activated localization microscopy (PALM) and (d) STORM described in Section 6, as well as cluster analysis software (Owen, Oddos, et al., 2010; Rossy et al., 2013). Surprisingly the size and number of Lck cluster following the T cell activation did not display any drastic changes, but the percentage of Lck molecules in these clusters did increase significantly after activation. They state signaling through the TCR pathway is associated with the appearance of denser Lck clusters. Casas et al. have also studied its spatiotemporal behavior following TCR engagement, and they have shown that the pool of Lck triggering the downstream signaling is not associated with CD8 coreceptors (Casas et al., 2014). Finally, it has been suggested that a conformational change of the TCR-CD3 complex geometry following TCR engagement, could explain the accessibility of ITAMs to Lck and therefore their phosphorylation (Xu et al., 2008).

Once phosphorylated, TCR-CD3 complex ITAMs allow the Syk-family kinase zeta-chain associated protein kinase of 70 kD (ZAP70) to dock (via tandem SH2s). In fact it has recently been shown that ZAP70 is partially present on TCR subunits even in resting cells, although it remains in an inactive (nonphosphorylated) form and therefore still has to change its conformation to allow the signal transduction downstream. It is evident that TCR-pathway nanoclusters are at least partially preformed in T cells, possibly to allow the rapid amplification of signaling after TCR engagement (Crites et al., 2014). Lck, in addition to phosphorylating the CD3 subunits on ITAMs, initiates ZAP70 phosphorylation. It has been shown that once the phosphorylation of the ZAP70 pool is initiated by Lck, the protein has the capacity to phosphorylate itself, allowing a very fast amplification of the signaling (Chan, Iwashima, Turck, & Weiss, 1992). It has been demonstrated that Lck must be active in order for ZAP70 to be active (Lovatt et al., 2006).

The ZAP70 phosphorylation mechanism results in the phosphorylation and recruitment of the scaffold protein LAT, found in nanoscale clusters in resting and activated T cells (Figure 3). LAT is also associated in nanoscale clusters with other signaling proteins such as SLP76 and VAV that will initiate the downstream cascade. It has been suggested by Williamson et al. that nanoscale clusters of LAT present at the plasma membrane do not take part in the early signaling process (Williamson et al., 2011). Instead a new pool of LAT is trafficked to the IS in vesicles and it is this pool that is primarily phosphorylated. As with preformed ZAP70-ITAM complexes, preformed clusters of LAT in subsynaptic vesicles could be an important mechanism for the bulk recruitment of proteins to the TCR and therefore

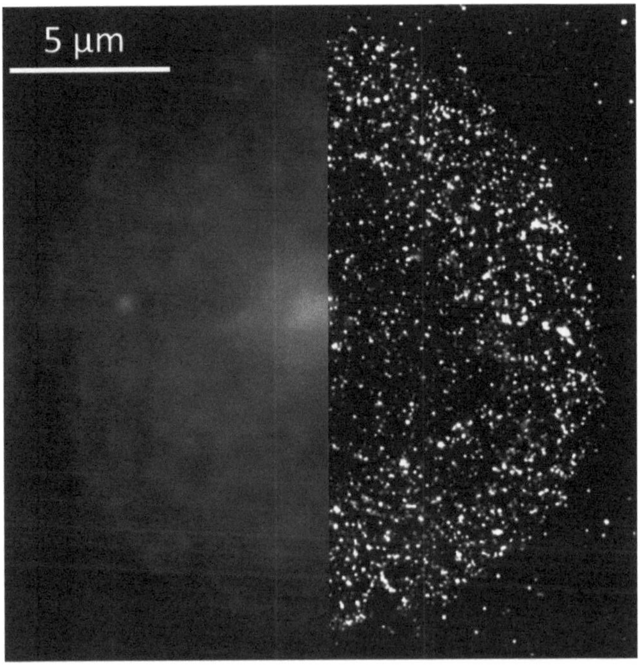

Figure 3 Linker for activation of T cells (LAT) clustering at the T cell immunological synapse. Primary human T cells activated on anti-CD3-coated glass coverslips for 10 min before fixation and labeling of LAT via immunostaining. Illustration of LAT clustering via conventional resolution TIRF microscopy (left) and super-resolution STORM imaging (right).

rapid signal amplification. However, this assumption remains a source of controversy in the field. In particular, other groups suggest that LAT nanoclusters present on the plasma membrane are the main pool mediating signaling (Balagopalan, Barr, Kortum, Park, & Samelson, 2013; Lillemeier et al., 2010). Lillemeier et al., for instance, suggest that LAT and TCR are located in different nanoscale islands at the plasma membrane in the steady state. Following activation, these membrane-embedded protein structures converge and colocalize in multiprotein signaling complexes. Thus, the key processes associated with T cell activation are the presence of signaling protein nanoclusters and the conformational modifications of the proteins via phosphorylation. LAT is a probing example of a protein facilitating the formation of multiprotein nanoclusters but which also plays a key role by interacting with lipid domains (Jin et al., 2008; Tanimura et al., 2003; Tanimura, Saitoh, Kawano, Kosugi, & Miyake, 2006).

The engagement of the TCR with a nonself peptide is therefore followed by a fast response and strong amplification of the signaling through key proximal signaling molecules such as Lck, leading to the TCR signaling cascade and eventually cell activation. The requirement for a fast and strong response even in the case of the engagement of a single TCR complex suggests the existence of at least a "partially ready" pool of the main components. Indeed, it is now accepted that the engagement of the TCR does not induce the switch from an OFF state to an ON state but rather the disruption of an equilibrium that leads to signal amplification. Lck clusters, for instance, have been shown to be present in its ON state (catalytically active) even in the case of resting cells (Nika et al., 2010).

It should be noted that the recruitment and clustering of activatory proteins to sites of TCR engagement occurs in parallel with the exclusion of negative-regulators of TCR signaling from the activation area. It has even been suggested that the increase in phosphorylated Lck, and therefore the activation itself was in fact principally resulting from the exclusion of one of these key regulatory proteins clusters away from the TCR: Csk, another member of the Srk family (Tan et al., 2014). Such a translocation allows an increase in activated Lck and therefore phosphorylation of ITAM sites and ZAP70. This segregation occurs through the dissociation of the Csk protein from the membrane-associated scaffold protein PAG on which it is docked in resting cells. Other complexes also have to be excluded from the signaling zone of the IS, demonstrating that the importance of Csk translocation is not anecdotal. The absence of Csk has been suggested to be sufficient to induce signaling (Tan et al., 2014). Other signaling complexes such as those containing the phosphatase CD45, have been shown to be associated to ordered lipid domains in the case of TCR activation, whereas they are preferentially found in disordered domains in nonstimulated T cells (Zhang et al., 2005).

As it appears that the spatial reorganization of signaling and regulatory nanodomains is necessary for the activation of T cells, it is interesting to speculate on how such a dynamic process might be regulated. Cytoskeletal reorganization after TCR engagement has been suggested to be the key mechanism in the spatial redistribution of these nanodomains as well as the formation of a mature synapse. It has been shown that actin could segregate negative regulators of TCR signaling away from the activation site, and could also be employed for the trafficking of vesicles to specific sites. In fact, it is known that retrograde actin flow is necessary for T cell activation (Yu, Smoligovets, & Groves, 2013). Ashdown, Cope, Wiseman, and Owen (2014) showed retrograde actin flow during synapse formation, which includes the presence of an extremely dense cortical actin meshwork. It has previously been demonstrated that changing the density of the underlying actin mesh can induce clustering of surface proteins in the membrane (Chaudhuri et al., 2011; Goswami et al., 2008; Gowrishankar et al., 2012). Such results therefore suggest that cortical actin may be an important regulator not only of protein nanoclustering itself, but also the dynamics of such clusters. In agreement with this, Babich et al. (2012) have shown that disrupting actin retrograde flow leads to the incapacity of the T cell to signal. These new discoveries explain the increasing interest in understanding the role of actin flow and meshwork in regulating signaling domain spatial organization and dynamics.

2.4 B and NK cell activation

As in the case of T cells, spatial reorganization and formation of nanoscale signaling domains, whether protein–protein complexes or protein-lipid domains, are key to the functioning of many other immune cells. For instance, lipid rafts are among the essential components of activation of B cells being used as recruitment platform for the BCR signaling pathway components (Flores-Borja, Kabouridis, Jury, Isenberg, & Mageed, 2005; Gupta & DeFranco, 2007; Pierce, 2002; Schmidt et al., 2009). As in the T cell, after BCR engagement with an antigenic protein, early signaling is initiated by a member of the Src family of kinases: Lyn. Lyn phosphorylates the B cell receptor complex, providing binding sites on which Syk kinase can dock, resulting in a downstream signaling cascade. Lyn, like other members of the Src family can be membrane targeted due to post-translational modifications and is specifically associated with ordered-phase lipid nanodomains (Casey, 1995). Thus, it has been suggested that the exclusion of the BCR complex from lipid rafts prior to antigen recognition is a regulatory process of B cell activation (Cheng et al., 1999). Also, Mattila et al., in agreement with current theory in T cell activation, suggest that cytoskeletal rearrangement following initial triggering allows BCR nanocluster spatial reorganization on the plasma membrane (Mattila et al., 2013). Actin is therefore an active component of BCR-mediated signaling (Treanor et al., 2010).

NK cells are cytotoxic lymphocytes of the innate immune system that are thought not to depend on MHC receptor complexes for their activation. It has been shown that they can be activated by other means such as the recognition of stress-induced ligand on neighboring cells, however, the impact of such triggering has not been yet observed in vivo (Vivier, Tomasello, Baratin, Walzer, & Ugolini, 2008). Therefore, instead of a unique receptor complex adapted to a given pathogen, NK cells present multiple different TLRs on their surface. These receptor complexes recruit signaling proteins forming multiprotein signaling domains (Treanor et al., 2006). The engagement of a TLR induces the production of interferon gamma (Adib-Conquy, Scott-Algara, Cavaillon, & Souza-Fonseca-Guimaraes, 2014). Recent research demonstrates that, in vivo, TLR engagement does not induce a strong response on cytotoxicity level but probably has another, yet to be discovered, effect on NK cell behavior. Interestingly, it has been suggested that the absence of MHC-1 on the targeted cell may very well be the main process for cell signaling. In fact, the "missing self-recognition" process based on the sensitivity of NK cells to the absence of MHC complexes on the plasma membrane of the target is an essential component of NK cell activation pathways. It results in NK cells being able to discriminate and attack MHC deficient self-cells (Raulet, 2006) meaning NK cells play an important role in combating tumors and infection (Grier et al., 2012).

3. CELL MIGRATION AND ADHESION

3.1 Introduction

Cell adhesion molecules (CAMs) play important and non-redundant roles in biology. The segregation of CAMs at the cell surface and the arrangement of signaling molecules strongly influence cell adhesion and migration. This implies a highly ordered and complex biological process that requires the concerted action of many different molecules. In this section we will explore integrins as an illustration of CAM nanoscale organization. However, it should be noted that many other molecules play important roles in the cellular machinery associated with migration and adhesion. Given the importance of such proteins in many different steady state and pathological processes, it follows that understanding how they function via their nanoscale organization might eventually help in the development of a new generation of therapeutic treatments for pathologies such as infection, autoimmunity, and cancer.

Cell adhesion, be it in the context of cell-matrix or cell–cell contacts, is pivotal in positioning and instructing cells. In its most basic form, cell adhesion refers to the process of the binding of a cell to another cell or to a surface via specific CAMs. For example, epithelia are surrounded or underlaid by an extracellular matrix of connective tissues like collagen fibers, proteoglycans, and multivalent matrix proteins such as fibronectin. This layer of matrix participates in many different biological processes that are often mediated through CAMs. These include the organization of cells into functional tissues with a specific architecture which in turn coordinates tissue structure–function relationships. The matrix also provides a substrate for migrating cells, and molecules in the matrix activate signal transduction pathways that can potentiate cell growth, differentiation, proliferation, and gene expression profiles. In addition to cell–matrix interactions, CAMs mediate many transient cell–cell interactions including that of the formation of a synapse between two immune cells as explored earlier in the chapter, by providing strong adhesive contacts that maintains an interface between two cells where information can be exchanged (Hogg, Laschinger, Giles, & McDowall, 2003; Montoya et al., 2002; Rothlein, Dustin, Marlin, & Springer, 1986). Alternatively, stable junctions can be formed between cells, a prime example being that of the endothelial cells that line blood vessels. Interestingly, even stable junctions can be regulated, promoting different states of permeability (leaky versus tight junctions). These operate under specific biological processes like inflammation to ensure the passage of immune cells out of the vasculature and into the tissue, and to allow for specific factors to enter the tissue via the blood. Virtually all of the aforementioned processes rely on the engagement of CAMs that act as information transducers and linkers between the cytoskeleton of the cell and the extracellular matrix or other cells (Evans et al., 2009; Hogg et al., 2003; Hogg, Patzak, & Willenbrock, 2011).

Broadly, adhesion receptors can be divided into calcium-dependent and calcium-independent families. The calcium-dependent CAMs include the integrins, cadherins, and selectins. The calcium-independent CAMs include the immunoglobulin superfamily CAMs and the addressins. CAMs consist of an extracellular portion (that can bind to specific counterligands or interact in a homo- or heterotypic fashion), a transmembrane spanning region (that positions the molecule appropriately within a lipid bilayer) and a cytoplasmic region (that can usually interact with cytoplasmic components), culminating in a specific response. Thus, adhesion receptors should not just be viewed as "sticking points" but rather as multifunctional receptors, operating as signaling domains on the plasma membrane (Hogg et al., 2003, 2011). As such, the nanoscale behavior of CAMs and their associated components has become a priority theme in research to understand how tissues are formed and maintained. Here, integrin biology is explored as these CAMs have been studied in the context of many different biological processes, focusing on how the cytoskeleton works in concert by binding cytoplasmic tails of integrins to form a structure known as a "focal adhesion" (Kanchanawong et al., 2010).

3.2 The integrin "adhesome" substructure

Integrins are heterodimeric proteins consisting of an α and β chain. 18 α integrins and 8 β integrins have been described in mammalian cells, which can form 24 different heterodimeric complexes (Hynes, 2002). Many of the components involved in integrin signaling have been identified (Zaidel-Bar, Itzkovitz, Ma'ayan, Iyengar, & Geiger, 2007), but to understand the mechanics and functional significance of these proteins, the spatial arrangement of these components must be carefully dissected. Microscopy has offered valuable insights into the arrangement of integrin receptors when complexed with their ligands, with the first descriptions of focal adhesion using interference microscopy being described in the 1970s (Abercrombie & Dunn, 1975; Abercrombie, Heaysman, & Pegrum, 1971). Subsequently, it was shown in the 1980s that integrin domains can link to the cytoskeleton via adaptor proteins like α-actinin and vinculin (Zaidel-Bar et al., 2007; Zamir & Geiger, 2001). Further structural insights into the make up of adhesion complexes remained incomplete due to constraints of light microscopy, as these complexes exist below the diffraction limit. However, with the development of super resolution microscopy capable of resolving domains below this limit (Hell & Wichmann, 1994; Hess, Girirajan, & Mason, 2006; Rust, Bates, & Zhuang, 2006), huge advances in understanding of the integrin adhesome have been made. In being able to localize single proteins, it has become increasingly clear that focal adhesions are highly ordered structures. They display features of a structure–function relationship between protein organization at the base of integrins and integrin attachment to the cytoskeleton which elicit signaling pathways associated with cell shape, adhesion, and locomotion.

In an elegant study presented by Waterman and colleagues, 3D super resolution microscopy revealed distinct "layers" of proteins that assemble at the cell membrane in clusters when an integrin is engaged by its counterligand (Kanchanawong et al., 2010). In this study, it was demonstrated that at the most membrane-proximal points (within 8–10 nm of the membrane), a layer of integrin associated signaling proteins, e.g., FAK and Paxillin, congregate. Both of these proteins have previously been shown to be important for cell adhesion and migration by instructing the underlying protein network that facilitates integrin–cytoskeleton interactions (Brown & Turner, 2004; Mitra, Hanson, & Schlaepfer, 2005). In keeping with this idea, the "signaling layer" is then followed by a layer of adaptor proteins like talin and vinculin that directly link to the integrin itself and interact with FAK and Paxillin (Brown et al., 2006; Galbraith, Yamada, & Sheetz, 2002; Hu, Ji, Applegate, Danuser, & Waterman-Storer, 2007; Jiang, Giannone, Critchley, Fukumoto, & Sheetz, 2003). This layer of adaptor proteins is positioned between 10 and 40 nm from the membrane. Finally, this layer of adaptor proteins is linked to the overlying cytoskeleton with which it interacts, providing the force required for cellular locomotion (Brown et al., 2006; Hu et al., 2007). Together, these data provide a blueprint of how signaling proteins, adaptor proteins and the cytoskeleton are arranged in a migrating cell, which in turn will allow for further investigation and a better understanding of how these components interact to form functional units.

3.3 Regulation of the spatial distribution of integrins

Given the importance of integrins in mediating cell adhesion, it is important to consider how a cell might regulate the activation of the associated signaling pathways. The idea that integrins can be constrained fits with a biological paradigm in which surface expression of integrin does not necessarily always lend itself to adherence, even in the presence of counterligand. It is thought that integrins might be segregated into domains at the cell surface independently of ligand engagement thus constraining signal or, integrin may be stored in vesicles that lie close to the membrane, and following membrane–vesicle fusion, are delivered to the cell surface. The segregation of integrin into specific lipid nanodomains is thought to occur, at least in part, by cytoskeletal constraint. In this model, the cytoskeleton constrains integrin diffusion and ensures that they do not enter ordered lipid domains which contain signaling intermediates that are able to initiate signal transduction (Hogg et al., 2002, 2011; Krauss & Altevogt, 1999; Leitinger & Hogg, 2002; Shamri et al., 2002). It is of interest that the cytoskeleton can participate in integrin biology in at least two ways, by participating in active signaling and force generation during migration, or the segregation of integrins out of microdomains during nonactive signaling.

Another interesting mechanism by which integrins might be spatially regulated is through vesicles. Studies have shown that integrins are constantly internalized and recycled to the cell surface (Lawson & Maxfield, 1995; Tohyama et al., 2003). In migrating cells, integrins have been shown to recycle from the back to the front via a vesicular transport system. Disrupting this system leads to disruption of cell migration. The vesicular transport system is therefore implicated in the spatial regulation of integrins into nanodomains that in turn regulate adhesion/deadhesion and signal transduction.

4. NANODOMAINS AS MEDIATORS OF INTER AND INTRACELLULAR COMMUNICATION

4.1 Introduction

We have discussed the importance of signaling domains in the activation of the immune system as well as their importance for migration and focal adhesion. Signaling protein nanoclusters and lipid domains are also suggested to be important components in intercellular communication and regulation of the cellular life cycle. They are involved in key cellular processes such as proliferation and cell division, as well as neurotransmission and embryonic development. Here, we will study nanodomains associated with cell proliferation, in particular Ras protein complexes. We will also focus on the role of nanodomains in the case of neurotransmission, as it is an illustration of the importance of nanoscale organization for endo- and exocytosis. Finally, we will understand how the spatial rearrangement of nanodomains plays a key role in early embryonic development. In fact, these topics have proven to be of real interest for the scientific community, profiting from latest technological developments in imaging and analysis. International scientific projects have even been created to try and tackle the issues associated with mapping nanodomains, such as the brain activity mapping (BAM) project for instance (Alivisatos et al., 2013).

4.2 Signaling nanodomains involved in cell proliferation

The Ras (rat sarcoma) protein family is a vivid illustration of the importance of nanoscale spatial organization, in particular clustering, for the activation and regulation of signaling pathways. Ras is a key protein group of the GTPase family which includes three isoforms: H, N, and K, expressed in mammalian cells. H- and N-Ras are targeted to the membrane due to posttranslational lipid modification (polyisoprenylation and palmitoylation) whereas K-Ras is only polyisoprenylated (Hancock, Magee, Childs, & Marshall, 1989). Ras protein signaling pathways are associated with cell survival and proliferation. More precisely, they are lipid-anchored G-proteins principally found in the inner leaflet of the plasma membrane in the form of signaling nanoclusters (Prior et al., 2001; Prior, Muncke, Parton, & Hancock, 2003; Prior, Parton, & Hancock, 2003). Prior et al. have shown that Ras proteins are organized in nanoclusters on the plasma membrane using immunogold electron microscopy. More precisely, their results suggest that Ras proteins form nonoverlapping isoform and conformation-specific nanoclusters. They are thought to exist on length scales of tens of nanometers, containing around six to eight molecules (Plowman, Muncke, Parton, & Hancock, 2005; Prior, Muncke, et al., 2003), with a lifetime highly dependent on their association with lipid-ordered nanodomains. For instance, the inactive form of the protein H-Ras (H-Ras GDP) is mainly found in lipid-ordered nanodomains and its clustering is therefore cholesterol dependant. Its active form (H-Ras GTP) on the contrary has been shown to cluster outside lipid-ordered domains (Harding & Hancock, 2008; Prior et al., 2001). Therefore, H-Ras GTP clustering is not cholesterol dependant but displays instead a strong dependency on the scaffold protein Galectin-1 (Abankwa, Gorfe, & Hancock, 2007). It has been suggested that H-Ras GDP clusters have a shorter lifetime than clusters of the activated conformation (Kusumi, Koyama-Honda, & Suzuki, 2004). Clusters of the activated form of H-Ras have been shown to be nearly immobile during the duration of their lifetime, whereas, unactivated clusters are highly dynamic (Murakoshi et al., 2004).

Harding et al. suggest that signaling networks, in particular those associated with Ras proteins, are dependent on the nanoscale organization of their main components (Harding & Hancock, 2008). Tian et al. have demonstrated that the Ras GTP- associated signaling cascades could only be activated by clustered proteins (Tian et al., 2007). They also suggest that the structure of these clusters (size and protein composition) is not dependent on Ras expression level. However, the number of Ras clusters that can be found in the plasma membrane depends on stimulation factors such as growth factor. Considering the importance of the cluster formation in Ras signal transduction, various models have been developed to try to understand and characterize this mechanism (Gurry, Kahramanoğulları, & Endres, 2009). Ras, because of its major role in cellular life cycle, in particular survival and proliferation, is frequently associated with cancer development. Indeed, it appears that Ras genes are found to be mutated in nearly one-third of human cancers, and an abnormal concentration of on-state Ras nanoclusters may result in cancer (Adjei, 2001).

Other signaling pathways such as epidermal growth factor receptors (EGFR) or G protein-coupled receptors (GPCRs) seem to be highly dependent on protein clustering. EGFRs for instance are transmembrane receptor tyrosine kinases, which have been shown associated with cell proliferation, differentiation, and migration, and form clusters at the plasma membrane (Ariotti et al., 2010; Wang et al., 2014). Wang et al. have demonstrated, using a super resolution fluorescence microscopy technique (d)STORM, the existence of these nanoclusters in healthy lung epithelial cells as well as cancerous lung cells. However, they also highlight an increase in the number of EGFR clusters and their size in the case of cancer cells. In fact, EGFR-associated pathways are thought to be regulated by the nanoscale organization of the receptor clusters at the plasma membrane, especially their association with a specific type of lipid domain (Galectin lattice and oligomerized Caveolin-1), in cancerous cells (Lajoie et al., 2007). More generally, it appears that protein clustering, as well as lipid nanodomains, play an active role in cancerous cell proliferation. Considering the importance of protein clustering and lipid nanodomains in cellular signaling machinery, it is understandable that their disorganization may result in disease.

4.3 Neurotransmission: a process relying on nanoscale organization

Research into neurotransmission, for decades, has been limited by the size of the intercellular communication platform: the synapse. The synapse is a cell—cell interface across which intercellular communication is triggered by electrical factors followed by the release of neurotransmitters. The development of advanced microscopy techniques such as electron microscopy or super resolution fluorescence microscopy, has unraveled the main components and mechanisms involved in nerve cell transmission. It is now thought that neuronal activity relies essentially on nanoscale events. Indeed, the spatial reorganization of signaling proteins based on mechanisms such as clustering, plays an essential role at the synapse (Dani, Huang, Bergan, Dulac, & Zhuang, 2010).

Neurotransmitters are the main mediators of signal transmission between neurons. Their delivery at the synaptic cleft that follows the opening of calcium channels assures their transportation from one neuron to the other via the synapse. Both the storage of these neurotransmitters in vesicles, called synaptic vesicles (SyV), and the mediation of exo- and endocytosis via lipid-protein domains at the synapse are key for neuron activity (Jin & Garner, 2008). A deficiency in the nanoscale organization of these domains could lead to impaired transmission. The pool of SyV is divided in three zones: the readily releasable pool at the synapse (preanchored at sites of protein nanodomains on the plasma membrane), the recycled pool and reserve pool. The SyVs are 40 nm in size and on average contain between 1500 and 2000 neurotransmitter molecules (Haucke, Neher, & Sigrist, 2011; Südhof, 2004; Willig, Rizzoli, Westphal, Jahn, & Hell, 2006).

The exo/endocytosis mechanism relies on specific nanodomains in the plasma membrane, with the ability to modify the lipid bilayer curvature, e.g., scaffold protein-based complexes or clathrin-coated domains. In particular, it has been shown that exocytosis is conditioned by the formation of a SNARE (soluble *N*-ethylmaleimide-sensitive factor attachment protein receptor) protein complexes on the inner leaflet of the plasma membrane. Mohrmann, de Wit, Verhage, Neher, and Sørensen (2010) suggest that at least three SNARE proteins are needed in such a cluster to allow rapid fusion of the vesicle. The opening of calcium channels activates SNAREs on the plasma membrane. Interestingly, Willig et al. (2006) have demonstrated that during exocytosis, the components (such as synaptotagmin) of the SyVs that have fused with the cell plasma membrane remain clustered, forming nanodomains on the surface. Note that endo and exocytosis happen quasisimultaneously to ensure the conservation of the number of SyVs, and is therefore described as a recycling process (Figure 4).

It has also been suggested that other key components of neurotransmission, such as brain-derived neurotrophic factor (BDNF), form clusters in the presynaptic zone. In particular, Blum and colleagues have imaged BDNF clusters in hippocampal neurons, using the super resolution technique (d) STORM, and estimated that these clusters are on average 60 nm in size (Andreska, Aufmkolk, Sauer, & Blum, 2014). On a larger scale, it is now

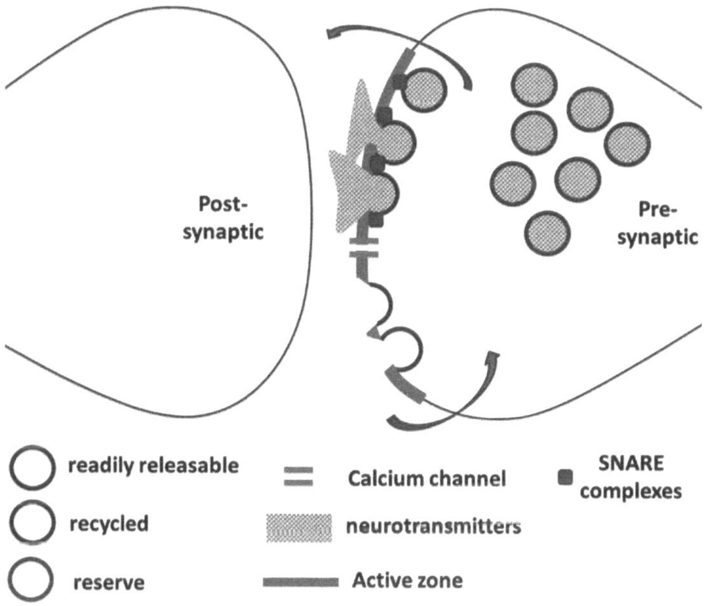

Figure 4 Synaptic cleft after opening of calcium channels showing the main three pools of synaptic vesicles. (See color plate)

suggested that the nanoscale organization of the cytoplasmic matrix of the active zone (where calcium channels have opened) is the main driver of intercellular communication in the case of neurons (Schoch & Gundelfinger, 2006).

4.4 The role of signaling domains in embryonic development

Signaling domain organization is now considered to be an essential component of embryonic development. The spatial organization of specific tissues or cellular structures relies on the ability of cells to communicate and reorganize themselves depending on the environment. Again, receptor tyrosine kinase based signaling domains have been shown to be essential in signaling events associated with embryonic development. Corson, Yamanaka, Lai, & Rossant (2003) highlight the existence of signaling events associated with Src family proteins by studying phosphorylated ERK organization as an indicator of signaling in mice.

Ordered-phase lipid domains in the plasma membrane play a key role in embryonic development (Yanagisawa, Nakamura, & Taga, 2004). Moreover, it has been suggested that these lipid domains, associated with key signaling proteins, display very specific spatial and temporal organization at critical steps of embryo development (Comiskey & Warner, 2007; Owen, Magenau, Majumdar, & Gaus, 2010). Therefore, signaling domains do not only play a role in neurotransmission and the immune system, but also are a key factor in embryonic development, appearing to be necessary for multicellular organism development. Studies of the importance of nanoscale organization of signaling domains in organism early development are still at an embryonic stage, but considering the key role of nanodomains in cellular lifecycle and intercellular communication, it is a very promising research topic.

5. SIGNALING NANODOMAINS IN DISEASE: THE CASE OF VIRUSES

5.1 Introduction

As well as playing a major role in the activation of immune cells, nanoscale signaling complexes have paradoxically been shown to represent a weakness in cellular defense against pathogens. Viruses and intracellular bacteria appear to target and exploit some specific signaling complexes to hijack the associated signaling pathways to their advantage. Furthermore, due to their importance in the cellular machinery the disruption of the spatial organization of these domains can have important implications for health. Indeed, studies suggest that a modification in the nanoscale organization of the PM could be implicated in neurodegenerative diseases, cancer, and pathogen infection (Michel & Bakovic, 2007; Simons & Ehehalt, 2002). It has, for instance, been shown that lipid rafts play a role in cancer cells abnormal migration and adhesion behavior (Hitosugi, Sato, Sasaki, & Umezawa, 2007; Patra, 2008; Zhuang, Lin, Lu, Solomon, & Freeman, 2002).

Being able to characterize these complexes in the context of disease or infection is necessary for the development of novel therapeutic strategies capable of readjusting the fragile equilibrium of this system. In particular, understanding how pathogens have adapted themselves to the nanoscale organization of the plasma membrane is at the forefront of a new generation of therapeutic strategies against viruses such as HIV (Barre-Sinoussi, Ross, & Delfraissy, 2013; Chojnacki et al., 2012; Ono, Waheed, & Freed, 2007; del Real et al., 2002). Interestingly, some new pharmaceutical drugs mimic the opportunistic strategy used by viruses to promote their entry into the cell. The aim of such drugs is to ensure successful delivery into the intracellular compartment. Pathogens have evolved to exploit the heterogeneous nature of the PM. Pathogens such as viruses specifically target signaling complexes such as lipid-ordered nanodomains or protein nanoclusters, to support their life cycle (in particular their entry, survival, and replication). Viruses depend on cellular machinery for their survival and proliferation. A successful entry in the host cell is therefore a key component of their lifecycle (Mañes, del Real, & Martinez-A, 2003).

5.2 Receptor complexes as an entry point for pathogens

Receptor complexes present on the PM have been shown to be a key element in a viral infection: the entry door to the intracellular components of the host cell. Viruses use the engagement with these receptor complexes to dock onto the cell surface. As these receptor complexes can be cell specific (such as CD4 on the surface of helper T lymphocytes), such a mechanism allows the pathogen to selectively dock on a given type of host cell. Moreover, virus entry into the host cell itself is a mechanism that relies essentially on the recycling of these signaling intermediates (via clathrin pathways, for instance) following engagement and activation of the associated downstream signaling cascade. Both these phenomena at least partially result from receptor (or associated machinery) rearrangement. Stuart, Eustace, McKee, & Brown (2002) for instance, demonstrate that some enteroviruses (the group that contains rhinoviruses, which are responsible for the common cold) are dependent on the presence of ordered-phase lipid domains in plasma membrane for their entry due to their interaction with CD55 (decay accelerating factor, DAF) which is an ordered phase-associated GPI-anchored protein. On the other hand, Ebola virus as well as hepatitis C virus, rely on clathrin-mediated endocytosis for their cellular entry (Bhattacharyya et al., 2010; Blanchard et al., 2006).

Although some viruses target receptor complexes that are excluded from lipid-ordered domains at the PM (such as MHC class 1—the target of Simian Virus 40 (SV40)), the majority of the targeted receptor complexes are principally found in rafts (including G_{M1}, also a target of SV40 (Ewers et al., 2010; Neu, Woellner, Gauglitz, & Stehle, 2008)). Takahashi and Suzuki (2011) underline the importance of ordered-phase domains in the viral life cycle. It appears that disrupting ordered-phase domains on the plasma membrane, for example, by using cholesterol depleting drugs, results in a decrease of virus entry into host cells without affecting virus binding to membrane receptors. These results suggest that receptor engagement is not affected by cholesterol depletion but that, instead, lipid domain coalescence and signaling are disrupted. While receptor engagement by a virus is the first stage in the infection process, Mudhakir and Harashima (2009) suggest that convergence of raft-associated signaling domains to the engagement site is necessary to amplify this first signaling event and therefore facilitate entry. Freitas et al. (2011) show using mammalian cells, that cholesterol depletion and therefore raft disruption decrease Ebola fusion capacity, for example. Also, it has been shown that lipid-ordered domains play an essential role in the influenza virus infection (Barman & Nayak, 2007; Carrasco, Amorim, & Digard, 2004). More surprisingly, it appears that membrane embedded nanodomains are necessary for some virus exit of the host cell (Khurana et al., 2007).

5.3 Pathogen-mediated signaling to facilitate entry, survival, and replication

If entry into the intracellular compartment is a critical step of pathogen infection, understanding the specificity of the signaling which actively promotes the internalization of the virus is essential. It has now been demonstrated that in many cases the interaction between a virus receptor and its associated plasma membrane receptor at the cell surface does not produce sufficient signal to assure virus entry. In fact, in addition to the primary receptor engagement, the pathogen recruits and activates other signaling domains (coreceptors complexes) to facilitate the internalization via an amplification of signaling. The binding of the virus to its specific receptor initiates the entry process, whereas coreceptor engagements boost signaling and facilitate entry (Marsh & Helenius, 2006). Most coreceptors are transmembrane embedded proteins, therefore, their activation and recruitment to the engagement site can be associated with raft coalescence and the creation of a lipid-ordered-phase-specific signaling structure at the docking site. Kinases, for instance, play the role of coreceptors for a range of pathogens. In particular, Pelkmans et al. among others, suggest the importance of caveolae, clathrin, and kinase pathways in endocytosis machinery for virus infection (Mudhakir & Harashima, 2009; Pelkmans et al., 2005). SV40, for example, has been shown to costimulate five kinase proteins in addition to its primary engagement targets (MHC I and G_{M1}). The amplification of signal via the activation of coreceptors as well as MHC association with caveolar domains after virus binding results in the pathogen entry (Mañes et al., 2003). Viruses display three main strategies to enter the intracellular compartment: fusion of the virus envelope with cellular plasma membrane, injection of genetic material through the plasma membrane, or uptake within endocytic vesicles recycled from the membrane (Chazal & Gerlier, 2003; Mudhakir & Harashima, 2009). We will particularly focus on this last mechanism, as it is an elegant illustration of the exploitation of nanoscale receptor clustering.

The ability to reorganize the nanoscale distribution of receptor complexes and signaling protein clusters on the cell surface allows a virus to control signaling pathways associated with membrane recycling, degradation pathways, cell division, and cell death. As said, the first advantage of causing the cell to signal via receptor-raft clustering is to facilitate the entry in the intracellular compartment, especially in the case of nonenveloped viruses, but it is also an efficient way for a virus to avoid detection and degradation. An example of this is the SV40 virus which binds MHC class 1 or G_{M1} on the cell surface followed by activation via engagement of kinase clusters. The targeting of coreceptors associated with lipid nanodomains induces raft convergence to the engagement site as well as signaling followed by endocytosis. By specifically targeting nanodomains-containing caveolae, SV40 avoids the degradation pathway by entering the intracellular compartment within ordered-phase vesicles (Chen & Norkin, 1999; Mudhakir & Harashima, 2009). The capacity of pathogens to be incorporated via vesicles is an efficient way to avoid the degradation pathway (a method a new generation of therapeutics is keen to mimic).

In the case of some bacteria on the other hand, triggering signaling also results in a cytoskeletal rearrangement which can facilitate pathogen entry. Lafont, Abrami, & van der Goot (2004) have shown that intracellular bacteria, such as *shigella*, rely on rafts and associated receptors as key components of their invasion process. In particular, *shigella* relies on its engagement with the transmembrane protein CD44, a clustered, raft-associated protein involved in cell migration through its association with the actin cytoskeleton. The activation of this receptor induces clustering of signaling proteins resulting in the rearrangement of the actin meshwork. Another example is the activation of the CD55 pathway by *Escherichia coli*, which mediates a reorganization of the F-actin meshwork (Ponta, Sherman, & Herrlich, 2003).

5.4 Pathogen-mediated signaling to facilitate proliferation: the example of HIV

The human immunodeficiency virus (HIV) is an illustration of the importance of signaling nanodomains, especially lipid-ordered nanodomain-associated receptor complexes, in virus infection strategy (Barre-Sinoussi et al., 2013). Virus invasion and proliferation result in the development of acquired immunodeficiency syndrome (AIDS), allowing opportunist infections to spread. HIV is an enveloped virus whose entry into the cell relies on the fusion of its membrane with the host cell plasma membrane. Therefore, the virus presents on its surface a receptor capable of binding with targeted host cells. In particular, gp120 will engage CD4 as well as a vast range of coreceptors now identified, such as the chemokine receptor complex family (Berger, Murphy, & Farber, 1999). By engaging these receptors, HIV triggers signaling through the clustering of these raft-associated domains. Thus, it also results in the convergence of lipid nanodomains, creating a stable platform for virus docking and the cellular stimulation that allows virus entry (Mañes et al., 2000; Nguyen & Hildreth, 2000; Nisole, Krust, & Hovanessian, 2002; Ono et al., 2007; del Real et al., 2002). It has been suggested by Hogue et al. that HIV is strongly dependant on the coalescence of lipid-ordered domains with tetraspanin-enriched domains (Hogue, Grover, Soheilian, Nagashima, & Ono, 2011). Similarly, del Real and colleagues have shown that the engagement of the virus with its receptor is not sufficient to ensure cell entry, but that HIV instead relies on CD4 associated with lipid-ordered domains (Mañes et al., 2000; del Real et al., 2002).

Once in the intracellular compartment, the virus must ensure its replication. To do so, HIV use its Nef protein to trigger the TCR signaling pathway and induce the signaling cascade described in Section 1 (Witte et al., 2008). Nef protein clusters are incorporated in host cell plasma membrane rafts, forming protein-lipid nanodomains. These domains, when in contact with TCR complexes, result in the activation of the T cells. Activation leads to the replication of the virus but also prevents apoptosis of the host cell. Indeed, Nef signaling clusters have been shown to have an inhibitory effect on key components of signaling pathways leading to the host cell apoptosis. These results underline the importance of controlling signaling nanodomains on the host cell membrane. On a wider scale, it has also been shown that HIV essentially relies on raft-associated signaling nanodomains for its spatial distribution around the human body, a key aspect of its life cycle. In particular, raft-associated protein signaling domains play a role in mechanisms such as the crossing of HIV through the host mucosa (Alfsen, Iniguez, Bouguyon, & Bomsel, 2001) and its dispersion in the vascular system once replicated and exited from the host cell (Liu et al., 2002).

6. NEW IMAGING TOOLS USED TO STUDY MEMBRANE NANODOMAINS

6.1 Nanoscale light microscopy

Although most biological questions have historically resulted from processes or symptoms visible at the scale of an organism or with repercussions on organisms such as diseases or development, the machinery on which these processes rely on takes place on a much smaller scale. Membrane-associated processes are an elegant illustration of the importance of the nanoscale organization of key molecules, with repercussions at a cellular level, but also at the level of an organism. The visualization of molecular distributions in key processes have been impossible until recently. Indeed, proteins, nanoclusters, membrane compartmentalization and vesicles are essentially subdiffraction-limited entities (<200 nm). Light microscopy has been limited for centuries by the so-called Abbe diffraction limit making it impossible to study with precision structures below this size-scale. Modern physics has nonetheless found ways to circumvent this limit, culminating in the development of a new field-bridging physics and biology: super resolution fluorescence microscopy, which won the 2014 Nobel Prize in Chemistry. This field has been initiated by the development of stimulated emission depletion (STED), localization microscopy and structured illumination microscopy (SIM) (Bates, Huang, Dempsey, & Zhuang, 2007; Gustafsson, 2000; Hein, Willig, & Hell, 2008; Hell & Wichmann, 1994; Huang, Wang, Bates, & Zhuang, 2008; Mueller et al., 2011; Rust et al., 2006, Shao, Kner, Rego, & Gustafsson, 2011; Wildanger, Rittweger, Kastrup, & Hell, 2008; Willig, Harke, Medda, & Hell, 2007). Although several techniques have been developed, we will focus on single molecule localization as it is probably the technique that has been used most extensively in the study of membrane compartmentalization and protein—protein signaling complexes.

Single molecule localization microscopy (SMLM) techniques such as PALM (Betzig et al., 2006; Hess et al., 2006) and (direct) (d)STORM (Bates et al., 2007; Huang et al., 2008; Rust et al., 2006) have allowed scientists to improve their understanding of the nanoscale organization of the plasma membrane. Indeed, the aim of these techniques is to be able to localize each fluorophore in the sample by temporally separating the fluorescence emission from each emitter. Therefore, for each frame, only a small subset of fluorophores are turned on and individual point spread functions (PSFs) are imaged, from which their localization can be estimated with nanometer precision. SMLM thus produces pointillist data sets of the coordinates of all localized fluorophores within the field of view. Careful processing is required to ensure the validity of such data sets (Annibale, Scarselli, Greco, & Radenovic, 2012; Annibale, Vanni, Scarselli, Rothlisberger, & Radenovic, 2011a, 2011b; Huang, Schwartz, Byars, & Lidke, 2011; Ovesný, Křížek, Borkovec, Švindrych, & Hagen, 2014). A large variety of analytical tools have also been developed, in particular cluster analysis tools, as clustering of proteins in complexes is a key process in cells (Griffié, Boelen, Burn, Cope, & Owen, 2015; Malkusch et al., 2012; Owen, Rentero, et al., 2010; Owen et al., 2013; Rossy, Cohen, Gaus, & Owen, 2014; Sengupta et al., 2011; Sengupta & Lippincott-Schwartz, 2012).

6.2 Other imaging methods

Fluorescence lifetime imaging microscopy (FLIM) presents a different approach as it does not provide subdiffraction limited images but remains nonetheless a very powerful tool in the context of nanodomain study. Indeed, by measuring lifetime decays in each pixel, it allows the acquisition of information that would not be available from standard microscopy (Elson et al., 2004; Festy, Ameer-Beg, Ng, & Suhling, 2007; Levitt, Matthews, Ameer-Beg, & Suhling, 2009). It is particularly powerful for sample with multiple species in each pixel. In the case of a membrane order-sensitive fluorescent dye for instance, the lifetime of the dye (as well as the emission wavelength) can be different depending on the order of the bilayer. It is therefore possible to visualize ordered lipid domains, or rafts, at the PM using FLIM (Demchenko, Mély, Duportail, & Klymchenko, 2009; Golfetto, Hinde, & Gratton, 2013; Haluska et al., 2008; Kwiatek et al., 2013; Owen et al., 2006). Interestingly, FLIM can detect the existence of more than one species per pixel making it possible to quantify the fraction of each phase using the phasor approach (Digman, Caiolfa, Zamai, & Gratton, 2008; Golfetto et al., 2013). Some interesting results concerning rafts have been acquired based on this methodology. In particular Owen, Williamson, Magenau, and Gaus (2012) have shown that the plasma membrane of live HeLa cells is composed of approximately 76% ordered and 24% disordered lipid domains. FLIM can also be used to detect FRET. Förster Resonance Energy Transfer (FRET) is a process that allows the study of protein–protein clustering and multiprotein complex formation (Bader, Hofman, Voortman, van Bergen en Henegouwen, & Gerritsen, 2009; Chichili, Cail, & Rodgers, 2012; Glebov & Nichols, 2004; Hashimoto-Tane et al., 2010; Jares-Erijman & Jovin, 2006; Loura, de Almeida, Silva, & Prieto,

2009; Pierce, 2004). In the case of structural studies, electron microscopy is routinely used due to its spatial precision. Choudhuri et al. (2014) for instance, have shown the presence of TCR complexes in extracellular vesicles at the center of the IS after engagement using this technique. New tools have also been developed to study dynamic associated with nanodomains FCS (and STED FCS), recovery after bleaching (FRAP), image correlation, and single molecule tracking (Ashdown et al., 2014; Eggeling et al., 2006, 2009; Favard, Wenger, Lenne, & Rigneault, 2011; Honigmann, Mueller, Hell, & Eggeling, 2013; Kenworthy et al., 2004; Kusumi et al., 2005; Manley et al., 2008; Ringemann et al., 2009; Ruan, Cheng, Levi, Gratton, & Mantulin, 2004; Umemura et al., 2008).

7. DISCUSSION

Nanodomains, or more precisely their formation and nanoscale organization at the plasma membrane, appear to be essential components of cellular activity. They appear in the form of nanoscale vesicles, protein–protein clusters or protein-lipid domains on the plasma membrane. Over the past decades, these domains have been identified as being necessary triggers and mediators in cellular signaling. More interestingly, it is now suggested that the existence and importance of these nanodomains are not confined to a specific cellular subset. On the contrary, they seem to be implicated in most cellular communication processes. This discovery explains the growing interest of the scientific community in their study, in a vast range of biology-related fields such as neurosciences, immunology, and therapeutic research. Recent advances in imaging techniques have increased this interest, as new tools are now available providing more detailed insight on the nanoscale organization of signaling domains. The specificity of this research topic, associating biology and physics-based instrumentation and analysis tools, justify the current emergence of pluridisciplinary research fields to try to answer biological issues. Moreover, the complexity of the field, bridging together different biology specialities as well as physics and even mathematics, has led to the development of international pluridisciplinary research projects, such as the BAM project. Therefore, understanding the nanoscale organization of signaling domains remains an ongoing process, although our insight has increased over the past decades due to the development of super resolution, electron microscopy and other forms of microscopy. The application of discoveries made on the nanoscale organization characteristics of these domains and on their function during key cellular events, could help in the development of new therapeutic strategies.

ACKNOWLEDGMENTS

We acknowledge funding from the ERC (grant # 337187) and a Marie-Curie CIG (grant #334303).

REFERENCES

Abankwa, D., Gorfe, A. A., & Hancock, J. F. (2007). Ras nanoclusters: molecular structure and assembly. *Seminars in Cell & Developmental Biology, 18*(5), 599–607.

Abercrombie, M., & Dunn, G. A. (1975). Adhesions of fibroblasts to substratum during contact inhibition observed by interference reflection microscopy. *Experimental Cell Research, 92*(1), 57–62.

Abercrombie, M., Heaysman, J. E. M., & Pegrum, S. M. (1971). The locomotion of fibroblasts in culture: IV. Electron microscopy of the leading lamella. *Experimental Cell Research, 67*(2), 359–367.

Adib-Conquy, M., Scott-Algara, D., Cavaillon, J.-M., & Souza-Fonseca-Guimaraes, F. (2014). TLR-mediated activation of NK cells and their role in bacterial/viral immune responses in mammals [Review] *Immunology & Cell Biology, 92*(3), 256–262.

Adjei, A. A. (2001). Blocking oncogenic Ras signaling for cancer therapy. *Journal of the National Cancer Institute, 93*(14), 1062–1074.

Alfsen, A., Iniguez, P., Bouguyon, E., & Bomsel, M. (2001). Secretory IgA specific for a conserved epitope on gp41 envelope glycoprotein inhibits epithelial transcytosis of HIV-1. *Journal of Immunology, 166*(10), 6257–6265.

Alivisatos, A. P., Andrews, A. M., Boyden, E. S., Chun, M., Church, G. M., Deisseroth, K., et al. (2013). Nanotools for neuroscience and brain activity mapping. *ACS Nano, 7*(3), 1850–1866.

Andreska, T., Aufmkolk, S., Sauer, M., & Blum, R. (2014). High abundance of BDNF within glutamatergic presynapses of cultured hippocampal neurons. *Frontiers in Cellular Neuroscience, 8*, 107.

Annibale, P., Scarselli, M., Greco, M., & Radenovic, A. (2012). Identification of the factors affecting co-localization precision for quantitative multicolor localization microscopy. *Optical Nanoscopy, 1*(9), 1–13.

Annibale, P., Vanni, S., Scarselli, M., Rothlisberger, U., & Radenovic, A. (2011a). Identification of clustering artifacts in photoactivated localization microscopy. *Nature Methods, 8*(7), 527–528.

Annibale, P., Vanni, S., Scarselli, M., Rothlisberger, U., & Radenovic, A. (2011b). Quantitative photo activated localization microscopy: unraveling the effects of photoblinking. *PLoS One, 6*(7), e22678.

Ariotti, N., Liang, H., Xu, Y., Zhang, Y., Yonekubo, Y., Inder, K., et al. (2010). Epidermal growth factor receptor activation remodels the plasma membrane lipid environment to induce nanocluster formation. *Molecular and Cellular Biology, 30*(15), 3795–3804.

Ashdown, G. W., Cope, A., Wiseman, P. W., & Owen, D. M. (2014). Molecular flow quantified beyond the diffraction limit by spatiotemporal image correlation of structured illumination microscopy data. *Biophysical Journal, 107*(9), L21–L23.

Babich, A., Li, S., O'Connor, R. S., Milone, M. C., Freedman, B. D., & Burkhardt, J. K. (2012). F-actin polymerization and retrograde flow drive sustained PLCγ1 signaling during T cell activation. *Journal of Cell Biology, 197*(6), 775–787.

Babiychuk, E. B., & Draeger, A. (2006). Biochemical characterization of detergent-resistant membranes: a systematic approach. *Biochemical Journal, 397*, 407–416.

Bader, A. N., Hofman, E. G., Voortman, J., van Bergen en Henegouwen, P. M. P., & Gerritsen, H. C. (2009). Homo-FRET imaging enables quantification of protein cluster sizes with subcellular resolution. *Biophysical Journal, 97*(9), 2613–2622.

Balagopalan, L., Barr, V. A., Kortum, R. L., Park, A. K., & Samelson, L. E. (2013). Cutting edge: cell surface linker for activation of T cells is recruited to microclusters and is active in signaling. *Journal of Immunology, 190*(8), 3849–3853.

Barman, S., & Nayak, D. P. (2007). Lipid raft disruption by cholesterol depletion enhances influenza a virus budding from MDCK cells. *Journal of Virology, 81*(22), 12169–12178.

Barre-Sinoussi, F., Ross, A., & Delfraissy, J. (2013). Past, present and future: 30 years of HIV research [Perspectives] *Nature Reviews Microbiology, 11*(12), 877–883.

Bates, M., Huang, B., Dempsey, G. T., & Zhuang, X. (2007). Multicolor super-resolution imaging with photo-switchable fluorescent probes. *Science, 317*, 1749–1753.

Berger, E. A., Murphy, P. M., & Farber, J. M. (1999). Chemokine receptors AS HIV-1 coreceptors: roles in viral entry, tropism, and disease. *Annual Review of Immunology, 17*(1), 657–700.

Betzig, E., Patterson, G. H., Sougrat, R., Lindwasser, O. W., Olenych, S., Bonifacino, J. S., et al. (2006). Imaging intracellular fluorescent proteins at nanometer resolution. *Science, 313*, 1642–1645.

Bhattacharyya, S., Warfield, K. L., Ruthel, G., Bavari, S., Aman, M. J., & Hope, T. J. (2010). Ebola virus uses clathrin mediated endocytosis as an entry pathway. *Virology, 401*(1), 18–28.

Blanchard, E., Belouzard, S., Goueslain, L., Wakita, T., Dubuisson, J., Wychowski, C., et al. (2006). Hepatitis C virus entry depends on clathrin-mediated endocytosis. *Journal of Virology, 80*(14), 6964–6972.

Brown, D. A. (2006). Lipid rafts, detergent-resistant membranes, and raft targeting signals. *Physiology, 21*(6), 430–439.

Brown, C. M., Hebert, B., Kolin, D. L., Zareno, J., Whitmore, L., Horwitz, A. R., et al. (2006). Probing the integrin-actin linkage using high-resolution protein velocity mapping. *Journal of Cell Science, 119*(24), 5204–5214.

Brown, M. C., & Turner, C. E. (2004). Paxillin: adapting to change. *Physiological Reviews, 84*.

Carrasco, M., Amorim, M. J., & Digard, P. (2004). Lipid raft-dependent targeting of the influenza a virus nucleoprotein to the apical plasma membrane. *Traffic, 5*(12), 979–992.

Casas, J., Brzostek, J., Zarnitsyna, V. I., Hong, J. S., Wei, Q., Hoerter, J. A., et al. (2014). Ligand-engaged TCR is triggered by Lck not associated with CD8 coreceptor. *Nature Communications, 5*, 5624.

Casey, P. J. (1995). Protein lipidation in cell signaling. *Science, 268*(5208), 221–225.

Cebecauer, M., Spitaler, M., Serge, A., & Magee, A. I. (2010). Signalling complexes and clusters: functional advantages and methodological hurdles. *Journal of Cell Science, 123*(3), 309–320.

Chan, A. C., Iwashima, M., Turck, C. W., & Weiss, A. (1992). ZAP-70: a 70 kd protein-tyrosine kinase that associates with the TCR ζ chain. *Cell, 71*(4), 649–662.

Chaudhuri, A., Bhattacharya, B., Gowrishankar, K., Mayor, S., & Rao, M. (2011). Spatiotemporal regulation of chemical reactions by active cytoskeletal remodeling. *Proceedings of the National Academy of Sciences of the United States of America, 108*(36), 14825–14830.

Chazal, N., & Gerlier, D. (2003). Virus entry, assembly, budding, and membrane rafts. *Microbiology and Molecular Biology Reviews, 67*(2), 226–237.

Cheng, P. C., Dykstra, M. L., Mitchell, R. N., & Pierce, S. K. (1999). A role for lipid Rafts in B Cell antigen receptor signaling and antigen targeting. *Journal of Experimental Medicine, 190*(11), 1549–1560.

Chen, Y., & Norkin, L. C. (1999). Extracellular simian virus 40 transmits a signal that promotes virus enclosure within caveolae. *Experimental Cell Research, 246*(1), 83–90.

Chichili, G. R., Cail, R. C., & Rodgers, W. (2012). Cytoskeletal modulation of lipid interactions regulates Lck kinase activity. *Journal of Biological Chemistry, 287*(29), 24186–24194.

Chojnacki, J., Staudt, T., Glass, B., Bingen, P., Engelhardt, J., Anders, M., et al. (2012). Maturation-dependent HIV-1 surface protein redistribution revealed by fluorescence nanoscopy. *Science, 338*(6106), 524–528.

Choudhuri, K., Llodra, J., Roth, E. W., Tsai, J., Gordo, S., Wucherpfennig, K. W., et al. (2014). Polarized release of T-cell-receptor-enriched microvesicles at the immunological synapse [Letter] *Nature, 507*, 118–123.

Comiskey, M., & Warner, C. M. (2007). Spatio-temporal localization of membrane lipid rafts in mouse oocytes and cleaving preimplantation embryos. *Developmental Biology, 303*(2), 727–739.

Corson, L. B., Yamanaka, Y., Lai, K. V., & Rossant, J. (2003). Spatial and temporal patterns of ERK signaling during mouse embryogenesis. *Development, 130*(19), 4527–4537.

Crites, T. J., Padhan, K., Muller, J., Krogsgaard, M., Gudla, P. R., Lockett, S. J., et al. (2014). TCR microclusters pre-exist and contain molecules necessary for TCR signal transduction. *Journal of Immunology, 193*(1), 56–67.

Dani, A., Huang, B., Bergan, J., Dulac, C., & Zhuang, X. (2010). Super-resolution imaging of chemical synapses in the brain. *Neuron, 68*(5), 843–856.

Davey, A. M., Walvick, R. P., Liu, Y., Heikal, A. A., & Sheets, E. D. (2007). Membrane order and molecular dynamics associated with IgE receptor cross-linking in mast cells. *Biophysical Journal, 92*(1), 343–355.

Davis, D. M., & Dustin, M. L. (2004). What is the importance of the immunological synapse? *Trends in Immunology, 25*(6), 323–327.

Davis, M. M., Krogsgaard, M., Huppa, J. B., Sumen, C., Purbhoo, M. A., Irvine, D. J., et al. (2003). Dynamics of cell surface molecules during T cell recognition. *Annual Review of Biochemistry, 72*(1), 717–742.

Demchenko, A. P., Mély, Y., Duportail, G., & Klymchenko, A. S. (2009). Monitoring biophysical properties of lipid membranes by environment-sensitive fluorescent probes. *Biophysical Journal, 96*(9), 3461–3470.

Denny, M. F., Patai, B., & Straus, D. B. (2000). Differential T-cell antigen receptor signaling mediated by the Src family kinases Lck and Fyn. *Molecular and Cellular Biology, 20*(4), 1426–1435.

Digman, M. A., Caiolfa, V. R., Zamai, M., & Gratton, E. (2008). The phasor approach to fluorescence lifetime imaging analysis. *Biophysical Journal, 94*(2), L14–L16.

Dustin, M. L., & Groves, J. T. (2012). Receptor signaling clusters in the immune synapse. *Annual Review of Biophysics, 41*(1), 543–556.

Dykstra, M., Cherukuri, A., Sohn, H. W., Tzeng, S. J., & Pierce, S. K. (2003). Location is everything: lipid rafts and immune cell signaling. *Annual Review of Immunology, 21*, 457–481.

Eggeling, C., Ringemann, C., Medda, R., Schwarzmann, G., Sandhoff, K., Polyakova, S., et al. (2009). Direct observation of the nanoscale dynamics of membrane lipids in a living cell. *Nature, 457*, 1159–1163.

Eggeling, C., Widengren, J., Brand, L., Schaffer, J., Felekyan, S., & Seidel, C. A. M. (2006). Analysis of photobleaching in single-molecule multicolor excitation and Forster resonance energy transfer measurements. *Journal of Physical Chemistry, 110*(9), 2979–2995.

Elson, D. S., Requejo-Isidro, J., Munro, I., Reavell, F., Siegel, J., Suhling, K., et al. (2004). Time-domain fluorescence lifetime imaging applied to biological tissue. *Photochemical & Photobiological Sciences, 8*, 795–801.

Evans, R., Patzak, I., Svensson, L., De Filippo, K., Jones, K., McDowall, A., et al. (2009). Integrins in immunity. *Journal of Cell Science, 122*(Pt 2), 215–225.

Ewers, H., Romer, W., Smith, A. E., Bacia, K., Dmitrieff, S., Chai, W., et al. (2010). GM1 structure determines SV40-induced membrane invagination and infection. *Nature Cell Biology, 12*(1), 11–18.

Favard, C., Wenger, J., Lenne, P.-F., & Rigneault, H. (2011). FCS diffusion laws in two-phase lipid membranes: determination of domain mean size by experiments and Monte Carlo simulations. *Biophysical Journal, 100*(5), 1242–1251.

Festy, F., Ameer-Beg, S. M., Ng, T., & Suhling, K. (2007). Imaging proteins in vivo using fluorescence lifetime microscopy. *Molecular BioSystems, 3*, 381–391.

Flores-Borja, F., Kabouridis, P. S., Jury, E. C., Isenberg, D. A., & Mageed, R. A. (2005). Decreased Lyn expression and translocation to lipid raft signaling domains in B lymphocytes from patients with systemic lupus erythematosus. *Arthritis & Rheumatism, 52*(12), 3955–3965.

Freitas, M. S., Follmer, C., Costa, L. T., Vilani, C., Bianconi, M. L., Achete, C. A., et al. (2011). Measuring the strength of interaction between the Ebola fusion peptide and lipid rafts: implications for membrane fusion and virus infection. *PLoS One, 6*(1), e15756.

Galbraith, C. G., Yamada, K. M., & Sheetz, M. P. (2002). The relationship between force and focal complex development. *Journal of Cell Biology, 159*(4), 695–705.

Garcia-Parajo, M. F., Cambi, A., Torreno-Pina, J. A., Thompson, N., & Jacobson, K. (2014). Nanoclustering as a dominant feature of plasma membrane organization. *Journal of Cell Science, 127*(23), 4995–5005.

Garner, A. E., Smith, D. A., & Hooper, N. M. (2008). Visualization of detergent solubilization of membranes: implications for the isolation of rafts. *Biophysical Journal, 94*(4), 1326–1340.

Glebov, O. O., & Nichols, B. J. (2004). Lipid raft proteins have a random distribution during localized activation of the T-cell receptor. *Nature Cell Biology, 6*(3), 238–243.

Goldrath, A., & Bevan, M. (1999). Selecting and maintaining a diverse T-cell repertoire. *Nature, 403*, 255–262.

Golfetto, O., Hinde, E., & Gratton, E. (2013). Laurdan fluorescence lifetime discriminates cholesterol content from changes in fluidity in living cell membranes. *Biophysical Journal, 104*(6), 1238–1247.

Gomez-Mouton, C., Lacalle, R. A., Mira, E., Jimenez-Baranda, S., Barber, D. F., Carrera, A. C., et al. (2004). Dynamic redistribution of raft domains as an organizing platform for signaling during cell chemotaxis. *Journal of Cell Biology, 194*(5), 759–768.

Goswami, D., Gowrishankar, K., Bilgrami, S., Ghosh, S., Raghupathy, R., Chadda, R., et al. (2008). Nanoclusters of GPI-anchored proteins are formed by cortical actin-driven activity. *Cell, 135*(6), 1085–1097.

Gowrishankar, K., Ghosh, S., Saha, S., C, R., Mayor, S., & Rao, M. (2012). Active remodeling of cortical actin regulates spatiotemporal organization of cell surface molecules. *Cell, 149*(6), 1353–1367.

Grakoui, A., Bromley, S. K., Sumen, C., Davis, M. M., Shaw, A. S., Allen, P. M., et al. (1999). The immunological synapse: a molecular machine controlling T cell activation *Science, 285*(5425), 221–227.

Grier, J. T., Forbes, L. R., Monaco-Shawver, L., Oshinsky, J., Atkinson, T. P., Moody, C., et al. (2012). Human immunodeficiency causing mutation defines CD16 in spontaneous NK cell cytotoxicity. *Journal of Clinical Investigation, 122*(10), 3769–3780.

Griffié, J., Boelen, L., Burn, G., Cope, A. P., & Owen, D. M. (2015). Topographic prominence as a method for cluster identification in single-molecule localisation data. *Journal of Biophotonics, 9999*(9999).

Gupta, N., & DeFranco, A. L. (2007). Lipid rafts and B cell signaling. *Seminars in Cell & Developmental Biology, 18*(5), 616–626.

Gurry, T., Kahramanoğulları, O., & Endres, R. G. (2009). Biophysical mechanism for Ras-nanocluster formation and signaling in plasma membrane. *PLoS One, 4*(7), e6148.

Gustafsson, M. G. L. (2000). Surpassing the lateral resolution limit by a factor of two using structured illumination microscopy. *Journal of Microscopy, 198*(2), 82–87.

Haluska, C. K., Schroder, A. P., Didier, P., Heissler, D., Duportail, G., Mely, Y., et al. (2008). Combining fluorescence lifetime and polarization microscopy to discriminate phase separated domains in giant unilamellar vesicles. *Biophysical Journal, 95*(12), 5737–5747.

Hancock, J. F., Magee, A. I., Childs, J. E., & Marshall, C. J. (1989). All Ras proteins are polyisoprenylated but only some are palmitoylated. *Cell, 57*(7), 1167–1177.

Harding, A. S., & Hancock, J. F. (2008). Using plasma membrane nanoclusters to build better signaling circuits. *Trends in Cell Biology, 18*(8), 364–371.

Hashimoto-Tane, A., Yokosuka, T., Ishihara, C., Sakuma, M., Kobayashi, W., & Saito, T. (2010). T-cell receptor microclusters critical for T-cell activation are formed independently of lipid raft clustering. *Molecular and Cellular Biology, 30*(14), 3421–3429.

Haucke, V., Neher, E., & Sigrist, S. J. (2011). Protein scaffolds in the coupling of synaptic exocytosis and endocytosis. *Nature Review Neurosciences, 12*(3), 127–138.

Hein, B., Willig, K. I., & Hell, S. W. (2008). Stimulated emission depletion (STED) nanoscopy of a fluorescent protein-labeled organelle inside a living cell. *Proceedings of the National Academy of Sciences of the United States of America, 105*(38), 14271–14276.

Hell, S. W., & Wichmann, J. (1994). Breaking the diffraction resolution limit by stimulated emission: stimulated-emission-depletion fluorescence microscopy. *Optics Letters, 19*(11), 780–782.

Hess, S. T., Girirajan, T. P. K., & Mason, M. D. (2006). Ultra-High resolution imaging by fluorescence photoactivation localization microscopy. *Biophysical Journal, 91*(11), 4258–4272.

Hitosugi, T., Sato, M., Sasaki, K., & Umezawa, Y. (2007). Lipid raft specific knockdown of Src family kinase activity inhibits cell adhesion and cell cycle progression of breast cancer cells. *Cancer Research, 67*(17), 8139–8148.

Hogg, N., Henderson, R., Leitinger, B., McDowall, A., Porter, J., & Stanley, P. (2002). Mechanisms contributing to the activity of integrins on leukocytes. *Immunological Reviews, 186*(1), 164–171.

Hogg, N., Laschinger, M., Giles, K., & McDowall, A. (2003). T-cell integrins: more than just sticking points. *Journal of Cell Science, 116*(23), 4695–4705.

Hogg, N., Patzak, I., & Willenbrock, F. (2011). The insider's guide to leukocyte integrin signalling and function. *Nature Reviews Immunology, 11*(6), 416–426.

Hogue, I. B., Grover, J. R., Soheilian, F., Nagashima, K., & Ono, A. (2011). Gag induces the coalescence of clustered lipid rafts and tetraspanin-enriched microdomains at HIV-1 assembly sites on the plasma membrane. *Journal of Virology, 85*(19), 9749–9766.

Honigmann, A., Mueller, V., Hell, S. W., & Eggeling, C. (2013). STED microscopy detects and quantifies liquid phase separation in lipid membranes using a new far-red emitting fluorescent phosphoglycerolipid analogue. *Faraday Discussions, 161*(0), 77–89.

Hu, K., Ji, L., Applegate, K. T., Danuser, G., & Waterman-Storer, C. M. (2007). Differential transmission of actin motion within focal adhesions. *Science, 315*(5808), 111–115.

Huang, B. (2010). Super-resolution optical microscopy: multiple choices. *Current Opinion in Chemical Biology, 14*(1), 10–14.

Huang, B., Wang, W., Bates, M., & Zhuang, X. (2008). Three-dimensional super-resolution imaging by stochastic optical reconstruction microscopy. *Science, 319*(5864), 810–813.

Huang, F., Schwartz, S. L., Byars, J. M., & Lidke, K. A. (2011). Simultaneous multiple-emitter fitting for single molecule super-resolution imaging. *Biomedical Optics Express, 2*(5), 1377–1393.

Hynes, R. O. (2002). Integrins: bidirectional, allosteric signaling machines. *Cell, 110*(6), 673–687.

Jacobson, K., & Dietrich, C. (1999). Looking at lipid rafts? *Trends in Cell Biology, 9*(3), 87–91.

Jacobson, K., Mouritsen, O. G., & Anderson, R. G. W. (2007). Lipid rafts: at a crossroad between cell biology and physics. *Nature Cell Biology, 9*(1), 7–14.

Jares-Erijman, E. A., & Jovin, T. M. (2006). Imaging molecular interactions in living cells by FRET microscopy. *Current Opinion in Chemical Biology, 10*(5), 409–416.

Jiang, G., Giannone, G., Critchley, D. R., Fukumoto, E., & Sheetz, M. P. (2003). Two-piconewton slip bond between fibronectin and the cytoskeleton depends on talin. *Nature, 424*(6946), 334–337.

Jin, Y., & Garner, C. C. (2008). Molecular mechanisms of presynaptic differentiation. *Annual Review of Cell and Developmental Biology, 24*(1), 237–262.

Jin, Z.-X., Huang, C.-R., Dong, L., Goda, S., Kawanami, T., Sawaki, T., et al. (2008). Impaired TCR signaling through dysfunction of lipid rafts in sphingomyelin synthase 1 (SMS1)-knockdown T cells. *International Immunology, 20*(11), 1427–1437.

Jury, E. C., Flores-Borja, F., & Kabouridis, P. S. (2007). Lipid rafts in T cell signalling and disease. *Seminars in Cell & Developmental Biology, 18*(5), 608–615.

Kanchanawong, P., Shtengel, G., Pasapera, A. M., Ramko, E. B., Davidson, M. W., Hess, H. F., et al. (2010). Nanoscale architecture of integrin-based cell adhesions. *Nature, 468*(7323), 580–584.

Kenworthy, A. K., Nichols, B. J., Remmert, C. L., Hendrix, G. M., Kumar, M., Zimmerberg, J., et al. (2004). Dynamics of putative raft-associated proteins at the cell surface. *Journal of Cell Biology, 165*(5), 735–746.

Ketchum, C., Miller, H., Song, W., & Upadhyaya, A. (2014). Ligand mobility regulates B cell receptor clustering and signaling activation. *Biophysical Journal, 106*(1), 26–36.

Khurana, S., Krementsov, D. N., de Parseval, A., Elder, J. H., Foti, M., & Thali, M. (2007). Human immunodeficiency virus type 1 and influenza virus exit via different membrane microdomains. *Journal of Virology, 81*(22), 12630–12640.

Krauss, K., & Altevogt, P. (1999). Integrin leukocyte function-associated antigen-1-mediated cell binding can be activated by clustering of membrane rafts. *Journal of Biological Chemistry, 274*(52), 36921–36927.

Kusumi, A., Fujiwara, T. K., Chadda, R., Xie, M., Tsunoyama, T. A., Kalay, Z., et al. (2012). Dynamic organizing principles of the plasma membrane that regulate signal transduction: commemorating the fortieth anniversary of Singer and Nicolson's fluid-mosaic model. *Annual Review of Cell and Developmental Biology, 28*(1), 215–250.

Kusumi, A., Koyama-Honda, I., & Suzuki, K. (2004). Molecular dynamics and interactions for creation of stimulation-induced stabilized rafts from small unstable steady-state rafts. *Traffic, 5*(4), 213–230.

Kusumi, A., Nakada, C., Ritchie, K., Murase, K., Suzuki, K., Murakoshi, H., et al. (2005). Paradigm shift of the plasma membrane concept from the two-dimensional continuum fluid to the partitioned fluid: high-speed single-molecule tracking of membrane molecules. *Annual Review of Biophysics and Biomolecular Structure, 34*(1), 351–378.

Kwiatek, J. M., Owen, D. M., Abu-Siniyeh, A., Yan, P., Loew, L. M., & Gaus, K. (2013). Characterization of a new series of fluorescent probes for imaging membrane order. *PLoS One, 8*(2), e52960.

Lafont, F., Abrami, L., & van der Goot, F. G. (2004). Bacterial subversion of lipid rafts. *Current Opinion in Microbiology, 7*(1), 4–10.

Lajoie, P., Partridge, E. A., Guay, G., Goetz, J. G., Pawling, J., Lagana, A., et al. (2007). Plasma membrane domain organization regulates EGFR signaling in tumor cells. *Journal of Cell Biology, 179*(2), 341–356.

Lawson, M. A., & Maxfield, F. R. (1995). Ca^{2+} and calcineurin-dependent recycling of an integrin to the front of migrating neutrophils. *Nature, 377*(6544), 75–79.

Leitinger, B., & Hogg, N. (2002). The involvement of lipid rafts in the regulation of integrin function. *Journal of Cell Science, 115*(5), 963–972.

Levitt, J. A., Matthews, D. R., Ameer-Beg, S. M., & Suhling, K. (2009). Fluorescence lifetime and polarization-resolved imaging in cell biology. *Current Opinion in Biotechnology, 20*(1), 28–36.

Lillemeier, B. F., Mortelmaier, M. A., Forstner, M. B., Huppa, J. B., Groves, J. T., & Davis, M. M. (2010). TCR and LAT are expressed on separate protein islands on T cell membranes and concatenate during activation. *Nature Immunology, 11*(1), 90–96.

Lingwood, D., & Simons, K. (2007). Detergent resistance as a tool in membrane research. *Nature Protocols, 2*(9), 2159–2165.

Lingwood, D., & Simons, K. (2010). Lipid rafts as a membrane-organizing principle. *Science, 327*(5961), 46–50.

Lippincott-Schwartz, J., & Manley, S. (2009). Putting super-resolution fluorescence microscopy to work. *Nature Methods, 6*(1), 21–23.

Liu, N. Q., Lossinsky, A. S., Popik, W., Li, X., Gujuluva, C., Kriederman, B., et al. (2002). Human immunodeficiency virus type 1 enters brain microvascular endothelia by macropinocytosis dependent on lipid rafts and the mitogen-activated protein kinase signaling pathway. *Journal of Virology, 76*(13), 6689–6700.

Loura, L. M. S., de Almeida, R. F. M., Silva, L. C., & Prieto, M. (2009). FRET analysis of domain formation and properties in complex membrane systems. *Biochimica et Biophysica Acta, 1788*(1), 209–224.

Lovatt, M., Filby, A., Parravicini, V., Werlen, G., Palmer, E., & Zamoyska, R. (2006). Lck regulates the threshold of activation in primary T cells, while both Lck and Fyn contribute to the magnitude of the extracellular signal-related kinase response. *Molecular and Cellular Biology, 26*(22), 8655–8665.

Malkusch, S., Endesfelder, U., Mondry, J., Gelléri, M., Verveer, P., & Heilemann, M. (2012). Coordinate-based colocalization analysis of single-molecule localization microscopy data. *Histochemistry and Cell Biology, 137*(1), 1–10.

Mañes, S., del Real, G., Lacalle, R. A., Lucas, P., Gómez-Moutón, C., Sánchez-Palomino, S., et al. (2000). Membrane raft microdomains mediate lateral assemblies required for HIV-1 infection. *EMBO Reports, 1*(2), 190–196.

Mañes, S., del Real, G., & Martinez-A, C. (2003). Pathogens: raft hijackers. *Nature Reviews Immunology, 3*(7), 557–568.

Manley, S., Gillette, J. M., Patterson, G. H., Shroff, H., Hess, H. F., Betzig, E., et al. (2008). High-density mapping of single-molecule trajectories with photoactivated localization microscopy. *Nature Methods, 5*(2), 155–157.

Marsh, M., & Helenius, A. (2006). Virus entry: open sesame. *Cell, 124*(4), 729–740.

Mattila, P. K., Feest, C., Depoil, D., Treanor, B., Montaner, B., Otipoby, K. L., et al. (2013). The actin and tetraspanin networks organize receptor nanoclusters to regulate B cell receptor-mediated signaling. *Immunity, 38*(3), 461–474.

Maxfield, F. R., & Tabas, I. (2005). Role of cholesterol and lipid organization in disease. *Nature, 438*(7068), 612–621.

Michel, V., & Bakovic, M. (2007). Lipid rafts in health and disease. *Biology of the Cell, 99*(3), 129–140.

Mitra, S. K., Hanson, D. A., & Schlaepfer, D. D. (2005). Focal adhesion kinase: in command and control of cell motility. *Nature Reviews Molecular Cell Biology, 6*(1), 56–68.

Mohrmann, R., de Wit, H., Verhage, M., Neher, E., & Sørensen, J. B. (2010). Fast vesicle fusion in living cells requires at least three SNARE complexes. *Science, 330*(6003), 502–505.

Montoya, M. C., Sancho, D., Bonello, G., Collette, Y., Langlet, C., He, H. T., et al. (2002). Role of ICAM-3 in the initial interaction of T lymphocytes and APCs. *Nature Immunology, 3*(2), 159–168.

Mossman, K. D., Campi, G., Groves, J. T., & Dustin, M. L. (2005). Altered TCR signaling from geometrically repatterned immunological synapses. *Science, 310*(5751), 1191–1193.

Mudhakir, D., & Harashima, H. (2009). Learning from the viral journey: how to enter cells and how to overcome intracellular barriers to reach the nucleus. *AAPS Journal, 11*(1), 65–77.

Mueller, V., Ringemann, C., Honigmann, A., Schwarzmann, G., Medda, R., Leutenegger, M., et al. (2011). STED nanoscopy reveals molecular details of cholesterol- and cytoskeleton-modulated lipid interactions in living cells. *Biophysical Journal, 101*(7), 1651–1660.

Murakoshi, H., Iino, R., Kobayashi, T., Fujiwara, T., Ohshima, C., Yoshimura, A., et al. (2004). Single-molecule imaging analysis of Ras activation in living cells. *Proceedings of the National Academy of Sciences of the United States of America, 101*(19), 7317–7322.

Neu, U., Woellner, K., Gauglitz, G., & Stehle, T. (2008). Structural basis of GM1 ganglioside recognition by simian virus 40. *Proceedings of the National Academy of Sciences of the United States of America, 105*(13), 5219–5224.

Nguyen, D. H., & Hildreth, J. E. K. (2000). Evidence for budding of human immunodeficiency virus type 1 selectively from glycolipid-enriched membrane lipid rafts. *Journal of Virology, 74*(7), 3264–3272.

Nika, K., Soldani, C., Salek, M., Paster, W., Gray, A., Etzensperger, R., et al. (2010). Constitutively active Lck Kinase in T Cells drives antigen receptor signal transduction. *Immunity, 32*(6), 766–777.

Nisole, S., Krust, B., & Hovanessian, A. G. (2002). Anchorage of HIV on permissive cells leads to coaggregation of viral particles with surface nucleolin at membrane raft microdomains. *Experimental Cell Research, 276*(2), 155–173.

Ono, A., Waheed, A. A., & Freed, E. O. (2007). Depletion of cellular cholesterol inhibits membrane binding and higher-order multimerization of human immunodeficiency virus type 1 Gag. *Virology, 360*(1), 27–35.

Ovesný, M., Křížek, P., Borkovec, J., Švindrych, Z., & Hagen, G. M. (2014). ThunderSTORM: a comprehensive ImageJ plugin for PALM and STORM data analysis and super-resolution imaging. *Bioinformatics, 30*(16), 2389–2390.

Owen, D. M., Lanigan, P. M. P., Dunsby, C., Munro, I., Grant, D., Neil, M. A. A., et al. (2006). Fluorescence lifetime imaging provides enhanced contrast when imaging the phase-sensitive dye di-4-ANEPPDHQ in model membranes and live cells. *Biophysical Journal, 90*(11), L80–L82.

Owen, D. M., Magenau, A., Majumdar, A., & Gaus, K. (2010). Imaging membrane lipid order in whole, living vertebrate organisms. *Biophysical Journal, 99*(1), L7–L9.

Owen, D. M., Magenau, A., Williamson, D., & Gaus, K. (2012). The lipid raft hypothesis revisited—new insights on raft composition and function from super-resolution fluorescence microscopy. *BioEssays, 34*(9), 739–747.

Owen, D. M., Oddos, S., Kumar, S., Davis, D. M., Neil, M. A. A., French, P. M. W., et al. (2010). High plasma membrane lipid order imaged at the immunological synapse periphery in live T cells. *Molecular Membrane Biology, 27*(4–6), 178–189.

Owen, D. M., Rentero, C., Rossy, J., Magenau, A., Williamson, D., Rodriguez, M., et al. (2010). PALM imaging and cluster analysis of protein heterogeneity at the cell surface. *Journal of Biophotonics, 3*(7), 446–454.

Owen, D. M., Williamson, J., Boelen, L., Magenau, A., Rossy, J., & Gaus, K. (2013). Quantitative analysis of three-dimensional fluorescence localization microscopy data. *Biophysical Journal, 105*(2), L05–L07.

Owen, D. M., Williamson, D. J., Magenau, A., & Gaus, K. (2012). Sub-resolution lipid domains exist in the plasma membrane and regulate protein diffusion and distribution. *Nature Communications, 3*, 1256.

Palacios, E. H., & Weiss, A. (2004). Function of the Src-family kinases, Lck and Fyn, in T-cell development and activation. *Oncogene, 23*(48), 7990–8000.

Patra, S. K. (2008). Dissecting lipid raft facilitated cell signaling pathways in cancer. *Biochimica et Biophysica Acta, 1785*(2), 182–206.

Pelkmans, L., Fava, E., Grabner, H., Hannus, M., Habermann, B., Krausz, E., et al. (2005). Genome-wide analysis of human kinases in clathrin- and caveolae/raft-mediated endocytosis. *Nature, 436*(7047), 78–86.

Pierce, S. K. (2002). Lipid rafts and B-cell activation. *Nature Reviews Immunology, 2*(2), 96–105.
Pierce, S. K. (2004). To cluster or not to cluster: FRETting over rafts. *Nature Cell Biology, 6*(3), 180–181.
Pike, L. J. (2003). Lipid rafts: bringing order to chaos [Review] *Journal of Lipid Research, 44*(4), 655–667.
Pike, L. J. (2006). Rafts defined: a report on the Keystone symposium on lipid rafts and cell function. *Journal of Lipid Research, 47*(7), 1597–1598.
Pizzo, P., & Viola, A. (2004). Lipid rafts in lymphocyte activation. *Microbes and Infection, 6*(7), 686–692.
Plowman, S. J., Muncke, C., Parton, R. G., & Hancock, J. F. (2005). H-Ras, K-Ras, and inner plasma membrane raft proteins operate in nanoclusters with differential dependence on the actin cytoskeleton. *Proceedings of the National Academy of Sciences of the United States of America, 102*(43), 15500–15505.
Ponta, H., Sherman, L., & Herrlich, P. A. (2003). CD44: from adhesion molecules to signalling regulators. *Nature Reviews Molecular Cell Biology, 4*(1), 33–45.
Prior, I. A., Harding, A., Yan, J., Sluimer, J., Parton, R. G., & Hancock, J. F. (2001). GTP-dependent segregation of H-Ras from lipid rafts is required for biological activity. *Nature Cell Biology, 3*(4), 368–375. http://dx.doi.org/10.1038/35070050.
Prior, I. A., Muncke, C., Parton, R. G., & Hancock, J. F. (2003). Direct visualization of Ras proteins in spatially distinct cell surface microdomains. *Journal of Cell Biology, 160*(2), 165–170.
Prior, I. A., Parton, R. G., & Hancock, J. F. (2003). Observing cell surface signaling domains using electron microscopy. *Science STKE, 2003*(177), l9.
Purbhoo, M. A., Liu, H., Oddos, S., Owen, D. M., Neil, M. A. A., Pageon, S. V., et al. (2010). Dynamics of subsynaptic vesicles and surface microclusters at the immunological synapse. *Science Signaling, 3*(121), ra36.
Raulet, D. H. (2006). Missing self recognition and self tolerance of natural killer (NK) cells. *Seminars in Immunology, 18*(3), 145–150.
del Real, G., Jimenez-Baranda, S., Lacalle, R. A., Mira, E., Lucas, P., Gomez-Mouton, C., et al. (2002). Blocking of HIV-1 infection by targeting CD4 to Nonraft membrane domains. *Journal of Experimental Medicine, 196*(3), 293–301.
Ringemann, C., Harke, B., von Middendorff, C., Medda, R., Honigmann, A., Wagner, R., et al. (2009). Exploring single-molecule dynamics with fluorescence nanoscopy. *New Journal of Physics, 11*(10), 103054.
Rocks, O., Gerauer, M., Vartak, N., Koch, S., Huang, Z.-P., Pechlivanis, M., et al. (2010). The palmitoylation machinery is a spatially organizing system for peripheral membrane proteins. *Cell, 141*(3), 458–471.
Rossy, J., Cohen, E., Gaus, K., & Owen, D. M. (2014). Method for co-cluster analysis in multichannel single-molecule localisation data. *Histochemistry and Cell Biology, 141*(6), 605–612.
Rossy, J., Owen, D. M., Williamson, D. J., Yang, Z., & Gaus, K. (2013). Conformational states of the kinase Lck regulate clustering in early T cell signaling. *Nature Immunology, 14*(1), 82–89.
Rossy, J., Williamson, D. J., & Gaus, K. (2012). The regulation of TCR signalling by the spatial organisation of membrane proteins [Mini Review] *Frontiers in Immunology, 3*.
Rothlein, R., Dustin, M. L., Marlin, S. D., & Springer, T. A. (1986). A human intercellular adhesion molecule (ICAM-1) distinct from LFA-1. *Journal of Immunology, 137*(4), 1270–1274.
Ruan, Q., Cheng, M. A., Levi, M., Gratton, E., & Mantulin, W. W. (2004). Spatial-temporal studies of membrane dynamics: scanning fluorescence correlation spectroscopy (SFCS). *Biophysical Journal, 87*(2), 1260–1267.

Rust, M. J., Bates, M., & Zhuang, X. (2006). Sub-diffraction-limit imaging by stochastic optical reconstruction microscopy (STORM). *Nature Methods, 3*(10), 793–796.

Schermelleh, L., Heintzmann, R., & Leonhardt, H. (2010). A guide to super-resolution fluorescence microscopy. *Journal of Cell Biology, 190*(2), 165–175.

Schmidt, C., Kim, D., Ippolito, G. C., Naqvi, H. R., Probst, L., Mathur, S., et al. (2009). Signalling of the BCR is regulated by a lipid rafts-localised transcription factor, Bright. *EMBO Journal, 28*(6), 711–724.

Schoch, S., & Gundelfinger, E. (2006). Molecular organization of the presynaptic active zone. *Cell and Tissue Research, 326*(2), 379–391.

Sengupta, P., Baird, B., & Holowka, D. (2007). Lipid rafts, fluid/fluid phase separation, and their relevance to plasma membrane structure and function. *Seminars in Cell & Developmental Biology, 18*(5), 583–590.

Sengupta, P., Jovanovic-Talisman, T., Skoko, D., Renz, M., Veatch, S. L., & Lippincott-Schwartz, J. (2011). Probing protein heterogeneity in the plasma membrane using PALM and pair correlation analysis. *Nature Methods, 8*(11), 969–975.

Sengupta, P., & Lippincott-Schwartz, J. (2012). Quantitative analysis of photoactivated localization microscopy (PALM) datasets using pair-correlation analysis. *BioEssays, 34*(5), 396–405.

Shamri, R., Grabovsky, V., Feigelson, S. W., Dwir, O., van Kooyk, Y., & Alon, R. (2002). Chemokine stimulation of lymphocyte α4-integrin avidity but not of leukocyte function-associated antigen-1 avidity to endothelial ligands under shear flow requires cholesterol membrane rafts. *Journal of Biological Chemistry, 277*(42), 40027–40035.

Shao, L., Kner, P., Rego, E. H., & Gustafsson, M. G. L. (2011). Super-resolution 3D microscopy of live whole cells using structured illumination. *Nature Methods, 8*(12), 1044–1046.

Sherman, E., Barr, V., Manley, S., Patterson, G., Balagopalan, L., Akpan, I., et al. (2011). Functional nanoscale organization of signaling molecules downstream of the T Cell antigen receptor. *Immunity, 35*(5), 705–720.

Simons, K., & Ehehalt, R. (2002). Cholesterol, lipid rafts, and disease. *Journal of Clinical Investigation, 110*(5), 597–603.

Simons, K., & Ikonen, E. (1997). Functional rafts in cell membranes. *Nature, 387*, 569–572.

Simons, K., & Vaz, W. L. C. (2004). Model systems, lipid rafts and cell membranes. *Annual Review of Biophysics and Biomolecular Structure, 33*(1), 269–295.

Singer, S. J., & Nicolson, G. L. (1972). The fluid mosaic model of the structure of cell membranes. *Science, 175*, 720–731.

Smotrys, J. E., & Linder, M. E. (2004). Palmitoylation of intracellular signaling proteins: regulation and function. *Annual Review of Biochemistry, 73*(1), 559–587.

Stuart, A. D., Eustace, H. E., McKee, T. A., & Brown, T. D. K. (2002). A novel cell entry pathway for a DAF-using human enterovirus is dependent on lipid rafts. *Journal of Virology, 76*(18), 9307–9322.

Südhof, T. C. (2004). The synaptic vesicle cycle. *Annual Review of Neuroscience, 27*(1), 509–547.

Takahashi, T., & Suzuki, T. (2011). Function of membrane rafts in viral lifecycles and host cellular response. *Biochemistry Research International, 2011*, 1–23.

Tanimura, N., Nagafuku, M., Minaki, Y., Umeda, Y., Hayashi, F., Sakakura, J., et al. (2003). Dynamic changes in the mobility of LAT in aggregated lipid rafts upon T cell activation. *Journal of Cell Biology, 160*(1), 125–135.

Tanimura, N., Saitoh, S.-i., Kawano, S., Kosugi, A., & Miyake, K. (2006). Palmitoylation of LAT contributes to its subcellular localization and stability. *Biochemical and Biophysical Research Communications, 341*(4), 1177–1183. http://dx.doi.org/10.1016/j.bbrc.2006.01.076.

Tan, Y. X., Manz, B. N., Freedman, T. S., Zhang, C., Shokat, K. M., & Weiss, A. (2014). Inhibition of the kinase Csk in thymocytes reveals a requirement for actin remodeling in the initiation of full TCR signaling [Article] *Nature Immunology, 15*(2), 186–194.

Tian, T., Harding, A., Inder, K., Plowman, S., Parton, R. G., & Hancock, J. F. (2007). Plasma membrane nanoswitches generate high-fidelity Ras signal transduction. *Nature Cell Biology, 9*(8), 905–914.

Tohyama, Y., Katagiri, K., Pardi, R., Lu, C., Springer, T. A., & Kinashi, T. (2003). The critical cytoplasmic regions of the αL/β2 integrin in Rap1-induced adhesion and migration. *Molecular Biology of the Cell, 14*(6), 2570–2582.

Treanor, B., Depoil, D., Gonzalez-Granja, A., Barral, P., Weber, M., Dushek, O., et al. (2010). The membrane skeleton controls diffusion dynamics and signaling through the B cell receptor. *Immunity, 32*(2), 187–199.

Treanor, B., Lanigan, P. M. P., Kumar, S., Dunsby, C., Munro, I., Auksorius, E., et al. (2006). Microclusters of inhibitory killer immunoglobulin-like receptor signaling at natural killer cell immunological synapses. *Journal of Cell Biology, 174*(1), 153–161.

Umemura, Y. M., Vrljic, M., Nishimura, S. Y., Fujiwara, T. K., Suzuki, K. G. N., & Kusumi, A. (2008). Both MHC class II and its GPI-anchored form undergo hop diffusion as observed by single-molecule tracking. *Biophysical Journal, 95*(1), 435–450.

Ventimiglia, L. N., & Alonso, M. A. (2013). The role of membrane rafts in Lck transport, regulation and signalling in T-cells. *Biochemical Journal, 454*(2), 169–179.

Vivier, E., Tomasello, E., Baratin, M., Walzer, T., & Ugolini, S. (2008). Functions of natural killer cells. *Nature Immunology, 9*(5), 503–510.

Wang, Y., Gao, J., Guo, X., Tong, T., Shi, X., Li, L., et al. (2014). Regulation of EGFR nanocluster formation by ionic protein-lipid interaction [Original Article]. *Cell Research, 24*(8), 959–976.

Weiss, A., & Littman, D. R. (1994). Signal transduction by lymphocyte antigen receptors. *Cell, 76*(2), 263–274.

Wildanger, D., Rittweger, E., Kastrup, L., & Hell, S. W. (2008). STED microscopy with a supercontinuum laser source. *Optics Express, 16*(13), 9614–9621.

Williamson, D. J., Owen, D. M., Rossy, J., Magenau, A., Wehrmann, M., Gooding, J. J., et al. (2011). Pre-existing clusters of the adaptor Lat do not participate in early T cell signaling events. *Nature Immunology, 12*(7), 655–662.

Willig, K. I., Harke, B., Medda, R., & Hell, S. W. (2007). STED microscopy with continuous wave beams. *Nature Methods, 4*(11), 915–918.

Willig, K. I., Rizzoli, S. O., Westphal, V., Jahn, R., & Hell, S. W. (2006). STED microscopy reveals that synaptotagmin remains clustered after synaptic vesicle exocytosis. *Nature, 440*(7086), 935–939.

Witte, V., Laffert, B., Gintschel, P., Krautkramer, E., Blume, K., Fackler, O. T., et al. (2008). Induction of HIV transcription by Nef involves Lck activation and protein kinase Cθ raft recruitment leading to activation of ERK1/2 but not NFκB. *Journal of Immunology, 181*(12), 8425–8432.

Xavier, R., Brennan, T., Li, Q., McCormack, C., & Seed, B. (1998). Membrane compartmentation is required for efficient T cell activation. *Immunity, 8*, 723–732.

Xu, C., Gagnon, E., Call, M. E., Schnell, J. R., Schwieters, C. D., Carman, C. V., et al. (2008). Regulation of T Cell receptor activation by dynamic membrane binding of the CD3[var epsilon] cytoplasmic tyrosine-based motif. *Cell, 135*(4), 702–713.

Yanagisawa, M., Nakamura, K., & Taga, T. (2004). Roles of lipid rafts in integrin-dependent adhesion and gp130 signalling pathway in mouse embryonic neural precursor cells. *Genes to Cells, 9*(9), 801–809.

Yu, Y., Smoligovets, A. A., & Groves, J. T. (2013). Modulation of T cell signaling by the actin cytoskeleton. *Journal of Cell Science, 126*(5), 1049–1058.

Zaidel-Bar, R., Itzkovitz, S., Ma'ayan, A., Iyengar, R., & Geiger, B. (2007). Functional atlas of the integrin adhesome. *Nature Cell Biology, 9*(8), 858–867. http://dx.doi.org/10.1038/ncb0807-858.

Zamir, E., & Geiger, B. (2001). Molecular complexity and dynamics of cell-matrix adhesions. *Journal of Cell Science, 114*(20), 3583–3590.
Zhang, M., Moran, M., Round, J., Low, T. A., Patel, V. P., Tomassian, T., et al. (2005). CD45 signals outside of lipid rafts to promote ERK activation, synaptic raft clustering, and IL-2 production. *Journal of Immunology, 174*(3), 1479–1490.
Zhuang, L., Lin, J., Lu, M. L., Solomon, K. R., & Freeman, M. R. (2002). Cholesterol-rich lipid rafts mediate akt-regulated survival in prostate cancer cells. *Cancer Research, 62*(8), 2227–2231.

CHAPTER FIVE

Mechanisms Underlying Anomalous Diffusion in the Plasma Membrane

Diego Krapf
Department of Electrical and Computer Engineering and School of Biomedical Engineering, Colorado State University, Fort Collins, CO, USA
E-mail: krapf@engr.colostate.edu

Contents

Current Topics in Membranes, Volume 75
ISSN 1063-5823
http://dx.doi.org/10.1016/bs.ctm.2015.03.002

Abstract

The plasma membrane is a complex fluid where lipids and proteins undergo diffusive motion critical to biochemical reactions. Through quantitative imaging analyses such as single-particle tracking, it is observed that diffusion in the cell membrane is usually anomalous in the sense that the mean squared displacement is not linear with time. This chapter describes the different models that are employed to describe anomalous diffusion, paying special attention to the experimental evidence that supports these models in the plasma membrane. We review models based on anticorrelated displacements, such as fractional Brownian motion and obstructed diffusion, and nonstationary models such as continuous time random walks. We also emphasize evidence for the formation of distinct compartments that transiently form on the cell surface. Finally, we overview heterogeneous diffusion processes in the plasma membrane, which have recently attracted considerable interest.

1. INTRODUCTION

The plasma membrane separates the interior of the cell from the outside world maintaining intracellular components inside a living cell. As the interface between the cell and the extracellular fluid, the plasma membrane is responsible for a myriad of functions essential for cell physiology. These functions include the uptake of nutrients, such as sugars or amino acids, the regulation of ionic intracellular concentrations via the combination of transporters and channels, and the delivery of materials to the extracellular space (exocytosis). One of the most important functions of the cell membrane is the initiation of complex signaling cascades by recognizing and internalizing small molecules, which enable cell communication. Signaling pathways through the plasma membrane also control metabolism, growth and differentiation of tissues, and synthesis of proteins (Alberts et al., 2010; Lodish et al., 2007).

Given that the cell membrane is fluid, signaling reactions therein are limited by the rate at which reactants find each other. Thus, reactions are governed by the diffusive transport of membrane components. One would expect that the molecules are allowed to rapidly diffuse laterally across the membrane so that reactions can occur rapidly. This idea is very far from the real motion that characterizes cell membranes. On the contrary, the motion of membrane proteins is highly restricted and deviations from normal free diffusion are commonplace across different time and length scales. Physiological reasons for this behavior may be linked to the necessity of biomolecular processes to be tightly regulated. Further, cell membranes are known to be densely crowded structures where the motion of macromolecules is inherently restricted. Unraveling the diffusion pattern of lipids and membrane proteins is critical to understanding the organization of many fundamental physiological processes. These processes include receptor recognition, transport to and from the cell membrane, and cell-to-cell communication. In general, diffusive transport on the cell membrane has key functions in signal transduction and in the way cells interact with their surroundings.

This review deals with the mechanisms underlying anomalies in the diffusion of membrane molecules. The remaining of Section 1 reviews general aspects of normal and anomalous diffusion. Section 2 overviews the different physical models that explain anomalous diffusion, with particular emphasis on how different models arise in the plasma membrane. Section 3 focuses on the experimental evidence that has accumulated over the last 30 years on membrane diffusion and its implications on the mechanisms underlying anomalous diffusion.

1.1 Brownian motion and anomalous diffusion

Brownian motion, also known as Wiener process, is probably the most typical stochastic process in the physical sciences and engineering. Rigorously speaking, it is defined in one dimension as a continuous motion with independent and identically distributed Gaussian random displacements (Kuo, 2006). This definition has direct consequences on its properties. In particular, the mean square displacement (MSD) is linear with time. Brownian motion can be easily expanded to two or three dimensions by considering two or three independent processes, respectively. In a more physical context, Brownian motion develops when a molecule is constantly hit by other particles (for example, water molecules) and allowed to diffuse unrestrictedly as a consequence of the momentum transferred by collisions (Einstein, 1905). As such, Brownian motion is the solution to the diffusion equation

$$\frac{\partial P}{\partial t} = D\left(\frac{\partial^2 P}{\partial x^2} + \frac{\partial^2 P}{\partial y^2}\right), \tag{1}$$

where D is the diffusion coefficient and P represents either the particle concentration or the probability of finding a particle at location (x, y) at time t. Equation (1) is written in two dimensions because we deal with diffusion in membranes, namely a two-dimensional environment.

Assuming the particle is at the origin at time zero, the solution to Eqn (1) is a Gaussian function

$$P(\mathbf{r}, t) = \frac{1}{4\pi Dt} e^{-r^2/(4Dt)}, \tag{2}$$

where $r^2 = x^2 + y^2$ and the space integral under $P(\mathbf{r}, t)$ at any given time is unity because we solve for a single particle. Thus $P(\mathbf{r}, t)$ is the probability of finding the particle at location $\mathbf{r}$ at time t after starting from the origin. $P(\mathbf{r}, t)$ is a propagator since it describes the temporal evolution and, given the fundamental solution (Eqn (2)), one can construct the general solution for any initial conditions. From Eqn (2), we see that the MSD is

$$\langle r^2 \rangle = \int_0^\infty (2\pi r) P(\mathbf{r}, t) r^2 dr = 4Dt. \tag{3}$$

Note that in one and three dimensions the solutions for the MSD are $2Dt$ and $6Dt$, respectively. Thus the hallmark of "normal" diffusion is a square displacement that on average increases linearly in time.

A system is typically considered to exhibit anomalous diffusion when Eqn (3) breaks down. In general, it is possible that the MSD increases faster or slower than the linear solution. The most common situation is

$$\langle r^2 \rangle = Kt^\alpha, \tag{4}$$

where α is the anomalous exponent, and K is a generalized diffusion coefficient with units of cm^2/s^α. The process is subdiffusive when $\alpha < 1$ and superdiffusive when $\alpha > 1$. When $\alpha = 1$, "normal" diffusion is recovered. Other functional forms have been theoretically derived, e.g., an exponential time dependence (Cherstvy & Metzler, 2014), but this chapter deals primarily with subdiffusive processes of the form $MSD = Kt^\alpha$ where $\alpha < 1$.

2. MODELS FOR ANOMALOUS DIFFUSION

Anomalous diffusion can be described with different models (Bouchaud & Georges, 1990; Ralf Metzler, Jeon, Cherstvy, & Barkai, 2014). Among these, we find fractional Brownian motion (fBM), continuous time random walks (CTRWs), and obstructed diffusion, which have been extensively used to model diffusion in living cells (Höfling & Franosch, 2013). One of the most challenging aspects of anomalous diffusion is that different processes yield the same MSD (Eqn (4)). Here we provide a survey for the most common subdiffusion models, by describing their fundamental properties and we attempt to explain how these models may arise in the plasma membrane environment of a living cell.

2.1 Fractional Brownian motion

fBM is the archetypal subdiffusion process due to the simplicity and probably the elegance of its mathematical definition. fBM is defined via a generalization of Brownian motion, where the increments are no longer independent but they depend on previous increments (Mandelbrot & Van Ness, 1968). In one dimension, the position obeys the autocorrelation function

$$\langle x(t)x(t+s)\rangle = \frac{K}{2}\left[t^{\alpha} + (t+s)^{\alpha} - s^{\alpha}\right], \tag{5}$$

where t and s are positive times and α is the anomalous exponent as described before (Figure 1). Again, $\alpha = 1$ yields independent increments, i.e., Brownian motion. Often fBM is expressed in terms of the Hurst exponent H, with $2H = \alpha$. As in standard Brownian motion, the propagator of this process is also Gaussian, but with a time-dependent diffusion coefficient (Sebastian, 1995). In other words, Eqn (2) is modified so that $\langle x^2\rangle = Kt^{\alpha}$,

$$P(x,t) = \frac{1}{\sqrt{2\pi K t^{\alpha}}} e^{-x^2/(2Kt^{\alpha})}. \tag{6}$$

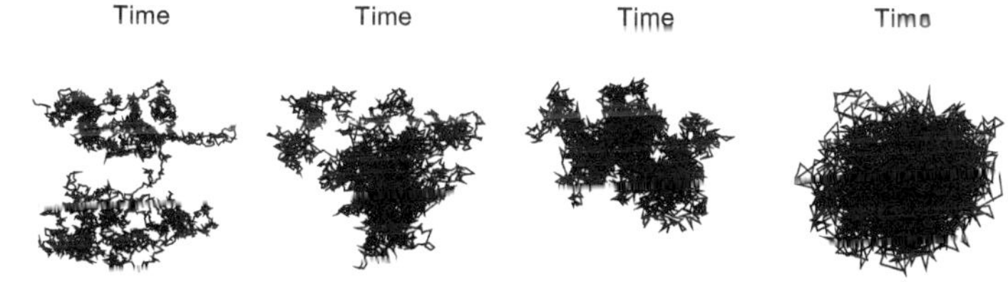

Figure 1 Numerical simulations of fractional Brownian motion in two dimensions for different α exponents between 0.4 and 1. $\alpha = 1$ is normal Brownian motion. Each trajectory is simulated for 5000 steps. The top panels are fractional Brownian motion in one dimension which is the x displacement of the trajectories seen in the bottom panels.

The extension of fBM to two dimensions is not as straightforward as the case of Brownian motion. To the best of our knowledge, there is currently no consensus as to what is the best way to define a 2D fBM, and typically it is described in terms of two independent fBM along the two dimensions.

In spite of the relative simplicity of the mathematical definition of fBM, describing it in terms of a physical mechanism is not trivial. A feasible origin for fBM is diffusion in a viscoelastic medium, which can arise from very densely crowded environments (Bronstein et al., 2009; Ernst, Hellmann, Kohler, & Weiss, 2012; Goychuk, Kharchenko, & Metzler, 2014; Szymanski & Weiss, 2009). In an elastic material such as a perfect spring, the stress is proportional to strain, i.e., the restoring force linearly increases with the deformation length. On the other hand, in a viscous material such as honey, the force exerted on an object, opposing its motion, is proportional to its velocity and independent of the distance it moves, with the proportionality constant termed the viscous drag coefficient ζ. The fluctuation-dissipation theorem tells us the diffusion coefficient in a viscous material is inversely proportional to the viscous drag coefficient, $D = k_B T/\zeta$. A viscoelastic material has both elastic and viscous properties; thus the restoring force depends on velocity and displacement. As a consequence, the material applies a restoring force against deformation that has memory in the sense that it depends on how much time passed since the initial deformation was applied. Eventually, the material relaxes into a new conformation and the restoring force decays to zero. In a viscous medium, a particle experiences a friction force in the opposite direction to its velocity independent of its past and thus it undergoes Brownian motion. However, a particle in viscoelastic medium experiences forces in both the direction opposite to the instantaneous velocity and in the direction to where it came from because the particle motion locally deforms the medium. As a consequence, the motion is antipersistent and it has a tendency to go back to the locations it visited in the past. This type of antipersistent motion can be modeled as fBM where the smaller the exponent α, the more compact is the motion. In other words, the higher α, the sparser the explored space (see Figure 1). An equivalent picture of elastic and viscous materials is that of solids and liquids, whereas a solid stores mechanical energy while a liquid dissipates it. A viscoelastic material both stores and dissipates energy.

Complex systems can exhibit rich viscoelastic behavior (Mason & Weitz, 1995). What specific phenomena may confer viscoelastic properties to the plasma membrane? At least three physical effects can induce viscoelasticity: (1) *Macromolecular crowding*. This is the best established mechanism for viscoelasticity in relation to anomalous diffusion (Guigas, Kalla, & Weiss, 2007;

Weiss, 2013). Let us envision the plasma membrane as a solvent densely packed with molecules of different sizes. As a tracer particle moves by diffusion in this medium, other molecules leave space for its path as dictated by thermal fluctuations. However, the sea of other molecules presses to close this space introducing an elastic effect. As time evolves, macromolecules relax in the new location and the elastic effect disappears. (2) *Transient interactions with cytoskeletal components.* In this scenario, interactions of different strengths maintain specific positions of the plasma membrane highly localized in space (Chichili & Rodgers, 2009; Luna & Hitt, 1992). Deformation of the plasma membrane introduces forces at these locations. Provided that these interactions are reversible, the membrane relaxes over time as the interactions break. (3) *Intermolecular interactions.* Interactions among membrane components can lead to short-range order in a fluid (Marguet, Lenne, Rigneault, & He, 2006). Thus, as the membrane is deformed, a short-range elastic effect gives rise to memory in the motion of tracer particles.

2.2 Continuous time random walk

A random walk consists of random steps forming a path, where, for example, each step occurs at equally spaced times. Thus, the waiting time between jumps is a constant. When the step size either involves a constant distance or is a random variable with finite mean and variance, the random walk converges in the long time limit to Brownian motion. In particular, we have that for a 1D random walk of mean step size L, $\langle x^2 \rangle = L^2 N$ after N steps, which is equivalent to Eqn (3).

Random walks can also be considered when steps take place at random times. We can then define a CTRW as a random walk with sojourn times between steps being random variables according to a distribution $\psi(\tau)$ for times $\tau > 0$. Again, when $\psi(\tau)$ has finite mean and variance, e.g., an exponential distribution with characteristic time τ_0, the average number of steps in time t is t/τ_0 and we recover normal diffusion at times $t \gg \tau_0$ with $\langle x^2 \rangle = (L^2/\tau_0)\, t$. Thus, we have a diffusion process, with a coefficient $D = L^2/2\tau_0$.

The situation changes dramatically when the distribution of waiting times between steps is scale free, such as a power-law distribution (Scher & Montroll, 1975; Scher, Shlesinger, & Bendler, 1991)

$$\psi(\tau) \sim \tau^{-(1+\alpha)}. \tag{7}$$

As we will see later, this type of power-law distributions has been measured in the immobilization times of membrane proteins (Weigel, Simon,

Tamkun, & Krapf, 2011). When $0 < \alpha < 1$, the mean of the waiting time diverges and thus we cannot define a characteristic time τ_0 as before. Instead, the only characteristic time of the process is the total time over which the experiment is conducted. As a consequence, experimental results depend on measurement time, a problem that will be discussed later.

Before describing the properties of random walks with power-law waiting times, let us discuss the reasons for such distributions to exist in the membrane of a living cell. First, we note that a CTRW is an excellent model for the motion of a particle that alternates between mobile and immobile periods. Here, the process may be described by random steps (mobile segments of the trajectory) that take place at random times (with waiting times defined as the time spent in the immobile phase). Immobilization can occur through binding to a cytoskeletal component or to a macromolecular complex, yielding long periods between steps. However, binding kinetics typically have a specific dissociation coefficient k, which yields a dissociation time with an exponential distribution

$$\psi(\tau) = k\exp(-k\tau). \tag{8}$$

Therefore, for times $t \gg 1/k$, this process exhibits normal diffusion.

Two situations can give rise to heavy-tailed (power law) CTRWs: nonstationary interactions and heterogeneous interactions.

2.2.1 Time evolution route to heavy-tailed CTRW

In the first scenario, a protein transiently binds to a macromolecular complex that grows over time and as a consequence, the dissociation coefficient changes. If we assume the escape probability decreases with time, we can postulate a model where the dissociation coefficient is $k(t) = \alpha/(t + t_0)$ and t_0 is a nucleation time for the macromolecular complex (Weigel, Tamkun, & Krapf, 2013). The survival probability P_s, i.e., the probability that the protein has not escaped after a time t obeys the rate equation:

$$\frac{dP_s}{dt} = -\frac{\alpha}{(t + t_0)}P_s, \tag{9}$$

which has a power-law solution $P_s(t) = t_0^\alpha(t + t_0)^{-\alpha}$. We can now obtain the residence time probability density function, i.e., the distribution of dissociation times,

$$\psi(\tau) = -\frac{dP_s(\tau)}{d\tau} \sim \tau^{-(1+\alpha)} \tag{10}$$

with the form of Eqn (7).

2.2.2 Heterogeneous interactions route to heavy-tailed CTRW

The second situation for which a heavy-tailed CTRW can arise consists of interactions with different complexes that translate into a distribution of dissociation coefficients (Burov, Jeon, Metzler, & Barkai, 2011; Klafter & Sokolov, 2005). In this case, we start from a narrow distribution of binding energies E_B such that

$$P(E_B) = \alpha \exp(-\,\alpha E_B),$$

where E_B is expressed in units of k_BT. Following Arrhenius rate law, the dissociation coefficient from an individual complex with binding energy E_B is $k = k_0 \exp(-E_B)$ and, therefore, the distribution of dissociation coefficients is $P(k) = \alpha\, k_0^{-\alpha} k^{\alpha-1}$ with $0 < k < k_0$. For a given dissociation coefficient k, the conditional waiting time probability is $P(\tau|k) = k\exp(-k\tau)$. By integration, we obtain the power law *a priori* distribution $P(\tau) = \int_0^{k_0} k \exp(-k\tau)P(k)dk$,

$$P(\tau) = \frac{\alpha}{k_0^{\alpha}}\gamma(\alpha+1, k_0\tau)\tau^{-(1+\alpha)}, \tag{11}$$

where $\gamma(s,t)$ is the lower incomplete gamma function $\gamma(\alpha+1, t) = \int_0^t \tau^{\alpha}e^{-\tau}d\tau$. For long times, $\gamma(\alpha+1, k_0\tau) \to \Gamma(\alpha+1)$ and $P(\tau)$ converges to the heavy-tailed power law $P(\tau) \sim \tau^{-(1+\alpha)}$.

Heavy-tailed CTRW have anomalous diffusion properties that greatly differ from those of fBM. Let us consider a measurement of a particle over a time t_1 that diffuses within the framework of a heavy-tailed CTRW. Given that all waiting times are possible, within this measurement time, we will likely observe the particle having at least one long waiting time of the same order of magnitude of the time t_1. However, if we measure over much longer times, the particle will probably display a waiting time within the observation time t_2 much longer than that observed within t_1. Three consequences stem from this concept.

2.2.3 Nonstationarity

The longer the observation time, the longer the effective average time the particle is immobilized and the process is not stationary (Metzler & Klafter, 2000). In other words, as measurement time progresses, the rate of jumps decreases and thus the apparent diffusion coefficient goes down, i.e., the system ages with time. This phenomenon is due to the fact that the expected value of the immobilization time diverges, $\langle\tau\rangle = \int_0^{\infty} \tau\psi(\tau)d\tau = \infty$. However, any physical measurement takes place in a finite time t_m and the observed effective immobilization time is $\langle\tau\rangle_{eff} = \int_0^{t_m} \tau\psi(\tau)d\tau$, which is bounded by t_m. This puzzling observation means that the measurement time affects the estimated diffusion coefficient.

2.2.4 Irreproducibility of results

Because one long waiting time of the order of the experimental time is typically observed, the measured parameters depend on a single random variable and the temporal averages remain random, even in the long time limit. As a consequence, different particles display a very broad distribution of estimated diffusion coefficients, no matter how long the observation time is. In terms of probability theory, because the variance diverges, the classical central limit theorem breaks down (Peebles, 1987).

2.2.5 Weak ergodicity breaking

In traditional systems, we estimate parameters by averaging measurements. For example, from measurements of MSD we can estimate the diffusion coefficient. Such estimates can be obtained by averaging over many different particles or alternatively by measuring a single trajectory that explores the whole phase space. Often, in single-particle tracking experiments, it is difficult to obtain a very large number of trajectories so that averages can be performed over an ensemble. Instead, averages are obtained for individual trajectories by averaging over time. For a random walk that visits the locations $\mathbf{r}_i$ with $i = 1,2,\ldots,N$ (here N−1 represents the number of steps), we define the time-averaged MSD for a lag time u as

$$\overline{\delta^2}(u) = \frac{1}{N-u}\sum_{i=1}^{N-u}(\mathbf{r}_{i+u} - \mathbf{r}_i)^2, \tag{12}$$

where δ^2 is the square displacement and the bar represents a temporal average. Note that u is the lag time expressed in number of steps, which is converted to real time units in a straightforward way $\Delta = ut_0$, where t_0 is the step time. The convergence of the temporal averages at long times to the ensemble average is known as the ergodic hypothesis. In a CTRW, the ergodic hypothesis breaks down and the time average does not converge to the ensemble average, independent of the measurement time. Further, for the power-law waiting time distribution, the time and ensemble mean-squared displacements have different dependence on lag time (He, Burov, Metzler, & Barkai, 2008; Lubelski, Sokolov, & Klafter, 2008). In this case, we obtain

$$\overline{\delta^2} \sim \Delta \tag{13}$$

but

$$\langle x^2 \rangle \sim t^{\alpha}. \tag{14}$$

We employ brackets as ensemble average and the bar as temporal average. Therefore, when the MSD is computed for an individual trajectory, the diffusion process appears "normal" in the sense that it scales linearly with lag time. Nevertheless, the diffusion is anomalous and the ensemble average of the temporal MSD explicitly depends on experimental time (aging behavior) due to its nonstationary properties (He et al., 2008)

$$\left\langle \overline{\delta^2} \right\rangle \sim \frac{\Delta}{T^{1-\alpha}}, \tag{15}$$

where T is the experimental time. Here, the temporal MSDs are averaged over many trajectories because the individual MSDs remain random variables even in the long time limit.

2.3 Obstructed diffusion

One of the simplest scenarios for anomalous diffusion we can conceptually envision is when the motion takes place in an environment that is densely packed with immobile objects. As the concentration of immobile obstacles increases, the available space decreases and diffusion becomes more and more restricted. At high concentrations, the obstacles form clusters that block the free motion of particles. Scaling relations have been derived for the average cluster size as a function of obstacle concentration (Ben-Avraham & Havlin, 2000) showing that the cluster size increases as a power law. At a critical concentration, a percolation transition takes place and a cluster that traverses the whole space forms. Thus the underlying medium is partitioned into disconnected regions. As the obstacle concentration approaches the percolation transition, a diffusing particle can explore the whole space, but its diffusion is constantly hindered. In this situation, the obstacles form an intricate maze where dead ends and cul de sacs appear over many length scales. Thus a particle diffusing in a percolating structure often has to retrace itself giving rise to anomalous diffusion with a compact exploration pattern (Figure 2(A–C)).

When the obstacle concentration is much smaller than criticality, the cluster sizes are small and transient anomalous diffusion is observed only over small length scales. Thus the MSD is sublinear over short times and linear at longer times (Figure 2(A)) (Saxton, 1996; Weigel, Ragi, et al., 2011). The subdiffusive regime grows as concentration increases and

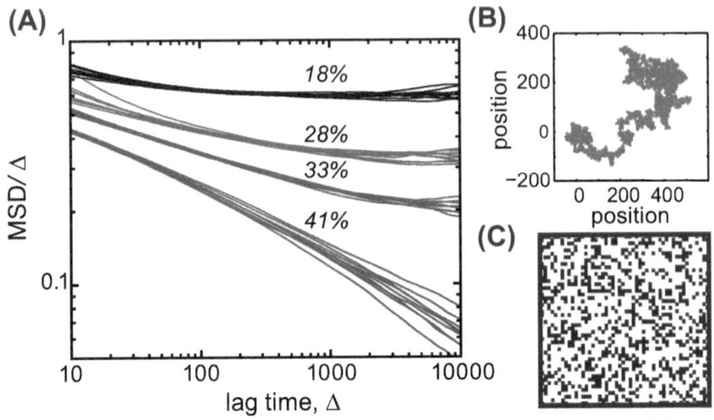

Figure 2 Simulations of obstructed diffusion. (A) Temporal MSD/Δ, i.e., $\overline{\delta^2}/\Delta$, as a function of Δ in a log–log plot for individual simulated tracers in the presence of immobile obstacles at different concentrations, where Δ is the lag time. Several simulations are presented for each obstacle concentration in order to show the small scatter of the mean square displacement (MSD). Simulations are performed in square lattices using a Monte Carlo blind ant algorithm (simulation details can be found in Weigel, Ragi, et al., 2011) so that 41% obstacles is right above the percolation threshold. The simulations with 41% obstacles show anomalous diffusion over the whole simulated time, but all other concentrations exhibit transient subdiffusion with a time upon which the process exhibits normal diffusion, i.e., MSD/Δ becomes flat, as expected from MSD $\sim$ Δ. (B) Example of a trajectory of 10^6 steps in a square lattice with 33% of sites obstructed. (C) Example of a 50 × 50 region of a square lattice where 33% of the sites are obstructed (dark pixels). Note that the region shown is only a small fraction of the lattice used to obtain trajectories with 10^6 steps.

when criticality is reached subdiffusion is observed at all length scales. At criticality, the obstacle mesh forms a random fractal with a noninteger dimension smaller than the dimension of the embedding space, i.e., $d_f < 2$. As mentioned above, the MSD is linear for any integer dimension (Eqn (3)), but this rule does not hold when diffusion takes place in a medium with fractal dimension (Ben-Avraham & Havlin, 2000).

Many similarities are shared between fBM and obstructed diffusion. Among them, the temporal averages converge to the ensemble average, the increments are stationary, and the exploration is more compact than in Brownian motion. None of these effects hold for the heavy-tailed CTRW. Nevertheless, the distinction between these stationary processes and a CTRW is not trivial because convergences can be extremely slow, a challenge that requires special attention (Burnecki et al., 2012; Deng & Barkai, 2009; Kepten, Bronshtein, & Garini, 2011; Magdziarz, Weron, Burnecki, & Klafter, 2009). Similarly, while subdiffusion in a percolation cluster at a concentration below criticality is transient, the convergence to normal diffusion can be slow, i.e., subdiffusion may be observed over several orders of magnitude. One clear distinction between fBM and obstructed diffusion is that while the propagator of fBM is Gaussian (Eqn (6)), the propagator for diffusion in a percolation cluster is not (see Eqn (4) in Weigel, Ragi, et al. (2011)).

2.4 Heterogeneous Landscapes

It has long been appreciated that beyond energy disorder, spatial heterogeneities can lead to complex dynamics (Havlin & Ben-Avraham, 1987). However, the extent of the complexity of motion in a heterogeneous environment has only been recently recognized. It was demonstrated that a heterogeneous diffusion process (HDP) with space-dependent diffusion coefficient can exhibit weak ergodicity breaking (EB) (Cherstvy, Chechkin, & Metzler, 2013). Cherstvy et al. (2013) showed that a Markovian HDP with power-law space dependence of the diffusion coefficient, $D(x) \sim D_0|x|^{\gamma}$, yields anomalous diffusion where the ensemble-averaged MSD scales as $\langle \delta^2 \rangle \sim t^{2/(2-\gamma)}$ but the time-averaged MSD remains linear in the lag time, $\overline{\delta^2} \sim \Delta$. Thus, the process is nonergodic in a similar fashion to a CTRW (see Eqns (13) and (14)), even though the mechanism is vastly different. This HDP leads to subdiffusion when $\gamma < 0$ and superdiffusion when $\gamma > 0$, but both cases give rise to nonergodic behavior with a large scatter in the time-averaged MSD. Cherstvy and Metzler (2014) studied in detail the amplitude scatter between individual realizations and the special case of the critical point when $\gamma = 2$. At this critical point, the fluctuations in the temporal MSD diverge and the ensemble-averaged MSD no longer follows a power law. Instead, for $\gamma = 2$, the MSD exhibits an exponential growth $\langle \delta^2 \rangle \sim \exp(2D_0 t)$.

A new model for HDP where the diffusivity varies randomly but it is constant within patches of random sizes was introduced by Massignan et al. (2014). This model was developed due to the resemblance of random patches to the rapidly varying diffusivity observed on the cell surface (Cutler et al., 2013; Giannone et al., 2010; Masson et al., 2014; Serge, Bertaux, Rigneault, & Marguet, 2008). Similar to the HDP with power-law space dependence of the diffusion coefficient (Cherstvy & Metzler, 2014), ergodicity breaking is found. The key component of this model is a heavy-tailed distribution of the residence time in different patches that can be achieved by a power-law distribution of patch sizes. The HDP model with random patches was found to be particularly relevant to the diffusion of DC-SIGN in the plasma membrane as discussed later (Manzo et al., 2015).

2.5 Subordination Schemes

So far, the discussed models for anomalous diffusion describe specific mechanisms. However, given that the plasma membrane is a highly complex environment (see Figure 3), different mechanisms can coexist. Therefore, different models should be combined into a stochastic process of increased complexity. Such a problem can be considered within a subordination scheme (Blumen, Klafter, White, & Zumofen, 1984; Fogedby, 1994; Klafter, Blumen, & Zumofen, 1984; Meroz, Sokolov, & Klafter, 2010; Sokolov & Klafter, 2005). For example, we can treat an fBM in the operational time as discussed before and then place it in a laboratory time where each step is drawn from a power-law distribution (Klafter & Sokolov, 2011; Tabei et al., 2013). Thus, we have fBM subordinated to a CTRW.

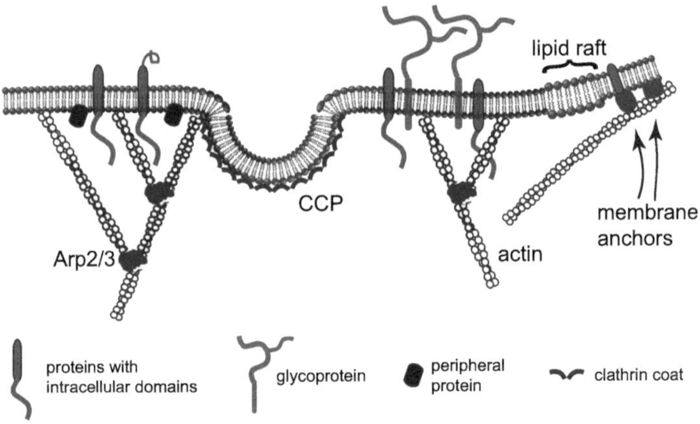

Figure 3 Largely simplified sketch of the plasma membrane. In this illustration, actin filaments can link to the plasma membrane and obstruct the motion of proteins. Two types of membrane anchors are shown, proteins that bind actin directly to the membrane (e.g., annexins) and to proteins, such as talin and vinculin, that connect actin to cell adhesion receptors, such as integrins (Winder & Ayscough, 2005). Membrane proteins may also have intracellular domains that contribute to interactions with cytoplasmic filaments such as the cortical actin. Extracellular carbohydrate chains (e.g., glycoproteins) can also contribute to interactions among themselves or with the extracellular matrix. A clathrin-coated pit (CCP) is shown where molecules with specific recognition sites can be captured. In addition, the high density of different membrane proteins contributes to its complexity. Transient functional domains such as cholesterol-rich lipid rafts also partition the membrane into regions with different diffusive behavior. (See color plate)

The statistics of subordinated processes are not trivial and each case requires special attention. Meroz et al. (2010) have considered the coexistence of diffusion on a fractal and a heavy-tailed CTRW. This is a special case of subordinating an ergodic anomalous process to a nonergodic one and it is particularly relevant to diffusion on the cell membrane as discussed later (Weigel, Simon, et al., 2011). The fractal structure models spatial obstruction to the motion of membrane proteins and the CTRW can represent immobilization due to trapping into macromolecular complexes. This process is found to exhibit aging and weak ergodicity breaking. The MSD is found to scale as

$$\langle x^2 \rangle \sim t^{\alpha\beta} \tag{16}$$

and

$$\left\langle \overline{\delta^2} \right\rangle \sim \frac{\Delta^{1-\alpha+\alpha\beta}}{T^{1-\alpha}}, \tag{17}$$

where the two constants α and β characterize the heavy-tailed CTRW and the fractal random walk subdiffusive exponents, respectively. As before, we denote with a bar $\overline{\delta^2}$ a moving temporal average over a single trajectory of time T and with brackets $\langle x^2 \rangle$ the ensemble average. Aging behavior is evident in the dependence of the MSD on the experimental time. Furthermore, the temporal MSD of individual trajectories $\overline{\delta^2}$ is found to exhibit a large scattering due to the nonergodic properties of the CTRW (Meroz et al., 2010).

A subordination scheme is an elegant way to solve for the simultaneous effect of two different subdiffusion models. However, this scheme has some limitations. For example, an fBM has memory, which is now considered in space but the temporal relaxation due to a particle remaining for a long time at a specific location is neglected.

3. EVIDENCE FOR ANOMALOUS DIFFUSION IN THE PLASMA MEMBRANE

The analysis of diffusion in living cells has to be placed in a specific length- and time-scale range. Diffusion in an amorphous material can be measured without tangible theoretical limitations beyond the time for a material to crystallize which is usually not within experimental reach. However, diffusion in living cells can only be reasonably measured for times smaller than its lifetime or the time until cell replication occurs. Thus, whenever we discuss power-law distributions, they are meant to occur within reasonable times. Obviously, these distributions will change for times upon which the system is altered. Further, the situation in a typical living cell is much more complex than in amorphous materials because it can involve several phases. As a result, different stochastic mechanisms govern diffusion in cellular environments at different timescales. For example, the motion of lipid granules in fission yeast was shown to be best described by CTRW subdiffusion with features of ergodicity breaking at short times (in the ms range) and by a mechanism closest to fBM at longer times (Jeon et al., 2011).

Anomalous diffusion in the membrane of living cells has been observed in a very broad variety of cell types. Both membrane proteins and lipids undergo anomalous diffusion in different cells including human embryonic kidney, neurons, Chinese hamster ovary, Cos7, fibroblasts, keratinocyte, mouse hepatoma (HEPA-OVA), sperm, and HeLa (James, Hennessy, Berge, & Jones, 2004; Kusumi et al., 2005; Manley et al., 2008; Murase et al., 2004; Ritchie, Iino, Fujiwara, Murase, & Kusumi, 2003). Further, anomalous diffusion of membrane proteins was also observed in living animals such as zebrafish (Schaaf et al., 2009). The purpose of this chapter is not the extensive discussion of the phenomena observed in different cell types and for different membrane molecules, but to discuss at a higher level the mechanisms responsible for the observed anomalous diffusion. For details on experimental work on different proteins and cell lines, the reader is referred to a recent review by Gal, Lechtman-Goldstein, and Weihs (2013).

As discussed above, some of the main physical mechanisms leading to subdiffusive motion involve intermolecular interactions within the membrane or with cytoskeletal components. These components include actin (Andrews et al., 2008; Heinemann, Vogel, & Schwille, 2013), cytoskeleton anchoring proteins (Bennett & Gilligan, 1993; Meier, Vannier, Serge, Triller, & Choquet, 2001; Winder & Ayscough, 2005), the extracellular matrix (Jiang, English, Hazan, Wu, & Ovryn, 2015; Torreno-Pina et al., 2014), clathrin-coated pits (CCP) (Weigel et al., 2013), lipid rafts (Eggeling et al., 2009), and the endoplasmic reticulum (Anderluh et al., 2014; Cui-Wang et al., 2012; Wu, Covington, & Lewis, 2014) (see Figure 3 for some examples).

3.1 Molecular dynamics simulations—nanoseconds to microseconds

The shortest timescales for diffusion in membranes have been studied by large-scale molecular dynamics simulations. These simulations investigate the behavior of lipids and proteins in small membrane patches with different lipid compositions (Javanainen et al., 2013). The membranes of live cells are extremely crowded with proteins. For example, the area occupied by proteins in the membrane of red blood cells was measured to be at least 23% (Dupuy & Engelman, 2008). It is generally estimated that the area fraction of proteins in the cell membrane can reach up to 50% (Casuso et al., 2012). Thus, simulations are also performed with different protein concentrations to identify and isolate protein crowding effects. Further, the packing of lipids in the plasma membrane is sensitive to lipid chemistry and cholesterol concentration. In particular, introducing cholesterol and sphingolipids into a lipid bilayer forms a liquid ordered phase with a higher packing density, thus the diffusivity in membranes with higher cholesterol concentrations decreases (Almeida, Vaz, & Thompson, 1992).

Jeon, Monne, Javanainen, and Metzler (2012) performed a detailed stochastic analysis of molecular dynamics simulations in lipid bilayers. Their work focused on the effect of cholesterol on subdiffusion and understanding whether the motion of lipids can be better modeled by a CTRW or by anticorrelated motion such as fBM or fractional Langevin equations. Strikingly, it was found that in the presence of cholesterol, lipids exhibit subdiffusive behavior up to times of the order of hundreds of nanoseconds before a transition to normal diffusion is observed. This timescale is much longer than what was previously estimated. In particular, these results are surprising because model membranes lack the complexity of the interactions with the cytoskeleton or with large protein assemblies. Thus, in native cell membranes, the subdiffusion regime probably stretches even further, into macroscopic timescales. The stochastic analysis described in Jeon et al. (2012) includes the scattering of the mean squared displacement, stationarity, the displacement autocorrelation function, and moment ratios. These tools show that the motions of lipids and cholesterol are ergodic processes, universally consistent with a fractional Langevin equation. Often, the ergodicity breaking parameter, which quantifies the fluctuations of the MSD, is used to characterize ergodicity breaking (He et al., 2008),

$$EB = \frac{\left\langle \left(\overline{\delta^2}\right)^2 \right\rangle - \overline{\langle \delta^2 \rangle}^2}{\overline{\langle \delta^2 \rangle}^2}. \tag{18}$$

Note that the ergodicity breaking parameter is the variance of the normalized dimensionless MSD, $\chi = \overline{\delta^2}/\langle\overline{\delta^2}\rangle$, thus it quantifies the fluctuations of the trajectories. If $EB = 0$ in the long time limit, we have ergodic behavior. It was found the ergodicity breaking parameter decays to zero as time increases and thus the process is ergodic (Jeon et al., 2012). Akimoto, Yamamoto, Yasuoka, Hirano, and Yasui (2011) analyzed a closely related metric for the relative fluctuations of the MSD in the trajectory of lipids from molecular dynamics simulations. The relative fluctuations are defined as

$$R = \frac{\left\langle \left| \overline{\delta^2} - \overline{\langle \delta^2 \rangle} \right| \right\rangle}{\overline{\langle \delta^2 \rangle}}. \tag{19}$$

Similar to the ergodicity breaking parameter, the relative fluctuations R decayed to zero in the long time limit which indicates ergodicity. However, it was found that the decay was slower than expected from a Gaussian process. Akimoto et al. (2011) suggested the non-Gaussianity is caused by trapping times distributed according to power laws. Intriguingly, such a behavior would cause a dependence of the time-averaged MSD on experimental time, but no such dependence was found (Jeon et al., 2012).

The analysis of lipid and protein lateral diffusion at different crowding levels was analyzed by Javanainen et al. (2013) to unravel the roles of lipid packing and protein crowding. In that work, simulations were performed in two different systems: one preferring protein aggregation (dipalmitoylphosphatidylcholine, DPPC) and another one avoiding it (dilinoleoylphosphatidylcholine, DLPC). In both systems, when proteins are sparse, lipids exhibit anomalous diffusion for times up to 100 ns. Nevertheless, for intermediate protein crowding (1:200 to 1:100 proteins per lipid), subdiffusion is observed for times up to 10 μs in aggregating protein systems (DPPC) and up to 1 μs in nonaggregating systems (DLPC). For high protein crowding levels similar to those estimated to be relevant to live cells (1:75 to 1:50 proteins per lipid), subdiffusion is observed for strikingly long times. In fact, the subdiffusion timescales exceed those enabled by molecular dynamics simulations even employing coarse graining. From the data collected in Javanainen et al. (2013), it is estimated that normal diffusion would only be achieved at times of the order of several hundred microseconds or even milliseconds for protein aggregating systems and between 10 and 100 μs for nonaggregating systems.

3.2 Obstructed (hop) diffusion—microseconds to seconds

3.2.1 *Fluorescence recovery after photobleaching and optical tweezers*

The concept of membrane proteins being corralled by cytoplasmic fences (cortical actin or actin binding proteins sketched in Figure 3) is one of the oldest and most studied mechanism for deviations from Brownian diffusion in the plasma membrane (Sheetz, 1983; Tsuji & Ohnishi, 1986). In this model, the part of the cytoskeleton in close proximity to the plasma membrane blocks the path of membrane proteins due to steric hindrance. One of the earliest experiments showing that cytoplasmic components introduce barriers to the lateral diffusion of membrane proteins involved measurements in cell blebs (Tank, Wu, & Webb, 1982). Blebs are spherical protrusions that form under specific conditions in eukaryotic cells. These structures resemble small balloons being inflated out of a plasma membrane patch and what makes them particularly useful for investigating the effect of the cytoskeleton on membrane diffusion is that they lack any actin cytoskeleton (Cunningham, 1995). Thus, differences between the diffusion properties of transmembrane proteins in blebs and those in membrane regions outside blebs can be attributed to the effect of the cytoskeleton in the vicinity of the plasma membrane, i.e., the cortical cytoskeleton. Using fluorescence recovery after photobleaching (FRAP), Tank et al. (1982) showed that in blebs, the lateral constraints to diffusion are released and, in turn, the diffusivity of receptors is enhanced.

Besides FRAP measurements, early evidence pointing to the existence of cytoplasmic barriers to the diffusion of membrane proteins came from optical tweezers measurements (Edidin, Kuo, & Sheetz, 1991; Edidin, Zuniga, & Sheetz, 1994; Kusumi & Sako, 1996). Proteins tagged with either gold or latex nanoparticles were dragged along the plasma membrane in a controlled manner using optical tweezers. The nanoparticle is captured in a three-dimensional optical trap and moved along the membrane until the protein bound to the nanoparticle encounters a barrier that causes it to escape from the trap. Thus, a measurement of the barrier free path (BFP) is obtained for each molecule. BFP and FRAP measurements indicate that the length upon which membrane proteins extend within the cytoplasm has dire consequences on the extent their path is obstructed (Edidin et al., 1994). These observations can be explained in terms of a cortical actin structure that lies close to the membrane but is not tightly bound. Thus, proteins with longer cytoplasmic domains encounter barriers more often and have increased difficulties in traversing such barriers.

Optical tweezers studies also shed light on the fluctuations and the mechanical properties of the cortical cytoskeleton (Sako & Kusumi, 1995). Thermal fluctuations introduce two different mechanisms by which membrane proteins can permeate through cytoplasmic barriers. On one hand, the actin cytoskeleton has a dynamic nature and it may exhibit fluctuations both in the lateral and in the normal directions. A barrier that retracts from the membrane due to thermal fluctuations may leave enough space for the protein to cross through it, or in other words, the barrier is temporarily lifted. A second mechanism by which proteins can cross through a barrier is by fluctuations in the conformation of their cytoplasmic domains. Many membrane proteins are believed to have disordered cytoplasmic tails which would resemble a random coil. Thus spontaneous fluctuations cause the tail to shrink and allow the protein to cross the barrier.

3.2.2 Single-particle tracking

The ultimate method for studying the motion of proteins in the plasma membrane is through single-molecule tracking. Here the protein of interest is tagged with a marker that can be localized with nanometer precision. Useful markers include gold nanoparticles, fluorescent proteins, organic fluorophores, and quantum dots (QD). Tracking individual fluorescent proteins is often termed single-molecule tracking, while tracking gold nanoparticles or QDs is usually termed single-particle tracking. Here we will use these terms interchangeable. Metal nanoparticles provide the advantage of allowing detection in times as short as microseconds using video-enhanced microscopy (de Brabander et al., 1991; Sheetz, Turney, Qian, & Elson, 1989). However, they are typically large (~40 nm) and interactions with the extracellular matrix are a concern. Fluorescent proteins can be genetically encoded for measurements in a live cell and are small, often yielding negligible interactions with other components. However, experiments show that the choice of fluorescent protein and expression level is important to avoid oligomerization, low affinity interactions between proteins that normally do not interact, and trafficking artifacts (Aguet, Antonescu, Mettlen, Schmid, & Danuser, 2013; Costantini, Fossati, Francolini, & Snapp, 2012; Snapp et al., 2003). The main disadvantage of single fluorescent proteins is that they photobleach and thus, long measurements are difficult to obtain. QDs provide the great advantage of allowing very long measurements, but they exhibit blinking, random fluctuations between bright and dark states causing arbitrary long periods where a molecule cannot be tracked (Michalet et al., 2005; Pinaud, Clarke, Sittner, & Dahan, 2010).

Single-particle tracking shows that membrane proteins explore small regions during periods of time that are statistically significantly longer than what would be expected from Brownian motion, before moving to explore a new region. These results are consistent with a model where proteins are temporarily confined by cytoplasmic barriers. Thus a protein is observed to reside within a region or compartment during a relatively long stochastic time before it hops to a new compartment, a process usually called hop diffusion (Figure 4(A) and (B)). As a consequence, proteins observed during periods of time much shorter than the time between collisions with the compartment perimeter are expected to display normal diffusion, i.e., MSD $\sim \Delta$ (Figure 4(C)). If a protein is confined within a compartment such that the escape probability is zero, as time increases the MSD can only increase to a finite value and thus it converges to a value that depends on corral size and geometry (Saxton & Jacobson, 1997). Specifically,

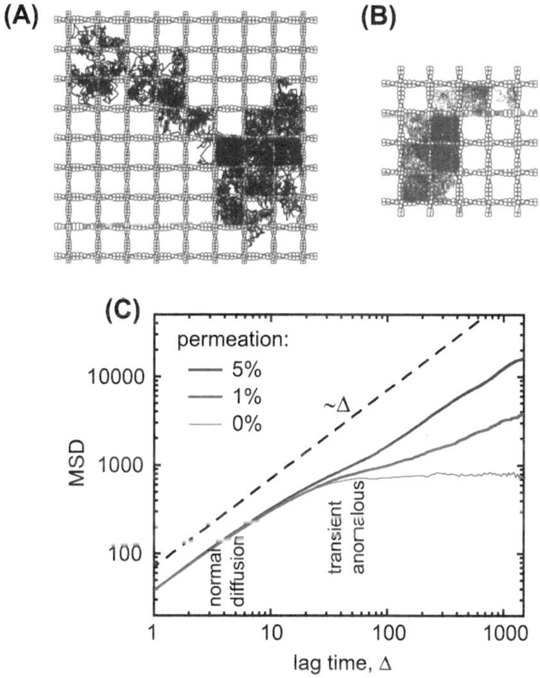

Figure 4 Fence model for transient anomalous diffusion. (A,B) Numerical simulations for random walks in the presence of permeable fences. The fences are sketched as actin filaments. These simulations are performed off-lattice with jumps drawn from a normal distribution with standard deviation $\sigma = 5$ in both x and y directions. The fences are placed a distance of 50 units from each other. Both trajectories are 10,000 jumps. The fences are permeable with a given probability of permeation: (A) 5% and (B) 1%. (C) Time-averaged mean square displacement (MSD) $(\overline{\delta^2})$ for the trajectories shown in the upper panels as a function of lag time. At lag times much shorter than the hopping time, the MSD is linear. At intermediate times, a sublinear regime is observed and diffusion becomes normal again at longer times. The recovery of linear MSD is much slower for 1% permeation probability than for 5% permeation probability. A trajectory with no permeation through the fences (0% permeation probability), so that the trajectory is confined to a compartment, is also shown as a thin line. The dashed line indicates linear behavior (free diffusion), i.e., MSD $\sim \Delta$. (See color plate)

$\text{MSD} = \langle r_c^2 \rangle [1 - A_1 \exp(-4A_2 Dt/\langle r_c^2 \rangle)]$ where r_c is the compartment radius and A1 and A2 are constants related to the shape of the compartment, as seen in Figure 4(C). Note that, even though the MSD is not linear, this behavior is considered normal diffusion within a confined region. However, when the tracked molecule is allowed to escape by hopping to a different compartment, the MSD will again become linear at times longer than the escape time, albeit with a reduced diffusion coefficient. The new apparent diffusion coefficient is $D_a = r_c^2/\tau_0$ with τ_0 being the characteristic escape time. Thus, diffusion is transiently anomalous for intermediate times between the characteristic confinement time and the escape time. In this model, we assume the confining compartments cover the whole membrane, so that the compartment size is also the distance between the centers of the compartments. A different model can also be proposed where the compartments are patches within the membrane, so that the molecule diffuses freely between periods of confined diffusion. This latter model shares some of the properties of CTRW arising from energetic disorder and will be discussed in Section 3.5. A detailed review of the current understanding of the organization of the plasma membrane is provided in Akihiro Kusumi et al. (2012).

Transient subdiffusion with an MSD of the form seen in Figure 4(C), i.e., with linear/sublinear/linear regimes, corresponds to a narrow distribution of compartment sizes. Such a distribution is expected to yield a distribution of escape times close to an exponential distribution

$$\psi(\tau) = \tau_0^{-1} \exp(-\tau/\tau_0). \tag{20}$$

However, many experiments fail to produce an MSD with the linear, sublinear, and again linear regimes. Instead, membrane protein subdiffusion is often observed over extended periods of time (Feder, Brust-Mascher, Slattery, Baird, & Webb, 1996; Notelaers et al., 2014, 2012; Smith, Morrison, Wilson, Fernandez, & Cherry, 1999). The hop diffusion model is consistent with these observations if the compartment size distribution is assumed to be broad. The escape time, of course, depends on compartment size and thus it would also have a very broad distribution. In a simplified picture, the escape time for a given compartment radius r_c is an exponential as Eqn (20) with $\tau_0 = r_c^2/D_a$ where D_a is the apparent diffusion coefficient. Thus,

$$\psi(\tau) = \int_0^\infty \frac{D_a}{r_c^2} \exp\left(-D_a\tau/r_c^2\right) P(r_c) dr_c \tag{21}$$

where $P(r_c)$ is the distribution of compartment radii.

A special and interesting case arises when the cytoskeletal barriers produce a self-similar meshwork. In this case, we expect a scale-free distribution of compartment sizes and in turn a distribution of escapes without a characteristic time. Some experiments agree with predictions of this type of anomalous diffusion where a linear MSD regime is not attained (Smith et al., 1999; Weigel, Ragi, et al., 2011; Weigel, Simon, et al., 2011). A different view of diffusion within a scale-free meshwork is that of nested compartmentalization. As a molecule escapes from a small compartment it remains confined within a larger one. Therefore, confinement can occur within many length scales. Evidence for such motion was observed in the plasma membrane of rat kidney fibroblasts (Fujiwara, Ritchie, Murakoshi, Jacobson, & Kusumi, 2002; Suzuki, Ritchie, Kajikawa, Fujiwara, & Kusumi, 2005). The neutral lipid dioleoylphosphatidylethanolamine (DOPE) and transferrin receptors (TfR) were observed to hop between small compartments of the order of 250 nm and remain confined for longer times inside 750 nm compartments, displaying an effective nested double compartment structure (Fujiwara et al., 2002). Fujiwara et al. (2002) reported residence times of the order of 11 ms for the small compartments and 0.6 s for the larger compartments for DOPE, and 55 ms and 1.8 s for TfR. Similarly, Suzuki et al. (2005) observed a nested compartmentalization of G protein-coupled receptors (GPCRs) in rat kidney fibroblasts with compartment sizes of 210 and 730 nm. As compartments become smaller, the observation of confinement becomes challenging. A scale-free compartmentalization of a membrane can also be obtained with a high concentration of immobile obstacles that generate a percolation cluster (Ben-Avraham & Havlin, 2000; Deverall et al., 2005; Saxton, 1994). Membrane proteins and lipids tethered to the cytoskeleton behave as immobile obstacles, hindering diffusion (Murase et al., 2004; Nakada et al., 2003). The effects of immobile obstacles on the motion of membrane proteins are described above in Section 2.3.

Direct observations of actin forming physical barriers to the motion of receptors over timescales of seconds were obtained from simultaneous single-particle tracking of QD-labeled IgE receptors and imaging cortical GFP-tagged actin in live RBL cells. Andrews et al. (2008) observed large actin bundles in the proximity of the plasma membrane and simultaneously tracked QD labeled receptors using both TIRF and confocal microscopy. TIRF measurements were performed at 33 frames/s and confocal measurements at 1 frame/s. Both measurements show that actin bundles create micrometer-sized dynamic compartments and that receptor motion remains limited to actin-poor regions. The actin structure is observed to reorganize over timescales of 1–10 s.

3.2.3 Fluorescence correlation spectroscopy

Fluorescence correlation spectroscopy (FCS) has provided evidence of anomalous diffusion of membrane proteins in live cells for more than a decade (Weiss, Hashimoto, & Nilsson, 2003). FCS yields local measurements with high temporal resolution and thus, it is an attractive method for the study of membrane organization but the reliable analysis of the modes of motion is challenging. Detailed reviews on its application to the study of diffusion in biological membranes are available (Chiantia, Ries, & Schwille, 2009; He & Marguet, 2011). Wawrezinieck, Rigneault, Marguet, and Lenne (2005) proposed an FCS analysis at multiple spatial scales, by varying the spot size, to study membrane organization and detect transient confinement within microdomains. Employing this analysis, it was observed that the motion of transferrin receptors is hindered by a meshwork, such as that sketched in Figure 4(A) (Lenne et al., 2006). However, the complex interplay of different mechanisms simultaneously affecting protein motion makes it difficult to characterize the effect of the cortical cytoskeleton on protein diffusion. Heinemann et al. (2013) studied the effect of membrane-bound actin on the lateral diffusion of proteins and lipids using an interesting in vitro system where biotinylated actin filaments were coupled to a free standing membrane, mimicking the cortical actin cytoskeleton. It was found that modulation of diffusion depended on the type and size of diffusing species and it was directly correlated to the bound actin density.

3.3 Macromolecular crowding and viscoelastic effects

The presence of mobile obstacles in the plasma membrane decreases the available space upon which molecules can move freely and as a consequence, the diffusion coefficient decreases. This intuitive effect is well established through both numerical simulations (Saxton, 1987) and experiments in reconstituted model membranes (Peters & Cherry, 1982; Vaz, Goodsaidzalduondo, & Jacobson, 1984). Thus, molecular crowding hinders mobility by reducing the diffusion coefficient, but the full extent of its effects on anomalous diffusion in membranes is not well understood (Banks & Fradin, 2005). The picture is clearer in three dimensions where macromolecular crowding has been observed to introduce viscoelasticity and subdiffusion (Banks & Fradin, 2005; Weiss, Guigas, & Kalla, 2007). As mentioned above, diffusion in a viscoelastic medium yields anomalous diffusion because the perturbations induced by a molecule on its surroundings display a slow relaxation to its new configuration. Thus, a memory effect pushes particles back to their past locations. Up to date, evidence for two-dimensional viscoelastic effects in live cells is indirect and further work is needed to quantify the implications of viscoelasticity on the diffusion of membrane proteins. In particular, better experiments need to be designed to probe the impact of the mechanical properties of the membrane on the diffusion pattern.

Some of the most clear experimental evidence for the effect of protein crowding on lateral diffusion was obtained by Horton, Hofling, Radler, and Franosch (2010) in synthetic model membrane systems with an adjustable degree of crowding. The study of protein anomalous diffusion was performed with FCS as a function of avidin content. Avidin was bound to biotinylated-anchored lipids and, in order to record an FCS signal, a fraction of the avidin was fluorescently labeled. For intermediate and high protein crowding levels, anomalous diffusion was observed for the whole measurement range, between 0.1 and 100 ms, in agreement with molecular dynamics simulations (Javanainen et al., 2013). The onset of anomalous dynamics due to molecular crowding was measured by varying the fraction of lipids that contained biotin anchors and thus, different protein concentrations were attained. Anomalous diffusion takes place at 0.05% anchors (1:2000 anchor-to-lipid ratio) and the anomalous diffusion exponent decreases monotonically as the anchor concentration increases from 0.05% to 10%.

The evidence for viscoelasticity in cell membranes comes from mechanical measurements. Several types of mechanical tests are performed on cellular environments showing that the plasma membrane is viscoelastic. These measurements typically employ atomic force microscopy (AFM), traction force spectroscopy, and tether formation by optical or magnetic tweezers. Traction force microscopy (TFM) is a method for probing the local forces exerted by adherent cells on a substrate by visualization of substrate deformations. Typically fluorescent beads (Beningo & Wang, 2002) or micropatterned fluorescent spots (Polio, Rothenberg, Stamenović, & Smith, 2012) are embedded in a polyacrylic acid (PAA) matrix and their displacement can be directly observed. Then, the mechanical deformation of the PAA substrate is inferred. TFM measurements show that the cytoskeleton exhibits local viscoelastic properties (Kumar et al., 2006). Given that cytoskeleton fibers confer tension to the plasma membrane, we expect the membrane to display viscoelastic properties as well. More direct evidence for the viscoelasticity of the plasma membrane is obtained via tether formation (Ermilov, Murdock,

Qian, Brownell, & Anvari, 2007; Jauffred, Callisen, & Oddershede, 2007; Schmitz, Benoit, & Gottschalk, 2008). In tether formation experiments, a functionalized microsphere is attached to the cell membrane and then it is pulled away using optical or magnetic tweezers. Pulling can be performed at different velocities and the membrane tension is measured. Alternatively, the membrane viscoelasticity can be probed by AFM. The advantages of AFM lie in their high spatiotemporal resolution and the ability to probe local forces (Bao & Suresh, 2003). AFM data provide clear evidence for viscoelasticity over broad length- and timescales (Cartagena & Raman, 2014). Measurements show that bacterial cell walls are also viscoelastic. Vadillo-Rodriguez and Dutcher (2011) review the viscoelastic properties of bacterial envelopes in terms of both experiments and modeling. Even though evidence for viscoelasticity of the cell membranes is irrefutable, the detailed mechanisms causing viscoelasticity are still poorly understood. In particular, deformations of the cytoskeleton are known to cause viscoelastic properties, but its impact on protein diffusion can be very different from those caused by protein crowding.

3.4 Transient binding—seconds to minutes

Transient immobilizations with power-law distributed sojourn times lead rich physical properties including anomalous diffusion, aging, and weak *ergodicity breaking*. Molecules in the plasma membrane can interact with other membrane or cytoplasmic complexes in such a way that they immobilize during the time they are bound to immobile complexes. We have shown that in mammalian cells, the voltage-gated potassium channels Kv2.1 and Kv1.4 immobilize in the plasma membrane with sojourn times distributed according to power laws (Weigel, Simon, et al., 2011). Further, these interactions introduce aging in the protein trajectories because the longer the measurement time, the higher the chances of finding a more stable trap with longer binding times (Figure 5(A)). Thus, the mobile fraction time of each trajectory decreases when the observation time increases and, in turn, the apparent diffusion coefficient decreases as well. A distribution of sojourn times $\psi(\tau) \sim \tau^{-(1+\alpha)}$ yields an ensemble- and time-averaged mean squared displacement that scales as $\langle\overline{\delta^2}\rangle \sim T^{-(1-\alpha)}$, where T is the experimental time. For Kv2.1, $\alpha = 0.9$ according to the scaling behavior of both the measured sojourn time distribution and the aging of the MSD (Figure 5(A) and (B)). This behavior is found in timescales ranging from seconds to minutes. Different Kv channels exhibit α values between 0.8 and 0.9.

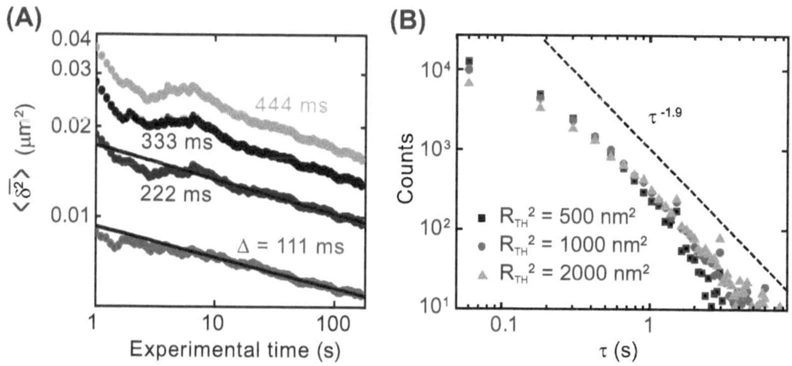

Figure 5 The ion channel Kv2.1 exhibits features of a heavy-tailed CTRW (Weigel, Simon, et al., 2011). (A) Aging. The ensemble average of the temporal mean square displacement (MSD) ($\langle\overline{\delta^2}\rangle$) is shown as a function of experimental time for four different lag times, namely $\Delta = 111, 222, 333,$ and 444 ms. The averaged MSD depends on the experimental time with $\langle\overline{\delta^2}\rangle \sim 1/T^{0.1}$. Following Eqn (15), this observation is consistent with a CTRW with $\alpha = 0.9$. (B) Distribution of immobilization times of the Kv2.1 channel. Immobilizations are identified as a particle being temporarily confined within a radius R_{TH} and analyses employing different thresholds are shown. The distribution exhibits a power-law dependence $\psi(\tau) \sim \tau^{-(1+\alpha)}$, also with $\alpha = 0.9$. (See color plate)

Systems with power-law sojourn times exhibit weak ergodicity breaking, that is, the time average does not converge to the ensemble average because the time averages remain a random variable in the long time limit (Bel & Barkai, 2005; He et al., 2008; Lubelski et al., 2008). The measurements of Kv channels show that the MSD when averaged over time exhibits a large scatter, i.e., it displays a very broad distribution, but the MSD averaged over an ensemble of molecules does not, i.e., they have a narrow distribution (Weigel, Simon, et al., 2011). Similar observations have been made for three-dimensional diffusion in cellular environments for lipid (Jeon et al., 2011) and insulin granules (Tabei et al., 2013). Other three-dimensional diffusion measurements have shown large scattering of the temporal MSD. These include RNA molecules in bacteria (Golding & Cox, 2006) and telomeres in the nucleus of eukaryotic cells (Bronstein et al., 2009), but rigorous analyses show that this motion cannot be described by a CTRW and an fBM is more suitable (Kepten et al., 2011; Magdziarz et al., 2009).

Recent studies have elucidated the mechanism by which power-law sojourn times are observed in the immobilization of Kv channels on the plasma membrane (Weigel et al., 2013). It was found that channels immobilize when they bind to CCP in a catch-and-release manner. Clathrin structures are responsible for the internalization of receptors from the plasma membrane (Doherty & McMahon, 2009; Edeling, Smith, & Owen, 2006). CCPs nucleate at locations next to junctions between the plasma membrane and the endoplasmic reticulum (Fox et al., 2013) and they grow as they mature and become ready for internalization. CCPs bind cargo such as Kv channels via interactions with an adaptor meshwork on the clathrin coat (Robinson, 2004). As a CCP grows, more adaptors become available for binding cargo and the escape rate from an individual coat decreases (Loerke, Mettlen, Schmid, & Danuser, 2011; Weigel et al., 2013). Thus the process is not Markovian because the longer a cargo molecule sits on a CCP, the smaller its probability of escape. As a result, sojourn times within CCPs have a power-law component that induces aging and anomalous diffusion. This type of motion is modeled as a heavy-tailed CTRW (see Section 2.2.1).

Different receptors have also been observed to display a similar catch-and-release behavior on CCPs as Kv channels (Torreno-Pina et al., 2014). Specifically, it was observed that transmembrane glycoprotein dendritic cell-specific intercellular adhesion molecule 3-grabbing nonintegrin (DC-SIGN) interacts with CCPs in a dynamic way. DC-SIGN interactions with clathrin coats appear to be tuned by the attachment of glycans. Interestingly, Torreno-Pina et al. (2014) showed that CCP-cargo interactions can be regulated by the localization of molecules to the vicinity of CCP regions. Specifically DC-SIGN localizes to nanodomains close to CCPs enhancing the probability of successful internalization. Therefore, antipersistent subdiffusion mechanisms, such as compartmentalization of the plasma membrane and fBM, enhance the interactions of cargo with CCP by maintaining reactants near their reaction centers. These observations are relevant for both Kv channels (Weigel, Simon, et al., 2011) and DC-SIGN (Torreno-Pina et al., 2014), where the exploration in the surrounding of CCPs is compact. We believe this is an important physiological role of anomalous diffusion on the plasma membrane and on the cell organization in general. Peculiarly, transient binding is not the sole mechanism for subdiffusion in either Kv channels or DC-SIGN. For the case of DC-SIGN, they are observed to exhibit clustering and confinement within nanoscale domains (de Bakker et al., 2007; Cambi et al., 2004). On the other hand, Kv channels are seen to exhibit anticorrelated type of subdiffusion while they are in the mobile phase and this subdiffusion persists for long times (Weigel, Simon, et al., 2011).

One key aspect of anomalous diffusion induced by transient immobilizations is that sojourn times exhibit very broad distributions. For interactions with CCP, this phenomenon is enhanced by the heterogeneity of clathrin-coated pits, which leads to a broad distribution in CCP dwell times on the plasma membrane. It was shown that for a single cell, different populations of CCP exist: short-lived populations correspond to abortive events that do not end in internalization and long-lived events lasting up to several minutes involve the production of a vesicle (Loerke et al., 2011, 2009; Mettlen, Loerke, Yarar, Danuser, & Schmid, 2010). In a recent work, it was shown that the distribution of dwell times of CCPs depends on the cargo molecules that bind to it (Flores-Otero et al., 2014). Flores-Otero et al. (2014) studied the signaling cascade of specific GPCR, one of the most important signal transduction systems in cell biology and pharmacology. It was found that different agonists bound to the same GPCR can introduce significant changes to CCP dwell times. Therefore, many sources lead to heterogeneity in the time that molecules remain bound to CCP. Importantly, longer CCP dwell times enhance G protein signaling; thus, the ability for molecules to remain immobilized during long times, such as those modeled with power-law distributions, has direct physiological implications.

3.5 Membrane heterogeneities

As proteins explore different microdomains on the cell surface, their diffusion coefficient depends on the local environment. For example, single-particle tracking revealed that glycine receptors alternate between diffusive and confined states due to interactions with gephyrin, which are believed to regulate glycine receptor clustering during the formation of synapses between neurons (Meier et al., 2001). As discussed above, a heterogeneous diffusion landscape can lead to complex dynamics involving anomalous diffusion, aging, and weak ergodicity breaking (Cherstvy et al., 2013; Massignan et al., 2014). These nontrivial aspects of diffusion in a heterogeneous landscape resemble a system with a broad distribution of energetic traps where transient immobilizations due to reversible chemical binding can be modeled by a CTRW (Condamin, Tejedor, Voituriez, Benichou, & Klafter, 2008; Klafter & Sokolov, 2011; Weigel, Simon, et al., 2011; Weigel et al., 2013). Notably, it was recently reported that the motion of DC-SIGN in the plasma membrane exhibits anomalous diffusion and weak ergodicity breaking (Manzo et al., 2015). Nevertheless, it is shown that in this case, ergodicity breaking arises from heterogeneous diffusivity such as that described by patches of random sizes (Massignan et al., 2014), rather than transient immobilizations.

3.5.1 Single-particle tracking—study of diffusion and energy landscapes

Several tools have been recently developed to probe diffusivity landscapes with high spatial resolution (Sibarita, 2014). These nanoscopy techniques comprise natural adaptations of single-particle tracking and superresolution imaging. Single-particle tracking photoactivated localization microscopy (sptPALM) (Manley et al., 2008) uses photoactivatable fluorescence proteins suitable for superresolution imaging (Betzig et al., 2006). In sptPALM, a small subset of the fluorescence proteins are photoactivated and imaged until they photobleach. By introducing a photoactivation violet laser beam, a steady-state density of photoactivated molecules are maintained on the cell surface and these molecules can be tracked (Manley et al., 2008). Short traces are typically obtained but it is possible to map the surface of the cell with very high trajectory densities. Alternatively, instead of employing photoactivatable proteins, it is possible to tag molecules with fluorescently labeled antibodies. Here, antibodies rapidly diffuse in solution and only when they are captured at the cell surface they slow down to a level that enables their individual detection (Giannone et al., 2010). Freely diffusing antibodies move too fast and they are only seen as a dim background. When an antibody binds to the target molecule, it is tracked until it either photobleaches or it dissociates from the cell surface. This technique has been termed universal point accumulation for imaging in nanoscale topographies (uPAINT). It provides single molecule trajectories of up to tens of seconds at densities up to 120 μm^{-2} (Giannone et al., 2010). A completely different approach to high-density tracking was developed by the Lidke group, by taking advantage of the different spectral properties of different QDs (Cutler et al., 2013). Individual QDs are identified by their spectral properties allowing the simultaneous tracking of multiple particles within a diffraction-limited spot. The advantage of this tool is that high densities are possible while individual trajectories can be traced for minutes due to the nonbleaching properties of QDs.

Using uPAINT, diffusion heterogeneities were analyzed and a superresolved two-dimensional spatial map of the displacements was built by Giannone et al. (2010). A correlation diagram of glycosylphosphatidylinisotol (GPI)-anchored protein displacement steps and surface density shows that molecules accumulate in micrometer-sized domains where they exhibit slower diffusivity. A similar analysis of a model transmembrane protein, TM-6His, shows reduced mobility at the cell edges of fibroblasts (Giannone et al., 2010). AMPA receptors have also been found to exhibit reduced mobility in regions of receptor accumulation such as synaptic sites (Giannone et al., 2010; Tardin, Cognet, Bats, Lounis, & Choquet, 2003). In order to accurately map both the diffusion and energy landscapes on the cell surface, a statistical approach using Bayesian inference was developed by the Masson group (Masson et al., 2009; Turkcan, Alexandrou, & Masson, 2012). This method was originally developed to analyze the motion of confined molecules and later expanded to a complete description of the heterogeneous diffusivity and energy landscapes (Masson et al., 2014). The starting point of this analysis is single-particle tracking data obtained by either uPAINT or sptPALM. It is found that diffusion maps exhibit large fluctuations at submicrometer length scales for different membrane proteins including glycine receptors (Masson et al., 2014), AMPA receptors (Hoze et al., 2012), and voltage-gated sodium channels (Akin et al., 2014).

3.5.2 Fluorescence correlation spectroscopy—clustering and lipid rafts

Membrane heterogeneities have been also recently studied with imaging FCS techniques such as imaging total internal reflection FCS (ITIR-FCS) (Kannan et al., 2007; Singh & Wohland, 2014). Imaging FCS is a multiplexed measurement that uses a fast camera to enable the simultaneous calculation of temporal correlations at each pixel. Di Rienzo, Gratton, Beltram, and Cardarelli (2013) analyzed the full space–time correlation function to probe population behavior in micrometer-sized plasma membrane regions and reconstructed the ensemble-averaged MSD for different lag times. With this approach the diffusion coefficient and confinement by the actin cytoskeleton were obtained for GFP-labeled transferrin receptors. Bag, Ali, Chauhan, Wohland, and Mishra (2013) used ITIR-FCS to study the dynamics of amylin-membrane interactions. Guo et al. extended this analysis by employing a Bayesian model selection and parameter estimation for resolving membrane heterogeneities and organization (Guo, Bag, Mishra, Wohland, & Bathe, 2014), a method originally developed for confocal FCS (Guo et al., 2012; He, Guo, & Bathe, 2012). The Bayesian ITIR-FCS approach was applied both to supported lipid bilayers and live cells, showing two-component phase separation in model membranes and the dynamic formation of amylin-induced domains in the plasma membrane of neuroblastoma cells (Guo et al., 2014).

A source of heterogeneity in the plasma membrane arises from the preferential aggregation between sphingolipids, sterols, and specific proteins. Therefore, dynamic complexes assemble in the form of isolated microdomains, termed lipid rafts, with different diffusion coefficient than the rest of the plasma membrane (Lingwood & Simons, 2010; Simons & Ikonen, 1997; Simons & Sampaio, 2011). However, the direct observation of lipid rafts in live cells is often elusive due to their transient nature and their small sizes, typically in the range 5–200 nm. Direct observations of the transient trapping of sphingolipids and GPI-anchored proteins within nanodomains were enabled by combining FCS with stimulated emission depletion (STED) fluorescence microscopy (Eggeling et al., 2009; Mueller et al., 2011). STED is a superresolution technique where the probe volume is confined to a region not constrained by the diffraction limit and thus, it permits diffusion measurements at length scales relevant to lipid rafts with high temporal resolution (Kastrup, Blom, Eggeling, & Hell, 2005). Billaudeau et al. (2013) review the use of spot variation FCS for the investigation of the nanoscale organization of the plasma membrane. These experiments reveal molecular details showing that constrained diffusion is often caused by cholesterol-assisted molecular complexes. Further STED-FCS allows the determination of on- and off-rates as well as the domain size (Veronika Mueller et al., 2012).

4. CONCLUSIONS

Many membrane proteins display anomalous diffusion. The physical mechanism of subdiffusion and its timescale depends on the specific type of molecule. Generally, it is observed that molecules with large cytoplasmic domains are strongly affected by collisions with the actin cytoskeleton that affect their motion up to minutes. Small molecules or lipids show subdiffusion on smaller timescales. The exact range of timescales for these molecules is still under debate and, depending on the specific molecule, can be from microseconds to milliseconds.

The picture that arises from the plasma membrane is that of a highly complex environment where interactions between many players lead to intriguing emerging phenomena. Early attempts to classify the mechanisms of anomalous diffusion on the cell surface appear to have underestimated cellular complexity. Instead, a picture with different mechanisms coexisting seems to be more realistic, with these mechanisms displaying strong coupling between them in such a way that it is difficult to decompose them into individual simpler phenomena. Furthermore, the complexity of diffusion in live cells continues to inspire new modeling tools which are needed to understand fundamental biophysical processes.

ACKNOWLEDGMENTS

I thank Eli Barkai for carefully reading the manuscript and for many productive discussions. I also thank Liz Akin, Jenny Higgins, Sanaz Sadegh, Mike Tamkun, and Aubrey Weigel for useful discussions.

I acknowledge the support from the National Science Foundation under grant number 1401432.

REFERENCES

Aguet, F., Antonescu, C. N., Mettlen, M., Schmid, S. L., & Danuser, G. (2013). Advances in analysis of low signal-to-noise images link dynamin and AP2 to the functions of an endocytic checkpoint. *Developmental Cell, 26*(3), 279–291.

Akimoto, T., Yamamoto, E., Yasuoka, K., Hirano, Y., & Yasui, M. (2011). Non-Gaussian fluctuations resulting from power-law trapping in a lipid bilayer. *Physical Review Letters, 107*(17), 178103.

Akin, E. J., Brown, K., Sadegh, S., Weigel, A. V., Masson, J.-B., Krapf, D., et al. (2014). Single-particle tracking PALM of Nav1.6 in hippocampal neurons demonstrates unique subcellular diffusion landscapes. *Biophysical Journal, 106*(2), 36a.

Alberts, B., Johnson, A., Lewis, J., Raff, M., Roberts, K., & Walter, P. (2010). *Molecular biology of the cell.* New York: Garland Science.

Almeida, P. F. F., Vaz, W. L. C., & Thompson, T. E. (1992). Lateral diffusion in the liquid phases of dimyristoylphosphatidylcholine cholesterol lipid bilayers: a free-volume analysis. *Biochemistry, 31*(29), 6739–6747.

Anderluh, A., Klotzsch, E., Ries, J., Reismann, A. W., Weber, S., Fölser, M., et al. (2014). Tracking single serotonin transporter molecules at the endoplasmic reticulum and plasma membrane. *Biophysical Journal, 106*(9), L33–L35.

Andrews, N. L., Lidke, K. A., Pfeiffer, J. R., Burns, A. R., Wilson, B. S., Oliver, J. M., et al. (2008). Actin restricts Fcepsilon RI diffusion and facilitates antigen-induced receptor immobilization. *Nature Cell Biology, 10*(8), 955–963.

Bag, N., Ali, A., Chauhan, V. S., Wohland, T., & Mishra, A. (2013). Membrane destabilization by monomeric hIAPP observed by imaging fluorescence correlation spectroscopy. *Chemical Communications, 49*(80), 9155–9157.

de Bakker, B. I., de Lange, F., Cambi, A., Korterik, J. P., van Dijk, E. M. H. P., van Hulst, N. F., et al. (2007). Nanoscale organization of the pathogen receptor DC-SIGN mapped by single-molecule high-resolution fluorescence microscopy. *ChemPhysChem, 8*(10), 1473–1480.

Banks, D. S., & Fradin, C. (2005). Anomalous diffusion of proteins due to molecular crowding. *Biophysical Journal, 89*(5), 2960–2971.

Bao, G., & Suresh, S. (2003). Cell and molecular mechanics of biological materials. *Nature Materials, 2*(11), 715–725.

Bel, G., & Barkai, E. (2005). Weak ergodicity breaking in the continuous-time random walk. *Physical Review Letters, 94*(24), 240602.

Ben-Avraham, D., & Havlin, S. (2000). *Diffusion and reactions in fractals and disordered systems.* Cambridge: Cambridge University Press.

Beningo, K. A., & Wang, Y. L. (2002). Flexible substrata for the detection of cellular traction forces. *Trends in Cell Biology, 12*(2), 79–84.

Bennett, V., & Gilligan, D. M. (1993). The spectrin-based membrane skeleton and micron-scale organization of the plasma membrane. *Annual Review of Cell Biology, 9*(1), 27–66.

Betzig, E., Patterson, G. H., Sougrat, R., Lindwasser, O. W., Olenych, S., Bonifacino, J. S., et al. (2006). Imaging intracellular fluorescent proteins at nanometer resolution. *Science, 313*(5793), 1642–1645.

Billaudeau, C., Mailfert, S., Trombik, T., Bertaux, N., Rouger, V., Hamon, Y., et al. (2013). Probing the plasma membrane organization in living cells by spot variation fluorescence correlation spectroscopy. *Methods in Enzymology, 519*, 277–302.

Blumen, A., Klafter, J., White, B. S., & Zumofen, G. (1984). Continuous-time random-walks on fractals. *Physical Review Letters, 53*(14), 1301–1304.

Bouchaud, J.-P., & Georges, A. (1990). Anomalous diffusion in disordered media: statistical mechanisms, models and physical applications. *Physics Reports, 195*(4), 127–293.

de Brabander, M., Nuydens, R., Ishihara, A., Holifield, B., Jacobson, K., & Geerts, H. (1991). Lateral diffusion and retrograde movements of individual cell-surface components on single motile cells observed with nanovid microscopy. *Journal of Cell Biology, 112*(1), 111–124.

Bronstein, I., Israel, Y., Kepten, E., Mai, S., Shav-Tal, Y., Barkai, E., et al. (2009). Transient anomalous diffusion of telomeres in the nucleus of mammalian cells. *Physical Review Letters, 103*(1), 018102.

Burnecki, K., Kepten, E., Janczura, J., Bronshtein, I., Garini, Y., & Weron, A. (2012). Universal algorithm for identification of fractional Brownian motion. A case of telomere subdiffusion. *Biophysical Journal, 103*(9), 1839–1847.

Burov, S., Jeon, J. H., Metzler, R., & Barkai, E. (2011). Single particle tracking in systems showing anomalous diffusion: the role of weak ergodicity breaking. *Physical Chemistry Chemical Physics, 13*(5), 1800–1812.

Cambi, A., de Lange, F., van Maarseveen, N. M., Nijhuis, M., Joosten, B., van Dijk, E. M., et al. (2004). Microdomains of the C-type lectin DC-SIGN are portals for virus entry into dendritic cells. *Journal of Cell Biology, 164*(1), 145–155.

Cartagena, A., & Raman, A. (2014). Local viscoelastic properties of live cells investigated using dynamic and quasi-static atomic force microscopy methods. *Biophysical Journal, 106*(5), 1033–1043.

Casuso, I., Khao, J., Chami, M., Paul-Gilloteaux, P., Husain, M., Duneau, J. P., et al. (2012). Characterization of the motion of membrane proteins using high-speed atomic force microscopy. *Nature Nanotechnology, 7*(8), 525–529.

Cherstvy, A. G., Chechkin, A. V., & Metzler, R. (2013). Anomalous diffusion and ergodicity breaking in heterogeneous diffusion processes. *New Journal of Physics, 15*, 083039.

Cherstvy, A. G., & Metzler, R. (2014). Nonergodicity, fluctuations, and criticality in heterogeneous diffusion processes. *Physical Review E, 90*(1), 012134.

Chiantia, S., Ries, J., & Schwille, P. (2009). Fluorescence correlation spectroscopy in membrane structure elucidation. *Biochimica et Biophysica Acta (BBA)-Biomembranes, 1788*(1), 225–233.

Chichili, G. R., & Rodgers, W. (2009). Cytoskeleton–membrane interactions in membrane raft structure. *Cellular and Molecular Life Sciences, 66*(14), 2319–2328.

Condamin, S., Tejedor, V., Voituriez, R., Benichou, O., & Klafter, J. (2008). Probing microscopic origins of confined subdiffusion by first-passage observables. *Proceedings of the National Academy of Sciences of the United States of America, 105*(15), 5675–5680.

Costantini, L. M., Fossati, M., Francolini, M., & Snapp, E. L. (2012). Assessing the tendency of fluorescent proteins to oligomerize under physiologic conditions. *Traffic, 13*(5), 643–649.

Cui-Wang, T., Hanus, C., Cui, T., Helton, T., Bourne, J., Watson, D., et al. (2012). Local zones of endoplasmic reticulum complexity confine cargo in neuronal dendrites. *Cell, 148*(1), 309–321.

Cunningham, C. C. (1995). Actin polymerization and intracellular solvent flow in cell surface blebbing. *Journal of Cell Biology, 129*(6), 1589–1599.

Cutler, P. J., Malik, M. D., Liu, S., Byars, J. M., Lidke, D. S., & Lidke, K. A. (2013). Multicolor quantum dot tracking using a high-speed hyperspectral line-scanning microscope. *PLoS One, 8*(5), e64320.

Deng, W. H., & Barkai, E. (2009). Ergodic properties of fractional Brownian-Langevin motion. *Physical Review E, 79*(1), 011112.

Deverall, M. A., Gindl, E., Sinner, E. K., Besir, H., Ruehe, J., Saxton, M. J., et al. (2005). Membrane lateral mobility obstructed by polymer-tethered lipids studied at the single molecule level. *Biophysical Journal, 88*(3), 1875–1886.

Di Rienzo, C., Gratton, E., Beltram, F., & Cardarelli, F. (2013). Fast spatiotemporal correlation spectroscopy to determine protein lateral diffusion laws in live cell membranes. *Proceedings of the National Academy of Sciences of the United States of America, 110*(30), 12307–12312.

Doherty, G. J., & McMahon, H. T. (2009). Mechanisms of endocytosis. *Annual Review of Biochemistry, 78*, 857–902.

Dupuy, A. D., & Engelman, D. M. (2008). Protein area occupancy at the center of the red blood cell membrane. *Proceedings of the National Academy of Sciences of the United States of America, 105*(8), 2848–2852.

Edeling, M. A., Smith, C., & Owen, D. (2006). Life of a clathrin coat: insights from clathrin and AP structures. *Nature Reviews Molecular Cell Biology, 7*(1), 32–44.

Edidin, M., Kuo, S. C., & Sheetz, M. P. (1991). Lateral movements of membrane-glycoproteins restricted by dynamic cytoplasmic barriers. *Science, 254*(5036), 1379–1382.

Edidin, M., Zuniga, M. C., & Sheetz, M. P. (1994). Truncation mutants define and locate cytoplasmic barriers to lateral mobility of membrane glycoproteins. *Proceedings of the National Academy of Sciences of the United States of America, 91*(8), 3378–3382.

Eggeling, C., Ringemann, C., Medda, R., Schwarzmann, G., Sandhoff, K., Polyakova, S., et al. (2009). Direct observation of the nanoscale dynamics of membrane lipids in a living cell. *Nature, 457*(7233), 1159–1162.

Einstein, A. (1905). The motion of elements suspended in static liquids as claimed in the molecular kinetic theory of heat. *Annalen Der Physik, 17*(8), 549–560.

Ermilov, S. A., Murdock, D. R., Qian, F., Brownell, W. E., & Anvari, B. (2007). Studies of plasma membrane mechanics and plasma membrane-cytoskeleton interactions using optical tweezers and fluorescence imaging. *Journal of Biomechanics, 40*(2), 476–480.

Ernst, D., Hellmann, M., Kohler, J., & Weiss, M. (2012). Fractional Brownian motion in crowded fluids. *Soft Matter, 8*(18), 4886–4889.

Feder, T. J., Brust-Mascher, I., Slattery, J. P., Baird, B., & Webb, W. W. (1996). Constrained diffusion or immobile fraction on cell surfaces: a new interpretation. *Biophysical Journal, 70*(6), 2767–2773.

Flores-Otero, J., Ahn, K. H., Delgado-Peraza, F., Mackie, K., Kendall, D. A., & Yudowski, G. A. (2014). Ligand-specific endocytic dwell times control functional selectivity of the cannabinoid receptor 1. *Nature Communications, 5*, 4589.

Fogedby, H. C. (1994). Langevin-equations for continuous-time Levy flights. *Physical Review E, 50*(2), 1657–1660.

Fox, P. D., Haberkorn, C. J., Weigel, A. V., Higgins, J. L., Akin, E. J., Kennedy, M. J., et al. (2013). Plasma membrane domains enriched in cortical endoplasmic reticulum function as membrane protein trafficking hubs. *Molecular Biology of the Cell, 24*(17), 2703–2713.

Fujiwara, T., Ritchie, K., Murakoshi, H., Jacobson, K., & Kusumi, A. (2002). Phospholipids undergo hop diffusion in compartmentalized cell membrane. *Journal of Cell Biology, 157*(6), 1071–1081.

Gal, N., Lechtman-Goldstein, D., & Weihs, D. (2013). Particle tracking in living cells: a review of the mean square displacement method and beyond. *Rheologica Acta, 52*(5), 425–443.

Giannone, G., Hosy, E., Levet, F., Constals, A., Schulze, K., Sobolevsky, A. I., et al. (2010). Dynamic superresolution imaging of endogenous proteins on living cells at ultra-high density. *Biophysical Journal, 99*(4), 1303–1310.

Golding, I., & Cox, E. C. (2006). Physical nature of bacterial cytoplasm. *Physical Review Letters, 96*(9), 098102.
Goychuk, I., Kharchenko, V. O., & Metzler, R. (2014). How molecular motors work in the crowded environment of living cells: coexistence and efficiency of normal and anomalous transport. *PLoS One, 9*(3), e91700.
Guigas, G., Kalla, C., & Weiss, M. (2007). The degree of macromolecular crowding in the cytoplasm and nucleoplasm of mammalian cells is conserved. *FEBS Letters, 581*(26), 5094–5098.
Guo, S.-M., Bag, N., Mishra, A., Wohland, T., & Bathe, M. (2014). Bayesian total internal reflection fluorescence correlation spectroscopy reveals hIAPP-induced plasma membrane domain organization in live cells. *Biophysical Journal, 106*(1), 190–200.
Guo, S.-M., He, J., Monnier, N., Sun, G., Wohland, T., & Bathe, M. (2012). Bayesian approach to the analysis of fluorescence correlation spectroscopy data II: application to simulated and in vitro data. *Analytical Chemistry, 84*(9), 3880–3888.
Havlin, S., & Ben-Avraham, D. (1987). Diffusion in disordered media. *Advances in Physics, 36*(6), 695–798.
He, Y., Burov, S., Metzler, R., & Barkai, E. (2008). Random time-scale invariant diffusion and transport coefficients. *Physical Review Letters, 101*(5), 058101.
He, J., Guo, S.-M., & Bathe, M. (2012). Bayesian approach to the analysis of fluorescence correlation spectroscopy data I: theory. *Analytical Chemistry, 84*(9), 3871–3879.
He, H.-T., & Marguet, D. (2011). Detecting nanodomains in living cell membrane by fluorescence correlation spectroscopy. *Annual Review of Physical Chemistry, 62*, 417–436.
Heinemann, F., Vogel, S. K., & Schwille, P. (2013). Lateral membrane diffusion modulated by a minimal actin cortex. *Biophysical Journal, 104*(7), 1465–1475.
Höfling, F., & Franosch, T. (2013). Anomalous transport in the crowded world of biological cells. *Reports on Progress in Physics, 76*(4), 046602.
Horton, M. R., Hofling, F., Radler, J. O., & Franosch, T. (2010). Development of anomalous diffusion among crowding proteins. *Soft Matter, 6*(12), 2648–2656.
Hoze, N., Nair, D., Hosy, E., Sieben, C., Manley, S., Herrmann, A., et al. (2012). Heterogeneity of AMPA receptor trafficking and molecular interactions revealed by superresolution analysis of live cell imaging. *Proceedings of the National Academy of Sciences of the United States of America, 109*(42), 17052–17057.
James, P. S., Hennessy, C., Berge, T., & Jones, R. (2004). Compartmentalisation of the sperm plasma membrane: a FRAP, FLIP and SPFI analysis of putative diffusion barriers on the sperm head. *Journal of Cell Science, 117*(26), 6485–6495.
Jauffred, L., Callisen, T. H., & Oddershede, L. B. (2007). Visco-elastic membrane tethers extracted from *Escherichia coli* by optical tweezers. *Biophysical Journal, 93*(11), 4068–4075.
Javanainen, M., Hammaren, H., Monticelli, L., Jeon, J. H., Miettinen, M. S., Martinez-Seara, H., et al. (2013). Anomalous and normal diffusion of proteins and lipids in crowded lipid membranes. *Faraday Discussions, 161*, 397–417.
Jeon, J. H., Monne, H. M. S., Javanainen, M., & Metzler, R. (2012). Anomalous diffusion of phospholipids and cholesterols in a lipid bilayer and its origins. *Physical Review Letters, 109*(18), 188103.
Jeon, J. H., Tejedor, V., Burov, S., Barkai, E., Selhuber-Unkel, C., Berg-Sorensen, K., et al. (2011). In vivo anomalous diffusion and weak ergodicity breaking of lipid granules. *Physical Review Letters, 106*(4), 048103.
Jiang, H., English, B. P., Hazan, R. B., Wu, P., & Ovryn, B. (2015). Tracking surface glycans on live cancer cells with single-molecule sensitivity. *Angewandte Chemie International Edition, 54*(6), 1765–1769.
Kannan, B., Guo, L., Sudhaharan, T., Ahmed, S., Maruyama, I., & Wohland, T. (2007). Spatially resolved total internal reflection fluorescence correlation microscopy using an electron multiplying charge-coupled device camera. *Analytical Chemistry, 79*(12), 4463–4470.

Kastrup, L., Blom, H., Eggeling, C., & Hell, S. W. (2005). Fluorescence fluctuation spectroscopy in subdiffraction focal volumes. *Physical Review Letters, 94*(17), 178104.
Kepten, E., Bronshtein, I., & Garini, Y. (2011). Ergodicity convergence test suggests telomere motion obeys fractional dynamics. *Physical Review E, 83*(4), 041919.
Klafter, J., Blumen, A., & Zumofen, G. (1984). Fractal behavior in trapping and reaction: a random-walk study. *Journal of Statistical Physics, 36*(5–6), 561–577.
Klafter, J., & Sokolov, I. M. (2005). Anomalous diffusion spreads its wings. *Physics World, 18*(8), 29–32.
Klafter, J., & Sokolov, I. M. (2011). *First steps in random walks: From tools to applications.* New York: Oxford University Press.
Kumar, S., Maxwell, I. Z., Heisterkamp, A., Polte, T. R., Lele, T. P., Salanga, M., et al. (2006). Viscoelastic retraction of single living stress fibers and its impact on cell shape, cytoskeletal organization, and extracellular matrix mechanics. *Biophysical Journal, 90*(10), 3762–3773.
Kuo, H.-H. (2006). *Introduction to stochastic integration.* New York: Springer.
Kusumi, A., Fujiwara, T. K., Chadda, R., Xie, M., Tsunoyama, T. A., Kalay, Z., et al. (2012). Dynamic organizing principles of the plasma membrane that regulate signal transduction: commemorating the fortieth anniversary of Singer and Nicolson's fluid-mosaic model. *Annual Review of Cell and Developmental Biology, 28*, 215–250.
Kusumi, A., Nakada, C., Ritchie, K., Murase, K., Suzuki, K., Murakoshi, H., et al. (2005). Paradigm shift of the plasma membrane concept from the two-dimensional continuum fluid to the partitioned fluid: high-speed single-molecule tracking of membrane molecules. *Annual Review of Biophysics and Biomolecular Structure, 34*, 351–378.
Kusumi, A., & Sako, Y. (1996). Cell surface organization by the membrane skeleton. *Current Opinion in Cell Biology, 8*(4), 566–574.
Lenne, P. F., Wawrezinieck, L., Conchonaud, F., Wurtz, O., Boned, A., Guo, X. J., et al. (2006). Dynamic molecular confinement in the plasma membrane by microdomains and the cytoskeleton meshwork. *EMBO Journal, 25*(14), 3245–3256.
Lingwood, D., & Simons, K. (2010). Lipid rafts as a membrane-organizing principle. *Science, 327*(5961), 46–50.
Lodish, H., Berk, A., Kaiser, C. A., Krieger, M., Scott, M. P., Bretscher, A., et al. (2007). *Molecular cell biology* (6th ed.). New York: W. H. Freeman
Loerke, D., Mettlen, M., Schmid, S. L., & Danuser, G. (2011). Measuring the hierarchy of molecular events during clathrin-mediated endocytosis. *Traffic, 12*(7), 815–825.
Loerke, D., Mettlen, M., Yarar, D., Jaqaman, K., Jaqaman, H., Danuser, G., et al. (2009). Cargo and dynamin regulate clathrin-coated pit maturation. *PLoS Biology, 7*(3), 628–639.
Lubelski, A., Sokolov, I. M., & Klafter, J. (2008). Nonergodicity mimics inhomogeneity in single particle tracking. *Physical Review Letters, 100*(25), 250602.
Luna, E. J., & Hitt, A. L. (1992). Cytoskeleton–plasma membrane interactions. *Science, 258*(5084), 955–964.
Magdziarz, M., Weron, A., Burnecki, K., & Klafter, J. (2009). Fractional Brownian motion versus the continuous-time random walk: a simple test for subdiffusive dynamics. *Physical Review Letters, 103*(18), 180602.
Mandelbrot, B., & Van Ness, J. W. (1968). Fractional Brownian motions fractional noises and applications. *SIAM Review, 10*(4), 422–437.
Manley, S., Gillette, J. M., Patterson, G. H., Shroff, H., Hess, H. F., Betzig, E., et al. (2008). High-density mapping of single molecule trajectories with photoactivated localization microscopy. *Nature Methods, 5*(2), 155–157.
Manzo, C., Torreno Pina, J. A., Massignan, P., Lapeyre, G. J., Lewenstein, M., & Garcia Parajo, M. F. (2015). Weak ergodicity breaking of receptor motion in living cells stemming from random diffusivity. *Physical Review X, 5*(1), 011021.

Marguet, D., Lenne, P. F., Rigneault, H., & He, H. T. (2006). Dynamics in the plasma membrane: how to combine fluidity and order. *EMBO Journal, 25*(15), 3446–3457.

Mason, T. G., & Weitz, D. A. (1995). Optical measurements of frequency-dependent linear viscoelastic moduli of complex fluids. *Physical Review Letters, 74*(7), 1250–1253.

Massignan, P., Manzo, C., Torreno-Pina, J. A., Garcia-Parajo, M. F., Lewenstein, M., & Lapeyre, G. J. (2014). Nonergodic subdiffusion from Brownian motion in an inhomogeneous medium. *Physical Review Letters, 112*(15), 150603.

Masson, J. B., Casanova, D., Turkcan, S., Voisinne, G., Popoff, M. R., Vergassola, M., et al. (2009). Inferring maps of forces inside cell membrane microdomains. *Physical Review Letters, 102*(4), 048103.

Masson, J. B., Dionne, P., Salvatico, C., Renner, M., Specht, C. G., Triller, A., et al. (2014). Mapping the energy and diffusion landscapes of membrane proteins at the cell surface using high-density single-molecule imaging and Bayesian inference: application to the multiscale dynamics of glycine receptors in the neuronal membrane. *Biophysical Journal, 106*(1), 74–83.

Meier, J., Vannier, C., Serge, A., Triller, A., & Choquet, D. (2001). Fast and reversible trapping of surface glycine receptors by gephyrin. *Nature Neuroscience, 4*(3), 253–260.

Meroz, Y., Sokolov, I. M., & Klafter, J. (2010). Subdiffusion of mixed origins: when ergodicity and nonergodicity coexist. *Physical Review E, 81*(1), 010101(R).

Mettlen, M., Loerke, D., Yarar, D., Danuser, G., & Schmid, S. L. (2010). Cargo- and adaptor-specific mechanisms regulate clathrin-mediated endocytosis. *Journal of Cell Biology, 188*(6), 919–933.

Metzler, R., Jeon, J.-H., Cherstvy, A. G., & Barkai, E. (2014). Anomalous diffusion models and their properties: non-stationarity, non-ergodicity, and ageing at the centenary of single particle tracking. *Physical Chemistry Chemical Physics, 16*(44), 24128–24164. http://dx.doi.org/10.1039/C4CP03465A.

Metzler, R., & Klafter, J. (2000). The random walk's guide to anomalous diffusion: a fractional dynamics approach. *Physics Reports, 339*(1), 1–77.

Michalet, X., Pinaud, F. F., Bentolila, L. A., Tsay, J. M., Doose, S., Li, J. J., et al. (2005). Quantum dots for live cells, in vivo imaging, and diagnostics. *Science, 307*(5709), 538–544.

Mueller, V., Honigmann, A., Ringemann, C., Medda, R., Schwarzmann, G., & Eggeling, C. (2012). FCS in STED microscopy: studying the nanoscale of lipid membrane dynamics. *Methods in Enzymology, 519*, 1–38.

Mueller, V., Ringemann, C., Honigmann, A., Schwarzmann, G., Medda, R., Leutenegger, M., et al. (2011). STED nanoscopy reveals molecular details of cholesterol-and cytoskeleton-modulated lipid interactions in living cells. *Biophysical Journal, 101*(7), 1651–1660.

Murase, K., Fujiwara, T., Umemura, Y., Suzuki, K., Iino, R., Yamashita, H., et al. (2004). Ultrafine membrane compartments for molecular diffusion as revealed by single molecule techniques. *Biophysical Journal, 86*(6), 4075–4093.

Nakada, C., Ritchie, K., Oba, Y., Nakamura, M., Hotta, Y., Iino, R., et al. (2003). Accumulation of anchored proteins forms membrane diffusion barriers during neuronal polarization. *Nature Cell Biology, 5*(7), 626–632.

Notelaers, K., Rocha, S., Paesen, R., Smisdom, N., De Clercq, B., Meier, J. C., et al. (2014). Analysis of alpha 3 GlyR single particle tracking in the cell membrane. *Biochimica Et Biophysica Acta-Molecular Cell Research, 1843*(3), 544–553.

Notelaers, K., Smisdom, N., Rocha, S., Janssen, D., Meier, J. C., Rigo, J. M., et al. (2012). Ensemble and single particle fluorimetric techniques in concerted action to study the diffusion and aggregation of the glycine receptor alpha 3 isoforms in the cell plasma membrane. *Biochimica Et Biophysica Acta-Biomembranes, 1818*(12), 3131–3140.

Peebles, P. Z. (1987). *Probability, random variables, and random signal principles* (3rd ed.). New York: McGraw-Hill.
Peters, R., & Cherry, R. J. (1982). Lateral and rotational diffusion of bacteriorhodopsin in lipid bilayers - experimental test of the Saffman-Delbruck equations. *Proceedings of the National Academy of Sciences of the United States of America, 79*(14), 4317–4321.
Pinaud, F., Clarke, S., Sittner, A., & Dahan, M. (2010). Probing cellular events, one quantum dot at a time. *Nature Methods, 7*(4), 275–285.
Polio, S. R., Rothenberg, K. E., Stamenović, D., & Smith, M. L. (2012). A micropatterning and image processing approach to simplify measurement of cellular traction forces. *Acta Biomaterialia, 8*(1), 82–88.
Ritchie, K., Iino, R., Fujiwara, T., Murase, K., & Kusumi, A. (2003). The fence and picket structure of the plasma membrane of live cells as revealed by single molecule techniques (review). *Molecular Membrane Biology, 20*(1), 13–18.
Robinson, M. S. (2004). Adaptable adaptors for coated vesicles. *Trends in Cell Biology, 14*(4), 167–174.
Sako, Y., & Kusumi, A. (1995). Barriers for lateral diffusion of transferrin receptor in the plasma membrane as characterized by receptor dragging by laser tweezers: fence versus tether. *Journal of Cell Biology, 129*(6), 1559–1574.
Saxton, M. J. (1987). Lateral diffusion in an archipelago. The effect of mobile obstacles. *Biophysical Journal, 52*(6), 989–997.
Saxton, M. J. (1994). Anomalous diffusion due to obstacles: a Monte-Carlo study. *Biophysical Journal, 66*(2), 394–401.
Saxton, M. J. (1996). Anomalous diffusion due to binding: a Monte Carlo study. *Biophysical Journal, 70*(3), 1250–1262.
Saxton, M. J., & Jacobson, K. (1997). Single-particle tracking: applications to membrane dynamics. *Annual Review of Biophysics and Biomolecular Structure, 26*, 373–399.
Schaaf, M. J. M., Koopmans, W. J. A., Meckel, T., van Noort, J., Snaar-Jagalska, B. E., Schmidt, T. S., et al. (2009). Single-molecule microscopy reveals membrane microdomain organization of cells in a living vertebrate. *Biophysical Journal, 97*(4), 1206–1214.
Scher, H., & Montroll, E. W. (1975). Anomalous transit-time dispersion in amorphous solids. *Physical Review B, 12*(6), 2455–2477.
Scher, H., Shlesinger, M. F., & Bendler, J. T. (1991). Time-scale invariance in transport and relaxation. *Physics Today, 44*(1), 26–34.
Schmitz, J., Benoit, M., & Gottschalk, K. E. (2008). The viscoelasticity of membrane tethers and its importance for cell adhesion. *Biophysical Journal, 95*(3), 1448–1459.
Sebastian, K. L. (1995). Path-integral representation for fractional Brownian-motion. *Journal of Physics. A: Mathematical and General, 28*(15), 4305–4311.
Serge, A., Bertaux, N., Rigneault, H., & Marguet, D. (2008). Dynamic multiple-target tracing to probe spatiotemporal cartography of cell membranes. *Nature Methods, 5*(8), 687–694.
Sheetz, M. P. (1983). Membrane skeletal dynamics: role in modulation of red-cell deformability, mobility of transmembrane proteins, and shape. *Seminars in Hematology, 20*(3), 175–188.
Sheetz, M. P., Turney, S., Qian, H., & Elson, E. L. (1989). Nanometer-level analysis demonstrates that lipid flow does not drive membrane glycoprotein movements. *Nature, 340*(6231), 284–288.
Sibarita, J. B. (2014). High-density single-particle tracking: quantifying molecule organization and dynamics at the nanoscale. *Histochemistry and Cell Biology, 141*(6), 587–595.
Simons, K., & Ikonen, E. (1997). Functional rafts in cell membranes. *Nature, 387*(6633), 569–572.
Simons, K., & Sampaio, J. L. (2011). Membrane organization and lipid rafts. *Cold Spring Harbor Perspectives in Biology, 3*(10), a004697.

Singh, A. P., & Wohland, T. (2014). Applications of imaging fluorescence correlation spectroscopy. *Current Opinion in Chemical Biology, 20*, 29–35.
Smith, P. R., Morrison, I. E. G., Wilson, K. M., Fernandez, N., & Cherry, R. J. (1999). Anomalous diffusion of major histocompatibility complex class I molecules on HeLa cells determined by single particle tracking. *Biophysical Journal, 76*(6), 3331–3344.
Snapp, E. L., Hegde, R. S., Francolini, M., Lombardo, F., Colombo, S., Pedrazzini, E., et al. (2003). Formation of stacked ER cisternae by low affinity protein interactions. *Journal of Cell Biology, 163*(2), 257–269.
Sokolov, I. M., & Klafter, J. (2005). From diffusion to anomalous diffusion: a century after Einstein's Brownian motion. *Chaos, 15*(2), 026103.
Suzuki, K., Ritchie, K., Kajikawa, E., Fujiwara, T., & Kusumi, A. (2005). Rapid hop diffusion of a G-protein-coupled receptor in the plasma membrane as revealed by single-molecule techniques. *Biophysical Journal, 88*(5), 3659–3680.
Szymanski, J., & Weiss, M. (2009). Elucidating the origin of anomalous diffusion in crowded fluids. *Physical Review Letters, 103*(3), 038102.
Tabei, S. M. A., Burov, S., Kim, H. Y., Kuznetsov, A., Huynh, T., Jureller, J., et al. (2013). Intracellular transport of insulin granules is a subordinated random walk. *Proceedings of the National Academy of Sciences of the United States of America, 110*(13), 4911–4916.
Tank, D. W., Wu, E. S., & Webb, W. W. (1982). Enhanced molecular diffusibility in muscle membrane blebs: release of lateral constraints. *Journal of Cell Biology, 92*(1), 207–212.
Tardin, C., Cognet, L., Bats, C., Lounis, B., & Choquet, D. (2003). Direct imaging of lateral movements of AMPA receptors inside synapses. *EMBO Journal, 22*(18), 4656–4665.
Torreno-Pina, J. A., Castro, B. M., Manzo, C., Buschow, S. I., Cambi, A., & Garcia-Parajo, M. F. (2014). Enhanced receptor-clathrin interactions induced by N-glycan-mediated membrane micropatterning. *Proceedings of the National Academy of Sciences of the United States of America, 111*(30), 11037–11042.
Tsuji, A., & Ohnishi, S. (1986). Restriction of the lateral motion of band-3 in the erythrocyte-membrane by the cytoskeletal network: dependence on spectrin association state. *Biochemistry, 25*(20), 6133–6139.
Turkcan, S., Alexandrou, A., & Masson, J. B. (2012). A Bayesian inference scheme to extract diffusivity and potential fields from confined single-molecule trajectories. *Biophysical Journal, 102*(10), 2288–2298.
Vadillo-Rodriguez, V., & Dutcher, J. R. (2011). Viscoelasticity of the bacterial cell envelope. *Soft Matter, 7*(9), 4101–4110.
Vaz, W. L. C., Goodsaidzalduondo, F., & Jacobson, K. (1984). Lateral diffusion of lipids and proteins in bilayer-membranes. *FEBS Letters, 174*(2), 199–207.
Wawrezinieck, L., Rigneault, H., Marguet, D., & Lenne, P.-F. (2005). Fluorescence correlation spectroscopy diffusion laws to probe the submicron cell membrane organization. *Biophysical Journal, 89*(6), 4029–4042.
Weigel, A. V., Ragi, S., Reid, M. L., Chong, E. K., Tamkun, M. M., & Krapf, D. (2011). Obstructed diffusion propagator analysis for single-particle tracking. *Physical Review E, 85*(4), 041924.
Weigel, A. V., Simon, B., Tamkun, M. M., & Krapf, D. (2011). Ergodic and nonergodic processes coexist in the plasma membrane as observed by single-molecule tracking. *Proceedings of the National Academy of Sciences of the United States of America, 108*(16), 6438–6443.
Weigel, A. V., Tamkun, M. M., & Krapf, D. (2013). Quantifying the dynamic interactions between a clathrin-coated pit and cargo molecules. *Proceedings of the National Academy of Sciences of the United States of America, 110*(48), E4591–E4600.
Weiss, M. (2013). Single-particle tracking data reveal anticorrelated fractional Brownian motion in crowded fluids. *Physical Review E, 88*(1), 010101.

Weiss, M., Guigas, G., & Kalla, C. (2007). Probing the nanoscale viscoelasticity of intracellular fluids in living cells. *Biophysical Journal, 93*(1), 316–323.
Weiss, M., Hashimoto, H., & Nilsson, T. (2003). Anomalous protein diffusion in living cells as seen by fluorescence correlation spectroscopy. *Biophysical Journal, 84*(6), 4043 4052.
Winder, S. J., & Ayscough, K. R. (2005). Actin-binding proteins. *Journal of Cell Science, 118*(4), 651–654.
Wu, M. M., Covington, E. D., & Lewis, R. S. (2014). Single-molecule analysis of diffusion and trapping of STIM1 and Orai1 at ER-plasma membrane junctions. *Molecular Biology of the Cell, 25*(22), 3672–3685.

CHAPTER SIX

N-3 Polyunsaturated Fatty Acids, Lipid Microclusters, and Vitamin E

Saame Raza Shaikh[1,2,3,*], Stephen R. Wassall[4], David A. Brown[3,5] and Rasagna Kosaraju[1,3]

[1]Department of Biochemistry & Molecular Biology, East Carolina University, Greenville, NC, USA
[2]Department of Microbiology and Immunology, East Carolina University, Greenville, NC, USA
[3]East Carolina Diabetes and Obesity Institute, East Carolina University, Greenville, NC, USA
[4]Department of Physics, Indiana University-Purdue University Indianapolis, Indianapolis, IN, USA
[5]Department of Physiology, East Carolina University, Greenville, NC, USA
*Corresponding author: E-mail: SHAIKHSA@ecu.edu

Contents

Abstract

Increased consumption of long-chain marine n-3 polyunsaturated fatty acids (PUFA) has potential health benefits for the general population and for select clinical populations. However, several key limitations remain in making adequate dietary recommendations on n-3 PUFAs in addition to translating the fatty acids into clinical trials for select diseases. One major constraint is an incomplete understanding of the underlying mechanisms of action of n-3 PUFAs. In this review, we highlight studies to show n-3 PUFA acyl chains reorganize the molecular architecture of plasma membrane sphingolipid—cholesterol-enriched lipid rafts and potentially sphingolipid-rich cholesterol-free domains and cardiolipin-protein scaffolds in the inner mitochondrial

Current Topics in Membranes, Volume 75
ISSN 1063-5823
http://dx.doi.org/10.1016/bs.ctm.2015.03.003

membrane. We also discuss the possibility that the effects of n-3 PUFAs on membrane organization could be regulated by the presence of vitamin E (α-tocopherol), which is necessary to protect highly unsaturated acyl chains from oxidation. Finally, we propose the integrated hypothesis, based predominately on studies in lymphocytes, cancer cells, and model membranes, that the mechanism by which n-3 PUFAs disrupt signaling microclusters is highly dependent on the type of lipid species that incorporate n-3 PUFA acyl chains. The current evidence suggests that n-3 PUFA acyl chains disrupt lipid raft formation by incorporating primarily into phosphatidylethanolamines but can also incorporate into other lipid species of the lipidome.

1. OVERVIEW OF LONG-CHAIN N-3 POLYUNSATURATED FATTY ACIDS IN HEALTH AND DISEASE

The long-chain n-3 polyunsaturated fatty acids (PUFA) eicosapentaenoic (EPA, 20:5) and docosahexaenoic (DHA, 22:6) acids are essential fatty acids enriched in marine oils that are not highly abundant in the Western diet (Simopoulos, 2008). As a consequence, increasing dietary intake of these fatty acids may have potential benefits for a range of normal physiological processes such as infant development and for diseases including neurological disorders, cancers, behavioral disorders, cardiovascular disease, and chronic inflammatory ailments (Berger et al., 2013; Janssen & Kiliaan, 2014; Jump, Depner, & Tripathy, 2012; Miles & Calder, 2012; Nelson & Munger, 2013; Schwab et al., 2014; Shewchuk, 2014). Indeed, n-3 PUFAs are now formulated as prescription supplements (i.e., Vascepa, Lovaza) for the treatment of elevated serum triglycerides, which is a common complication associated with metabolic disorders such as obesity and insulin resistance (Nelson & Munger, 2013). However, several key limitations need to be addressed in order to make better dietary recommendations on n-3 PUFAs, particularly for specific disease populations. These include a limited understanding of the differences between various n-3 PUFAs, doses, duration of treatment, and underlying cellular and molecular mechanisms.

The objective of this review is to summarize data showing that n-3 PUFA acyl chains have a fundamental role in reorganizing membrane architecture. We first provide a brief general overview of n-3 PUFAs and then focus on particularly on how n-3 PUFAs could disrupt the physical organization of various lipid microclusters including plasma membrane sphingolipid—cholesterol-enriched lipid rafts, cholesterol-free sphingolipid domains, and potentially cardiolipin-protein scaffolds in mitochondrial membranes. We predominately use lymphocytes and cancer cells as examples given the wealth of information on how n-3 PUFAs modulate membrane activity in inflammatory diseases and cancer models. A key point from these studies is that n-3 PUFAs reorganize the membrane based on which lipid species the fatty acids are associated with, which requires further investigation in order to develop a comprehensive mechanism by which n-3 PUFA disrupt lipid microcluster organization. We also propose that n-3 PUFA acyl chains may exert their effects on membrane architecture in unison with interactions with vitamin E.

2. THE PLASMA MEMBRANE IS A MOLECULAR TARGET OF N-3 PUFAs

The molecular mechanisms by which long-chain marine n-3 PUFAs exert their biological effects across numerous cell types are highly pleiotropic and there is increasing evidence that EPA and DHA are not mechanistically equivalent (Calder, 2012, 2015; Mozaffarian & Wu, 2012). N-3 PUFAs, as nonesterified fatty acids, enter cells with the use of transporters, and undergo activation by select enzymes to become fatty acyl-CoAs. The fatty acyl-CoAs have multiples fates, which include esterification into membrane polar phospholipids and neutral lipids, as well as serving as an ATP source through oxidative pathways (Jump, 2002). Here we focus on how n-3 PUFA acyl chains, upon incorporation into membrane phospholipids, influence the formation of membrane microdomains.

Upon incorporation into membrane polar lipids, n-3 PUFA acyl chains of phospholipids disrupt the formation of cholesterol-dependent signaling microdomains known as lipid rafts and thereby downstream signaling (Shaikh, 2012). Lipid rafts are operationally defined as sphingolipid–cholesterol-enriched microdomains that compartmentalize signaling proteins, particularly in the plasma membrane (Lingwood & Simons, 2010). These highly packed domains are generally small nanoscale assemblies that can form more distinct micron scale domains upon receptor ligation. N-3 PUFAs may also influence signaling events by directly binding the G-protein coupled receptor (GPR) 120, which leads to anti-inflammatory effects in macrophages, which has been shown to be central toward regulating insulin signaling in obesity (Oh et al., 2010, 2014; Talukdar, Olefsky, & Osborn, 2011). N-3 PUFA acyl chains, upon enzymatic release from the membrane, also regulate eicosanoid production and can serve as precursors for the synthesis of lipid mediators known as resolvins, protectins, and maresins (Buckley, Gilroy, & Serhan, 2014; Malkowski et al., 2001; Serhan, 2014).

Downstream of the plasma membrane, n-3 PUFAs exert their effects by targeting transcription factors to modify gene expression. For instance, n-3 PUFAs have been shown to bind peroxisome proliferator-activated receptors (PPARs) and liver x receptors. Numerous studies illustrate that n-3 PUFAs activate PPARs, which leads to suppression of the production of pro-inflammatory cytokines in addition to stimulating fatty acid oxidation genes (Anderson et al., 2014). The stimulation of fatty acid oxidation is of particular interest for metabolic diseases such as obesity, cardiovascular disease, and type II diabetes.

3. UNIQUE STRUCTURAL FEATURES OF N-3 PUFA ACYL CHAINS

In order to establish how long-chain n-3 PUFAs influence the molecular organization of plasma membrane microdomains, it is essential to understand their unique physical properties at a structural level. The structures of EPA and DHA are depicted in Figure 1. The majority of research in the area has focused on DHA acyl chains, with less emphasis on the structural properties of EPA (Shaikh, Kinnun, Leng, Williams, & Wassall, 2015; Williams et al., 2012). Even less is known about other n-3 PUFA acyl chains such as stearidonic acid (18:4) or α-linolenic acid (18:3), which is abundant in many food products. Furthermore, most of our understanding of DHA's

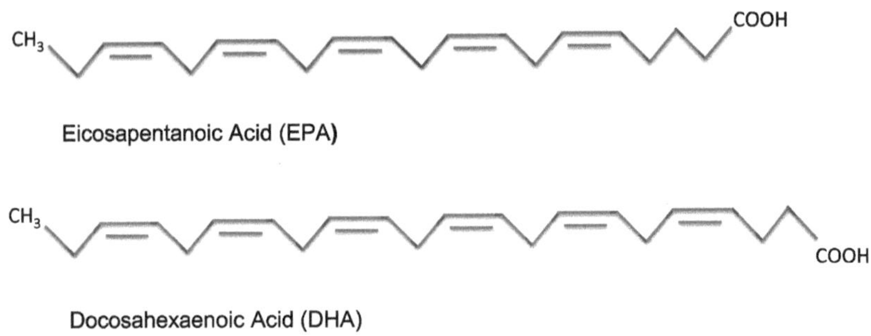

Figure 1 ***Molecular structures of the essential long-chain n-3 polyunsaturated fatty acids.*** Structures of eicosapentaenoic (EPA) and docosahexaenoic (DHA) acids that are the major bioactive fatty acids found in marine fish oils. The presence of multiple double bonds in n-3 PUFA acyl chains promotes a high degree of conformational freedom. This influences numerous membrane properties such as microviscosity, packing, and formation of signaling lipid microdomains.

structure comes from biophysical studies on model membranes by experimental methods including ^{2}H NMR and computational approaches such as molecular dynamics (MD) simulations. In these studies, DHA is typically esterified to phosphatidylcholine (PC) or phosphatidylethanolamine (PE) at the *sn*-2 position, which is where PUFAs primarily accumulate in membrane phospholipids, while saturated palmitic or stearic acid occupies the *sn*-1 position (Stillwell & Wassall, 2003).

Wassall and coworkers have reviewed the physical properties of DHA extensively (Stillwell & Wassall, 2003). The fundamental property that distinguishes n-3 PUFA from less unsaturated fatty acids is their tremendously high disorder. A repeating $=CH-CH_2-CH=$ unit that accompanies the multiple double bonds is the origin of the disorder. Because the energy barrier to rotation about C—C bonds within this recurring motif is shallow, an extremely flexible structure is produced (Feller, 2008). Indicative of low microviscosity, order parameters measured by solid state ^{2}H NMR throughout almost all of the DHA chain at the *sn*-2 position in 1-stearoyl-2-docosahexaenoylphosphatidylcholine are very small ($S_{CD} \approx 0.01$) and comparable to the highly mobile terminal methyl group (Gawrisch & Soubias, 2008). Order within the stearic acid chain at the *sn*-1 position, in turn, is substantially reduced by ~10% relative to monounsaturated 1-stearoyl-2-oleoylphosphatidylcholine with an oleic acid chain at the *sn*-2 position (Shaikh et al., 2015). The reduction is more marked in the lower portion of the saturated chain with the profile of order parameter along the chain retaining an overall signature shape in which there is a plateau region of slowly varying order in the top half followed by a drop-off in the bottom half (Holte, Peter, Sinnwell, & Gawrisch, 1995). Analysis of ^{13}C NMR relaxation times indicates that isomerization within the DHA chain is so rapid that the chain explores its entire conformational space in 10s of nanoseconds (Soubias & Gawrisch, 2007). The multitude of rapidly changing conformations adopted include, as visualized in MD simulations (Feller, Gawrisch, & MacKerell, 2001), bent configurations in which the terminal methyl group approaches the hydrophilic—hydrophobic interface which leads to a redistribution of the lower portion of DHA chains toward the interface (Eldho, Feller, Tristram-Nagle, Polozov, & Gawrisch, 2003; Mihailescu & Gawrisch, 2006).

The key point is that DHA acyl chains have a very large degree of conformational flexibility. They generally decrease membrane microviscosity, albeit at different magnitudes in differing cell types (Shaikh, 2012). The high disorder of DHA has also been shown to influence many other physical properties of bilayers such as reducing membrane thickness and increasing membrane compressibility, permeability, fusion, and phospholipid flip-flop (Wassall & Stillwell, 2008). Furthermore, of particular relevance to the lateral segregation of lipids into membrane domains, DHA has poor affinity for cholesterol. The rapid interconversion between torsional states undergone by PUFAs pushes the rigid steroid moiety away, reflected in measurements of low solubility (Shaikh et al., 2006) and binding coefficients (Niu & Litman, 2002; Williams, Wassall, Kemple, & Wassall, 2013) for cholesterol in polyunsaturated membranes. The PUFA-induced reduction in affinity is greater for PE than PC because, due to tighter packing associated with a smaller headgroup, sterol and PUFA are unfavorably brought closer together. It is the complete opposite of the situation for raft-forming sphingolipids that are predominantly saturated and adopt a largely linear conformation that is structurally better suited for close proximity to cholesterol. In the extreme case of bilayers comprised of phospholipids with a PUFA chain at both *sn*-1 and -2 positions, the sterol can even be forced to reside in an unusual location deep within the interior of the bilayer (Harroun, Katsaras, & Wassall, 2006, 2008). The aversion DHA has for cholesterol has led to numerous studies addressing its interactions with cholesterol-enriched microdomains, as briefly described below.

A series of experiments with model membranes address how sphingomyelin (SM) and cholesterol influenced the acyl chain and headgroup dynamics of phospholipids (either PC or PE) that had n-3 PUFA acyl chains esterified in the *sn*-2 position (Shaikh, Brzustowicz, Gustafson, Stillwell, & Wassall, 2002; Shaikh et al., 2004; Williams et al., 2012). Solid state ^{2}H NMR spectroscopy experiments, in combination with differential scanning calorimetry (DSC) and detergent extraction methods, showed that DHA remodels lipid raft-like domains that are nano-sized in a manner that is dependent upon the phospholipid headgroup. The majority of 1-palmitoyl-2-docosahexaenoylphosphatidylethanolamine (PDPE) was found to segregate into domains depleted in SM and cholesterol in 1:1:1 mol mixed membranes (Shaikh et al., 2004). Compared to a control system containing 1-palmitoyl-2-oleoylphosphatidylethanolamine (POPE), the introduction of cholesterol was seen to increase the order of PDPE ($\Delta\overline{S}_{CD} = 0.039$) by less than half compared to POPE ($\Delta\overline{S}_{CD} = 0.100$). These results were based on analyzing ^{2}H NMR spectra recorded for analogs of PE perdeuterated in the *sn*-1 chain (at 40 °C) (Shaikh et al., 2004). In detergent extraction experiments (conducted at 4 °C), more than 3× as much PDPE (~70%) as POPE (~22%) was isolated in detergent soluble ("nonraft") fractions while almost all of the cholesterol and SM (>90%) were isolated in detergent resistant ("raft") fractions.

More recent studies have revealed that the DHA-containing PC, 1-palmitoyl-2-docosahexaenoylphosphatidylcholine (PDPC) perdeuterated in the *sn*-1 chain, can directly infiltrate into SM-rich/cholesterol-rich domains based on an analysis of ^{2}H NMR spectra (Williams et al., 2012). A signal ascribed to PDPC in the more ordered raft-like domain was resolved that constituted most of the DHA-containing PC (~70%) in the mixed membrane. The greater affinity cholesterol has for PC versus PE offers an explanation. PC may also pack better with SM than PE, which due to its small headgroup is wedge shaped in conformation and has a tendency to form inverted hexagonal (H_{II}) phase that becomes accentuated with increasing levels of acyl chain unsaturation (Dekker, Geurts van Kessel, Klomp, Pieters, & De Kruijff, 1983). The sensitivity to details of lipid structure of the molecular interactions of n-3 PUFA with cholesterol-rich domains is further established by results obtained with 1:1:1 mol mixtures of 1-palmitoyl-2-eicosapentaenoylphospahtidylcholine (PEPC), SM, and cholesterol. On the basis of the intensity of signals in the ^{2}H NMR spectrum for a palmitoyl-perdeuterated analog of PEPC, it was determined that just 40% of the EPA-containing PC partitions into lipid raft-like domains. Altogether, these results suggest that DHA-containing PCs are more likely to incorporate into lipid rafts and disrupt their distribution compared to EPA-containing phospholipids (Williams et al., 2012).

4. PLASMA MEMBRANE LIPID—PROTEIN MICRODOMAIN DISTRIBUTION IS DISRUPTED BY N-3 PUFA ACYL CHAINS—EVIDENCE FROM BIOCHEMICAL AND IMAGING EXPERIMENTS

Studies in varying cell types (e.g., cancer cells, neuronal cells, hepatocytes, lymphocytes, etc.), have established that n-3 PUFA acyl chains are incorporated into detergent resistant (DRM) and detergent soluble membranes (DSM) (Briolay, Jaafar, Nemoz, & Bessueille, 2013; Fan, McMurray, Ly, & Chapkin, 2003; Li et al., 2005; Ravacci et al., 2013; Stulnig et al., 2001). DRMs and DSMs are isolated from sucrose gradient fractions of cells subjected to low levels of detergent at cold temperature and respectively are used to represent native rafts and nonrafts (Lingwood & Simons, 2007). Of course, these fractions are a very crude representation of membrane domains and do not represent native lipid rafts and nonrafts (Heerklotz, 2002; Lingwood & Simons, 2007). Therefore, our discussion of these fractions is only meant to provide some predictions about native membrane microdomains in cells. Nevertheless, this approach does still have proponents (Pathak & London, 2011) and has been used extensively to understand how n-3 PUFA influences the biochemical composition of lipid rafts (Shaikh, 2012).

The most common approach to studying n-3 PUFAs and lipid rafts with DRM and DSM fractions is to either treat immortal cells with EPA or DHA or to provide animals diets enriched in differing levels of EPA and DHA (Shaikh, 2012). As a consequence of the uptake of n-3 PUFAs into the differing membrane fractions, several key events take place. The most prevalent change is that arachidonic acid levels are lowered, which likely has an impact on raft and nonraft structure. Given that arachidonic acid itself is a disordered PUFA, its loss could make the membrane more packed, which has indeed been reported in a few studies but has not been thoroughly investigated (Rajamoorthi, Petrache, McIntosh, & Brown, 2005). Furthermore, the loss of arachidonic acid itself is highly relevant to disease pathology given that numerous lipid mediators are derived from this fatty acid and can have potent immunomodulatory properties (Kuhn, Banthiya, & van Leyen, in press).

A second consequence of n-3 PUFA uptake into the plasma membrane is that cholesterol molecules are displaced between DRM and DSM fractions. However, there is no clear consensus on which way cholesterol molecules move between DRMs and DSMs. For example, it is suggested that cholesterol levels decrease in the DRM fraction accompanied by an increase in the DSM fraction in response to DHA treatment (Grimm et al., 2011). A recent study also showed this effect in response to treatment with a diol of DHA in a neuronal cell line (Hu et al., 2014). Furthermore, characterization of DHA-treated MCF-7 breast cancer cells with Fourier transform infrared spectroscopy showed a decrease in both cholesterol and sphingolipids in the DRM fraction (Mereghetti et al., 2014).

There is also evidence to show that n-3 PUFAs can displace cholesterol from DSM to DRMs. Aliche-Djoudi et al. (2013) reported that treatment of hepatocytes with EPA promoted a displacement of cholesterol from nonraft-like to raft-like membrane fractions. In addition, further clouding the issue, n-3 PUFAs can selectively lower total cholesterol levels in just DRMs with no influence on the DSM fraction. For instance, cholesterol levels are lowered in DRM fractions of MDA-MB-231 breast cancer cells upon n-3 PUFA treatment, which may simply be driven by increased internalization of lipid rafts (Lee et al., 2014; Schley, Brindley, & Field, 2007). Overall, there is no clear consensus with biochemical studies on what n-3 PUFAs do to cholesterol lateral distribution or total levels, which is critical to establish in order to understand how lipid raft domains are modified with n-3 PUFAs. As described in sections below, one useful approach may be to understand how n-3 PUFA acyl chains in each major phospholipid of the lipidome interact with sphingolipids and cholesterol in order to establish a unified mode of action by which these fatty acids target the function of lipid microdomains.

Imaging studies provide more compelling evidence that n-3 PUFA acyl chains disrupt the molecular organization of lipid microdomains. Several studies have demonstrated that clustering of GM1 molecules is diminished in cell culture upon treatment with DHA, but strikingly not with EPA (Rockett et al., 2011, 2012; Shaikh, Rockett, Salameh, & Carraway, 2009). The clustering of GM1 molecules was induced by cholera toxin subunit b, an artificial means of inducing lipid rafts to form on a micron scale. Translational studies with splenic $B220^+$ B cells show that dietary administration of n-3 PUFAs to animals also diminishes GM1 clustering in a dose-dependent manner (Shaikh & Teague, 2012). Supporting polarization microscopy experiments to assess membrane packing, a key parameter of lipid raft organization, revealed that n-3 PUFAs enhanced membrane packing in both B and T lymphocytes. The increase in membrane packing with n-3 PUFAs was initially demonstrated by a study using fat-1 transgenic mice, which have enhanced levels of endogenous n-3 PUFAs due to the presence of a desaturase that endogenously converts n-6 to n-3 PUFAs (Kim, Barhoumi, McMurray, & Chapkin, 2013; Kim et al., 2008). In this study, it was shown that lipid raft accumulation in the immunological synapse, the organized junction between antigen-presenting cells and $CD4^+$ T cells, was enhanced with the fat-1 mouse relative to control mice (Kim et al., 2008). The consequence of disrupting the immunological synapse was suppressed helper $CD4^+$ Th1 cell activation, as measured by cytokine release and cellular proliferation. The suppression in helper Th1 activation with n-3 PUFAs has potential health benefits, at least in murine models, for the development of colitis and colon cancer (Jia et al., 2008).

Subsequent work in B cells and EL4 lymphomas also revealed a similar effect to $CD4^+$ T cells, where cross-linking of GM1 microdomains with cholera toxin subunit b enhanced membrane packing with dietary n-3 PUFAs, when compared to cells that were not cross-linked with cholera toxin (Rockett et al., 2012). More recent studies in a model of diet-induced obesity also revealed that EPA and DHA ethyl esters, modeling prescription supplements, differentially enhanced membrane packing upon formation of lipid microdomains, induced by lowering the temperature of cells to 4 °C (Teague et al., 2014). The changes in B cell membrane microdomain organization with n-3 PUFAs were associated with enhanced antibody production, which has implications for infection associated with obesity (Teague et al., 2013b, 2014).

Several unique concepts have emerged from imaging studies of lipid rafts. First, lipid raft spatial distribution and packing are clearly disrupted with n-3 PUFAs but the underlying mechanism is unknown. The results, when taking into consideration the biochemical measurements, suggest that n-3 PUFAs may be redistributing cholesterol in a manner that promotes the formation of lipid rafts. Second, it is conceivable that the effects of n-3 PUFAs are independent of cholesterol (Teague et al., 2013a). Furthermore, it is possible that EPA and DHA do not exert the same effects. This has been demonstrated in EL4 lymphomas in which DHA more effectively than EPA enhanced raft formation (Rockett et al., 2012). In contrast, a recent in vivo study showed that DHA had little effect on the accumulation of lipid rafts in the immunological synapse but raft packing was enhanced when both EPA and DHA were together in the diet (Kim et al., 2013). Clearly, more work is needed in this area to resolve differences between EPA and DHA. Furthermore, an area for future study about which little is currently known is how EPA's elongation product, docosapentaenoic acid (22:5), influences raft clustering and protein lateral organization.

Disrupting lipid raft microdomains with n-3 PUFAs has consequences for protein distribution and function. Again, in the case of B and T lymphocytes, key proteins are displaced upon a change in the molecular distribution of lipid microclusters, which can suppress the function of these cells. This has potential health benefits for diseases that are characterized by excessive inflammation such as rheumatoid arthritis, renal transplant, or metabolic diseases (Calder, 2015, Itariu et al., 2012; Sabbatini et al., 2013). In the case of B cells, major histocompatibility complex class (MHC) II molecules are displaced from raft-enriched regions of the immunological synapse, as demonstrated with confocal and acceptor photobleaching FRET studies (Rockett, Melton, Harris, Bridges, & Shaikh, 2013). Similar observations have also been made with MHC class I molecules (Shaikh et al., 2009). Analogously, in $CD4^+$ T cells, several key signaling molecules such as protein kinase C theta and phospholipase C gamma-1 were displaced from raft-like membranes, as demonstrated by biochemical DRM/DSM fractionations and confocal imaging (Fan, Ly, Barhoumi, McMurray, & Chapkin, 2004; Kim et al., 2008; McMurray, Ly, & Chapkin, 2003). Comparable results have been obtained with other proteins such as the toll-like receptor 4 and membrane-bound transducer glycoprotein 130 signaling in inflammatory diseases, which has downstream consequences for suppressing Akt and Erk signaling in cancer cells (Allen et al., 2014; Ravacci et al., 2013; Schley et al., 2007; Wong et al., 2009).

There remains a gap in knowledge on how the disruption in lipid rafts impacts protein molecular organization. This gap is largely driven by studies that have relied heavily on DRM/DSM isolation. More research, particularly at the nanometer scale with appropriate tools (i.e., STED, PALM imaging), is needed to understand how remodeling lipid rafts influences protein activity. Nevertheless, it is hypothesized that by disrupting lipid rafts with n-3 PUFAs, there is a direct consequence on downstream signaling and thereby gene activation, which ultimately influences health outcomes. For instance, n-3 PUFA mediated reorganization of lipid rafts can influence G-protein signaling cascades, which may improve clinical depression (Czysz & Rasenick, 2013). Similarly, n-3 PUFAs suppress or resolve inflammation in preclinical models of autoimmunity and metabolic disorders such as type II diabetes and obesity, potentially by targeting lipid raft organization (Kalupahana, Claycombe, & Moustaid-Moussa, 2011).

5. REMODELING OF LIPID MICRODOMAINS IS DEPENDENT ON THE INCORPORATION OF N-3 PUFAs INTO DIFFERENT PHOSPHOLIPIDS OF THE LIPIDOME—A POTENTIAL HYPOTHESIS TO EXPLAIN DISCREPANT DATA IN DIFFERING CELL TYPES

It is unknown how n-3 PUFAs promote a change in the distribution of lipid—protein microdomains. As described above, in some cases n-3 PUFAs promote cholesterol to move into DRMs whereas in other cell types cholesterol levels are lowered in DRMs or even displaced out of these crude membrane fractions. Thus, there is no clear consensus on how n-3 PUFAs are exerting their effects. Furthermore, developing a mechanism becomes more complicated when taking into account imaging studies that generally reveal that n-3 PUFAs are increasing the packing of lipid rafts, which leads to changes in protein localization and function (Kim et al., 2013). Thus, key questions emerge—Are n-3 PUFAs promoting the formation of lipid rafts in some cell types by incorporating into nonrafts and displacing cholesterol into rafts? Alternatively, are n-3 PUFAs fundamentally making nonraft regions of the membrane more raft-like by forcing cholesterol into nonrafts? An additional confounding variable to also consider when attempting to understand how n-3 PUFAs target cholesterol distribution in cells is the differences between the doses of fatty acids, the length of treatment, and the functional endpoints.

We propose that one approach toward developing a comprehensive mechanism by which n-3 fatty acids disrupt lipid rafts would be to establish how n-3 PUFA acyl chains associated with each major lipid species influences the formation of sphingolipid—cholesterol-enriched domains. Figure 2 presents potential models by which n-3 PUFAs could disrupt lipid spatial distribution. We integrate findings from model membranes with studies in cellular systems. The first possibility is that n-3 PUFAs may incorporate predominately into PCs, which are an abundant fraction of lipid rafts (Figure 2(A)). Thus, in this case, n-3 PUFAs may force cholesterol to move from rafts to nonrafts and make nonraft regions more raft-like. Although there is evidence that n-3 PUFAs incorporate into PCs in select cell types (Rockett et al., 2012), this is likely not the major mechanism by which n-3 PUFAs target lipid rafts given that in many of the cell types studied, n-3 PUFAs have a preference for incorporation into PEs (Fan et al., 2003). Thus, the second possibility is that n-3 PUFAs incorporate into PEs, which are generally not enriched in lipid rafts (Figure 2(B)). Indeed, as described above, DRM/DSM analysis in multiple cell types treated with n-3 PUFAs reveals that a significant fraction of n-3 PUFAs are localized into PEs. In this case, the n-3 PUFA acyl chains may force cholesterol to move into rafts and make them more packed. The hypothesis is supported by extensive solid state NMR and DSC experiments, described above, to show that n-3 PUFAs phase separate from liquid-ordered lipid raft-like domains when incorporated into PEs but incorporate directly into liquid-ordered rafts when esterified to PCs (Shaikh et al., 2004; Soni et al., 2008; Williams et al., 2012).

There are several other possibilities by which n-3 PUFAs may target lipid-protein spatial distribution in the plasma membrane. A third option is that n-3 PUFA acyl chains incorporate into sphingolipids (Figure 2(C)). This would directly modify the ability of lipid raft domains to pack and cluster (as well as influencing the formation of cholesterol-free sphingolipid domains, as discussed below). Overall, there is little evidence to show that sphingolipids can incorporate n-3 PUFA acyl chains. One recent in vivo

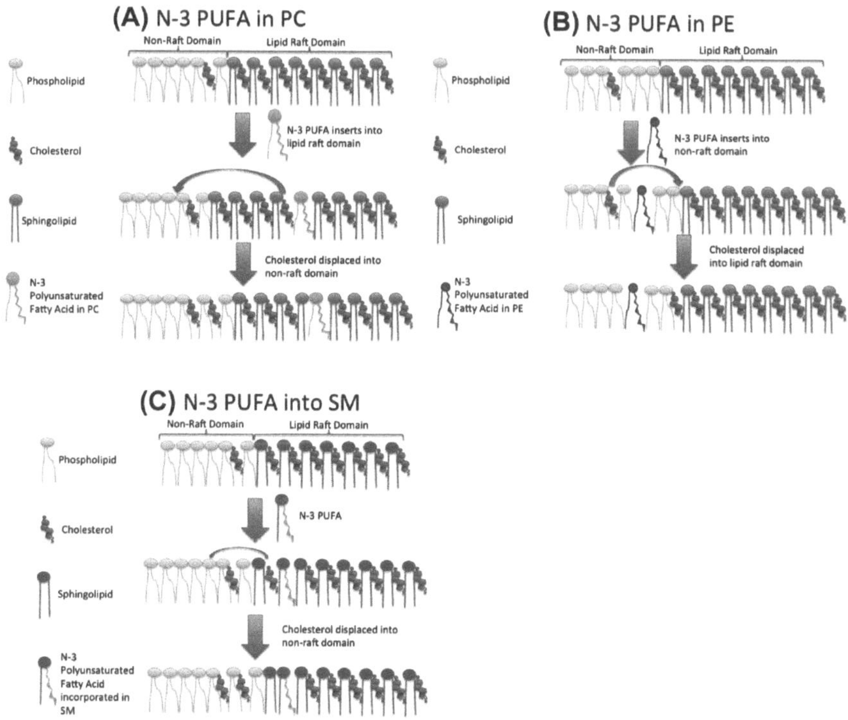

Figure 2 ***Potential mechanisms by which n-3 PUFAs could disrupt lipid microdomain distribution.*** Based on the existing literature, we propose that n-3 PUFA acyl chains esterified to (A) phosphatidylcholines (PCs) will incorporate largely into raft domains, which will displace cholesterol into nonrafts. In contrast, if n-3 PUFA acyl chains are esterified into (B) phosphatidylethanolamines (PEs), then these molecules will infiltrate nonrafts to promote cholesterol displacement into rafts. (C) There is some suggestion that n-3 PUFAs could incorporate as acyl chains of sphingomyelin (SM), which would prevent the appropriate packing of sphingolipid–cholesterol-enriched and sphingolipid–cholesterol-free domains. Finally, it is entirely plausible that n-3 PUFAs can esterify to cholesterol (not depicted), which could influence lipid raft packing but would have more pronounced effects on intracellular lipid pools (not shown). For simplicity, we only present a monolayer and proteins are not shown. The overall notion is that changes in lipid microcluster organization with n-3 PUFAs will fundamentally change the environment of lipid raft associated proteins and thereby downstream cellular function, potentially explaining some of the effects of the fatty acids on human health. (See color plate)

study demonstrated that a small fraction of n-3 PUFAs from dietary fish oil incorporated into sphingolipids of murine B cells (Rockett et al., 2012). These results were consistent with similar findings in cell culture with immortal T cells (Stulnig et al., 2001). A fourth possibility we propose is that n-3 PUFA acyl chains could become esterified to cholesterol to modify lipid raft packing. In this case, cholesterol esters with n-3 PUFA may prevent raft packing in an entirely different manner. Of course, cholesterol esters have very low membrane solubility and are generally localized to intracellular pools so this mechanism may be of less importance to the plasma membrane but we hypothesize that it could influence the organization of intracellular lipid pools. There is evidence to show that n-3 PUFAs can incorporate, albeit at low levels, into cholesterol esters (Rockett et al., 2012; Vidgren et al., 1997); however, there are no data available on how this may influence membrane molecular organization.

A key area for future investigation is to determine how n-3 PUFA acyl chains interact with raft domains when they are incorporated into other phospholipids. For instance, n-3 PUFA acyl chains are also incorporated into phosphatidylserines (PS) (Fan et al., 2003). Experiments with model membranes in this area would be of great value, which would inform us of how n-3 PUFA acyl chains of PS molecules interact with sphingolipids and cholesterol. Of course, these studies will also have to account for differences in lipid composition between the inner and outer leaflets of the plasma membrane. Furthermore, elucidating how incorporation of n-3 PUFA acyl chains into different lipid species is regulated will provide essential knowledge on how rafts are influenced by n-3 PUFAs. Finally, the differences between EPA and DHA on raft structure have yet to be ascertained with each of the major phospholipids of the mammalian plasma membrane. Select measurements in model membranes and cells show that DHA appears to be more effective than EPA in disrupting lipid raft organization (Rockett et al., 2012; Williams et al., 2012). This adds another layer of complexity toward understanding how n-3 PUFA acyl chains influence membrane organization.

6. VITAMIN E AS A REGULATOR OF N-3 PUFA MICRODOMAIN ORGANIZATION?

It is also important to consider other variables associated with n-3 PUFAs that may have a role in disrupting raft distribution. One recent idea is that n-3 PUFA acyl chains are not exerting their effects alone, but are doing so with vitamin E (α-tocopherol), the major lipid-soluble antioxidant that is found in low abundance in membranes (Atkinson, Epand, & Epand, 2008). This idea is highly relevant given that dietary formulations of fish oil often have vitamin E as an antioxidant (Valk & Hornstra, 2000). It has been proposed that vitamin E and polyunsaturated phospholipids in the bilayer may form distinct clusters (Atkinson, Harroun, Wassall, Stillwell, & Katsaras, 2010; Wassall & Stillwell, 2009), in some ways akin to cholesterol with sphingolipids in lipid rafts. Although vitamin E has been reported to reside with lipid rafts (Lemaire-Ewing, Desrumaux, Néel, & Lagrost, 2010), others have argued for its affinity for n-3 PUFA acyl chains (Atkinson et al., 2010; Wassall & Stillwell, 2009). The advantage of n-3 PUFA-vitamin E clusters could explain how low levels of the vitamin can protect relatively large amounts of n-3 PUFA from oxidation. The experimental evidence for these distinct domains is lacking and is an area for future investigation. Recent neutron scattering data confirm that the chromanol group on vitamin E, which is responsible for antioxidant activity, generally resides near the hydrophobic–hydrophilic interface of bilayers although a notable exception is DMPC where the vitamin sits at the center (Marquardt et al., 2014) (Marquardt et al., 2013). The location of the vitamin at the water interface is independent of the type of headgroup and acyl chains (Marquardt, Kučerka, Katsaras, & Harroun, in press). Nevertheless, the tremendous flexibility of n-3 PUFAs means that the entire chain regularly visits the hydrophilic–hydrophobic interface. It is intriguing to speculate that vitamin E may cluster with n-3 PUFA acyl chains to prevent oxidative damage and, in turn, this "cluster" could impact lipid raft organization (Atkinson et al., 2010). Overall, this is an area of investigation that is critical to address, which has broad implications for numerous diseases associated with oxidative stress in addition to ailments in which vitamin E levels are low and contribute toward disease pathology.

7. MOVING BEYOND LIPID RAFTS—THE POTENTIAL ROLE OF N-3 PUFAs IN REORGANIZING CHOLESTEROL-FREE SPHINGOLIPID DOMAINS, THE ACTIN CYTOSKELETON, AND MITOCHONDRIAL CARDIOLIPIN–PROTEIN SCAFFOLDING MICRODOMAINS

While there has been considerable work with n-3 PUFAs and sphingolipid/cholesterol-enriched lipid domains, there is far less information on how n-3 PUFAs remodel other microdomains and regulators of membrane domain organization. Recently, there is emerging evidence for the formation of sphingolipid-enriched cholesterol-free domains based on elegant studies using high resolution mass spectrometry imaging (Frisz, Klitzing, et al., 2013; Frisz, Lou, et al., 2013). These studies in cell culture reveal that labeled sphingolipids form distinct micron-sized membrane domains that are not disrupted upon cholesterol depletion but are sensitive to the disruption of the cytoskeleton. Therefore, the question arises if n-3 PUFAs can disrupt the formation of these domains. One mechanism by which n-3 PUFAs could disrupt cholesterol-free sphingolipid domains would be by incorporating directly into sphingolipids. As described above, evidence for this is limited but is an area for future investigation.

Two other areas of membrane research that are not well investigated with n-3 PUFAs are cytoskeletal dynamics and intracellular membrane domain formation and function. A few literature reports have emerged to show that n-3 PUFAs influence the cortical actin cytoskeleton, which is a key regulator of membrane microdomains (Hou et al., 2012; Yog, Barhoumi, McMurray, & Chapkin, 2010). For instance, Hou et al. (2012) demonstrated that helper T cells isolated from fat-1 transgenic mice or from mice fed DHA displayed a 50% reduction in phosphatidylinositol 4,5-bisphosphate, which contributed toward diminished actin remodeling and thereby T cell activation. An additional possibility is that n-3 PUFAs could target cardiolipin microdomain spatial distribution and function (Shaikh & Brown, 2013). Feeding studies in rats and humans have shown that DHA in particular remodels the abundance and acyl chain composition of mitochondrial phospholipids, which occurs at the expense of predominately arachidonic acid (Dabkowski et al., 2013; Herbst et al., 2014; Khairallah et al., 2010, 2012). There are vast functional consequences of remodeling the mitochondrial lipidome with n-3 PUFAs. For example, opening of the mitochondrial permeability transition pore in the rat myocardium has a pathological role in heart failure and EPA and DHA delay the opening of this pore in response to calcium in hypertrophied hearts (Khairallah et al., 2010).

It is unknown if n-3 PUFA can specifically disrupt the formation of multiprotein complexes that particularly depend on cardiolipin, and thereby influence mitophagy, apoptosis, and oxidative phosphorylation. As an example, proteins of the stomatin/prohibitin/flotillin/HflK/C family function as scaffolds in the inner mitochondrial membrane that regulates the formation of laterally distinct cardiolipin and/or phosphatidylethanolamine microdomains (Chicco & Sparagna, 2007; Christie et al., 2011; Renner & Weibel, 2011). These scaffolds and domains are critical for the morphology and function of mitochondria. How n-3 PUFAs may influence these structures is completely unknown. This is relevant from a mechanistic point of view given that n-3 PUFAs can improve select aspects of cardiovascular disease in which mitochondrial function is key in the etiology of the disease (Kromhout & de Goede, 2014).

8. CONCLUSIONS

Biochemical, spectroscopic, and imaging studies over the past decade have established that n-3 PUFA acyl chains remodel the architecture of plasma membrane sphingolipid–cholesterol-enriched microclusters in model membranes and in cells treated in cell culture or isolated from fish oil fed rodents. However, there remain considerable gaps in knowledge in the field of n-3 PUFAs and membrane microdomains. One major unknown is how EPA and DHA (and other n-3 PUFAs) influence lipid–protein clustering by targeting cholesterol lateral organization. Second, we need a better understanding of how n-3 PUFAs may indirectly influence lipid microdomains through disruptions in the actin cytoskeleton and the potential involvement of vitamin E as a regulator of n-3 PUFA oxidation. Finally, we propose that extensive lipidomic analysis coupled with studies in model membranes will significantly advance our understanding of how n-3 PUFAs influence lipid–protein microcluster formation in various cell types. Overall, there is a need for establishing specific mechanisms by which n-3 PUFA acyl chains disrupt lipid microclusters. This will aid our understanding in how these fatty acids influence normal physiological processes and will enhance our ability to design clinical trials for targeting select cell types in differing diseases.

REFERENCES

Aliche-Djoudi, F., Podechard, N., Collin, A., Chevanne, M., Provost, E., Poul, M., et al. (2013). A role for lipid rafts in the protection afforded by docosahexaenoic acid against ethanol toxicity in primary rat hepatocytes. *Food and Chemical Toxicology, 60*(0), 286–296.

Allen, M. J., Fan, Y.-Y., Monk, J. M., Hou, T. Y., Barhoumi, R., McMurray, D. N., et al. (2014). n–3 PUFAs reduce T-helper 17 cell differentiation by decreasing responsiveness to interleukin-6 in isolated mouse splenic CD4^{+} T cells. *Journal of Nutrition, 144*(8), 1306–1313.

Anderson, E. J., Thayne, K. A., Harris, M., Shaikh, S. R., Darden, T. M., Lark, D. S., et al. (2014). Do fish oil omega-3 fatty acids enhance antioxidant capacity and mitochondrial fatty acid oxidation in human atrial myocardium via PPARgamma activation? *Antioxidants & Redox Signaling, 21*(8), 1156–1163.

Atkinson, J., Epand, R. F., & Epand, R. M. (2008). Tocopherols and tocotrienols in membranes: a critical review. *Free Radical Biology & Medicine, 44*, 739–764.

Atkinson, J., Harroun, T., Wassall, S. R., Stillwell, W., & Katsaras, J. (2010). The location and behavior of α-tocopherol in membranes. *Molecular Nutrition & Food Research, 54*(5), 641–651.

Berger, M. M., Delodder, F., Liaudet, L., Tozzi, P., Schlaepfer, J., Chiolero, R. L., et al. (2013). Three short perioperative infusions of n-3 PUFAs reduce systemic inflammation induced by cardiopulmonary bypass surgery: a randomized controlled trial. *American Journal of Clinical Nutrition, 97*(2), 246–254.

Briolay, A., Jaafar, R., Nemoz, G., & Bessueille, L. (2013). Myogenic differentiation and lipid-raft composition of L6 skeletal muscle cells are modulated by PUFAs. *Biochimica et Biophysica Acta, 1828*(2), 602–613.

Buckley, C. D., Gilroy, D. W., & Serhan, C. N. (2014). Proresolving lipid mediators and mechanisms in the resolution of acute inflammation. *Immunity, 40*(3), 315–327.

Calder, P. C. (2012). Mechanisms of action of (n-3) fatty acids. *Journal of Nutrition, 142*(3), 592S–599S.

Calder, P. C. (2015). Marine omega-3 fatty acids and inflammatory processes: effects, mechanisms and clinical relevance. *Biochimica et Biophysica Acta, 1851*(4), 469–484.

Chicco, A. J., & Sparagna, G. C. (2007). Role of cardiolipin alterations in mitochondrial dysfunction and disease. *American Journal of Physiology — Cell PhysiologyJournal of Physiology, 292*(1), C33–C44.

Christie, D. A., Lemke, C. D., Elias, I. M., Chau, L. A., Kirchhof, M. G., Li, B., et al. (2011). Stomatin-like protein 2 binds cardiolipin and regulates mitochondrial biogenesis and function. *Molecular and Cellular Biology, 31*(18), 3845–3856.

Czysz, A. H., & Rasenick, M. M. (2013). G-protein signaling, lipid rafts and the possible sites of action for the antidepressant effects of n-3 polyunsaturated fatty acids. *CNS & Neurological Disorders: Drug Targets, 12*(4), 466–473.

Dabkowski, E. R., O'Connell, K. A., Xu, W., Ribeiro, R. F., Jr., Hecker, P. A., Shekar, K. C., & Stanley, W. C. (2013). Docosahexaenoic acid supplementation alters key properties of cardiac mitochondria and modestly attenuates development of left ventricular dysfunction in pressure overload-induced heart failure. *Cardiovascular Drugs and Therapy, 27*(6), 499–510.

Dekker, C. J., Geurts van Kessel, W. S. M., Klomp, J. P. G., Pieters, J., & De Kruijff, B. (1983). Synthesis and polymorphic phase behaviour of polyunsaturated phosphatidylcholines and phosphatidylethanolamines. *Chemistry and Physics of Lipids, 33*(1), 93–106.

Eldho, N. V., Feller, S. E., Tristram-Nagle, S., Polozov, I. V., & Gawrisch, K. (2003). Polyunsaturated docosahexaenoic vs docosapentaenoic acid-differences in lipid matrix properties from the loss of one double bond. *Journal of the American Chemical Society, 125*(21), 6409–6421.

Fan, Y.-Y., Ly, L. H., Barhoumi, R., McMurray, D. N., & Chapkin, R. S. (2004). Dietary docosahexaenoic acid suppresses T cell protein kinase C theta lipid raft recruitment and IL-2 production. *Journal of Immunology, 173*(10), 6151–6160.

Fan, Y.-Y., McMurray, D. N., Ly, L. H., & Chapkin, R. S. (2003). Dietary (n-3) polyunsaturated fatty acids remodel mouse T-cell lipid rafts. *Journal of Nutrition, 133*(6), 1913–1920.

Feller, S. E. (2008). Acyl chain conformations in phospholipid bilayers: a comparative study of docosahexaenoic acid and saturated fatty acids. *Chemistry and Physics of Lipids, 153*(1), 76–80.

Feller, S. E., Gawrisch, K., & MacKerell, A. D. (2001). Polyunsaturated fatty acids in lipid bilayers: intrinsic and environmental contributions to their unique physical properties. *Journal of the American Chemical Society, 124*, 318–326.

Frisz, J. F., Klitzing, H. A., Lou, K., Hutcheon, I. D., Weber, P. K., Zimmerberg, J., et al. (2013). Sphingolipid domains in the plasma membranes of fibroblasts are not enriched with cholesterol. *Journal of Biological Chemistry, 288*(23), 16855–16861.

Frisz, J. F., Lou, K., Klitzing, H. A., Hanafin, W. P., Lizunov, V., Wilson, R. L., et al. (2013). Direct chemical evidence for sphingolipid domains in the plasma membranes of

fibroblasts. *Proceedings of the National Academy of Sciences of the United States of America, 110*(8), E613–E622.
Gawrisch, K., & Soubias, O. (2008). Structure and dynamics of polyunsaturated hydrocarbon chains in lipid bilayers-significance for GPCR function. *Chemistry and Physics of Lipids, 153*, 64–75.
Grimm, M. O. W., Kuchenbecker, J., Grösgen, S., Burg, V. K., Hundsdörfer, B., Rothhaar, T. L., et al. (2011). Docosahexaenoic acid reduces amyloid β production via multiple pleiotropic mechanisms. *Journal of Biological Chemistry, 286*(16), 14028–14039.
Harroun, T. A., Katsaras, J., & Wassall, S. R. (2006). Cholesterol hydroxyl group is found to reside in the center of a polyunsaturated lipid membrane. *Biochemistry, 45*(4), 1227–1233.
Harroun, T. A., Katsaras, J., & Wassall, S. R. (2008). Cholesterol is found to reside in the center of a polyunsaturated lipid membrane. *Biochemistry, 47*(27), 7090–7096.
Heerklotz, H. (2002). Triton promotes domain formation in lipid raft mixtures. *Biophysical Journal, 83*(5), 2693–2701.
Herbst, E. A. F., Paglialunga, S., Gerling, C., Whitfield, J., Mukai, K., Chabowski, A., et al. (2014). Omega-3 supplementation alters mitochondrial membrane composition and respiration kinetics in human skeletal muscle. *Journal of Physiology, 592*(6), 1341–1352.
Holte, L. L., Peter, S. A., Sinnwell, T. M., & Gawrisch, K. (1995). 2H nuclear magnetic resonance order parameter profiles suggest a change of molecular shape for phosphatidylcholines containing a polyunsaturated acyl chain. *Biophysical Journal, 68*(6), 2396–2403.
Hou, T. Y., Monk, J. M., Fan, Y. Y., Barhoumi, R., Chen, Y. Q., Rivera, G. M., et al. (2012). n-3 polyunsaturated fatty acids suppress phosphatidylinositol 4,5-bisphosphate-dependent actin remodelling during $CD4^+$ T-cell activation. *Biochemical Journal, 443*(1), 27–37.
Hu, J., Popp, R., Frömel, T., Ehling, M., Awwad, K., Adams, R. H., et al. (2014). Müller glia cells regulate notch signaling and retinal angiogenesis via the generation of 19,20-dihydroxydocosapentaenoic acid. *Journal of Experimental Medicine, 211*(2), 281–295.
Itariu, B. K., Zeyda, M., Hochbrugger, E. E., Neuhofer, A., Schindler, K., Bohjalian, A., et al. (2012). Long-chain n-3 PUFAs reduce adipose tissue and systemic inflammation in severely obese nondiabetic patients: a randomized controlled trial. *American Journal of Clinical Nutrition, 96*(5), 1137–1149.
Janssen, C. I. F., & Kiliaan, A. J. (2014). Long-chain polyunsaturated fatty acids (LCPUFA) from genesis to senescence: the influence of LCPUFA on neural development, aging, and neurodegeneration. *Progress in Lipid Research, 53*(0), 1–17.
Jia, Q., Lupton, J. R., Smith, R., Weeks, B. R., Callaway, E., Davidson, L. A., et al. (2008). Reduced colitis-associated colon cancer in fat-1 (n-3 fatty acid desaturase) transgenic mice. *Cancer Research, 68*(10), 3985–3991.
Jump, D. B. (2002). The biochemistry of n-3 polyunsaturated fatty acids. *Journal of Biological Chemistry, 277*(11), 8755–8758.
Jump, D. B., Depner, C. M., & Tripathy, S. (2012). Omega-3 fatty acid supplementation and cardiovascular disease. *Journal of Lipid Research, 53*, 2525–2545.
Kalupahana, N. S., Claycombe, K. J., & Moustaid-Moussa, N. (2011). (n-3) Fatty acids alleviate adipose tissue inflammation and insulin resistance: mechanistic insights. *Advances in Nutrition, 2*(4), 304–316.
Khairallah, R. J., Kim, J., O'Shea, K. M., O'Connell, K. A., Brown, B. H., Galvao, T., et al. (2012). Improved mitochondrial function with diet induced increase in either docosahexaenoic acid or arachidonic acid in membrane phospholipids. *PLoS One*, 7(3), e34402.
Khairallah, R. J., O'Shea, K. M., Brown, B. H., Khanna, N., Des Rosiers, C., & Stanley, W. C. (2010). Treatment with docosahexaenoic acid, but not eicosapentaenoic acid, delays Ca^{2+}-induced mitochondria permeability transition in normal and

hypertrophied myocardium. *Journal of Pharmacology and Experimental Therapeutics, 335*(1), 155–162.

Kim, W., Barhoumi, R., McMurray, D. N., & Chapkin, R. S. (2013). Dietary fish oil and DHA down-regulate antigen-activated $CD4^+$ T-cells while promoting the formation of liquid-ordered mesodomains. *British Journal of Nutrition, 111*(2), 254–260.

Kim, W., Fan, Y. Y., Barhoumi, R., Smith, R., McMurray, D. N., & Chapkin, R. S. (2008). n-3 polyunsaturated fatty acids suppress the localization and activation of signaling proteins at the immunological synapse in murine $CD4^+$ T cells by affecting lipid raft formation. *Journal of Immunology, 181*(9), 6236–6243.

Kromhout, D., & de Goede, J. (2014). Update on cardiometabolic health effects of ω-3 fatty acids. *Current Opinion in Lipidology, 25*(1), 85–90.

Kuhn, H., Banthiya, S., & van Leyen, K. Mammalian lipoxygenases and their biological relevance. *Biochimica et Biophysica Acta,* in press.

Lee, E. J., Yun, U.-J., Koo, K. H., Sung, J. Y., Shim, J., Ye, S.-K., et al. (2014). Down-regulation of lipid raft-associated onco-proteins via cholesterol-dependent lipid raft internalization in docosahexaenoic acid-induced apoptosis. *Biochimica et Biophysica Acta, 1841*(1), 190–203.

Lemaire-Ewing, S., Desrumaux, C., Néel, D., & Lagrost, L. (2010). Vitamin E transport, membrane incorporation and cell metabolism: is α-tocopherol in lipid rafts an oar in the lifeboat? *Molecular Nutrition & Food Research, 54,* 631–640.

Li, Q., Wang, M., Tan, L., Wang, C., Ma, J., Li, N., & Li, J. (2005). Docosahexaenoic acid changes lipid composition and interleukin-2 receptor signaling in membrane rafts. *Journal of Lipid Research, 46*(9), 1904–1913.

Lingwood, D., & Simons, K. (2007). Detergent resistance as a tool in membrane research. *Nature Protocols, 2*(9), 2159–2165.

Lingwood, D., & Simons, K. (2010). Lipid rafts as a membrane-organizing principle. *Science, 327*(5961), 46–50.

Malkowski, M. G., Thuresson, E. D., Lakkides, K. M., Rieke, C. J., Micielli, R., Smith, W. L., et al. (2001). Structure of eicosapentaenoic and linoleic acids in the cyclooxygenase site of prostaglandin endoperoxide H synthase-1. *Journal of Biological Chemistry, 276*(40), 37547–37555.

Marquardt, D., Kučerka, N., Katsaras, J., & Harroun, T.A. α-Tocopherol's location in membranes is not affected by their composition. *Langmuir,* in press. http://dx.doi.org/10.1021/la502605c.

Marquardt, D., Williams, J. A., Kučerka, N., Atkinson, J., Wassall, S. R., Katsaras, J., et al. (2013). Tocopherol activity correlates with its location in a membrane: a new perspective on the antioxidant vitamin E. *Journal of the American Chemical Society, 135*(20), 7523–7533.

Marquardt, D., Williams, J. A., Kinnun, J. J., Kučerka, N., Atkinson, J., Wassall, S. R., et al. (2014). Dimyristoyl phosphatidylcholine: A remarkable exception to α-tocopherol's membrane presence. *Journal of the American Chemical Society, 136,* 203–210.

Mereghetti, P., Corsetto, P. A., Cremona, A., Rizzo, A. M., Doglia, S. M., & Ami, D. (2014). A fourier transform infrared spectroscopy study of cell membrane domain modifications induced by docosahexaenoic acid. *Biochimica et Biophysica Acta, 1840*(10), 3115–3122.

Mihailescu, M., & Gawrisch, K. (2006). The structure of polyunsaturated lipid bilayers important for rhodopsin function: a neutron diffraction study. *Biophysical Journal, 90,* L4–L6.

Miles, E. A., & Calder, P. C. (2012). Influence of marine n-3 polyunsaturated fatty acids on immune function and a systematic review of their effects on clinical outcomes in rheumatoid arthritis. *British Journal of Nutrition, 107*(Suppl. 2), S171–S184.

Mozaffarian, D., & Wu, J. H. Y. (2012). (n-3) Fatty acids and cardiovascular health: are effects of EPA and DHA shared or complementary? *Journal of Nutrition, 142*(3), 614S–625S.

Nelson, S. D., & Munger, M. A. (2013). Icosapent ethyl for treatment of elevated triglyceride levels. *Annals of Pharmacotherapy, 47*(11), 1517–1523.

Niu, S.-L., & Litman, B. J. (2002). Determination of membrane cholesterol partition coefficient using a lipid vesicle-cyclodextrin binary system: effect of phospholipid acyl chain unsaturation and headgroup composition. *Biophysical Journal, 83*, 3408–3415.

Oh, D. Y., Talukdar, S., Bae, E. J., Imamura, T., Morinaga, H., Fan, W., et al. (2010). GPR120 is an omega-3 fatty acid receptor mediating potent anti-inflammatory and insulin-sensitizing effects. *Cell, 142*(5), 687–698.

Oh, D. Y., Walenta, E., Akiyama, T. E., Lagakos, W. S., Lackey, D., Pessentheiner, A. R., et al. (2014). A GPR120-selective agonist improves insulin resistance and chronic inflammation in obese mice. *Nature Medicine, 20*(8), 942–947.

Pathak, P., & London, E. (2011). Measurement of lipid nanodomain (raft) formation and size in sphingomyelin/POPC/cholesterol vesicles shows TX-100 and transmembrane helices increase domain size by coalescing preexisting nanodomains but do not induce domain formation. *Biophysical Journal, 101*, 2417–2425.

Rajamoorthi, K., Petrache, H. I., McIntosh, T. J., & Brown, M. F. (2005). Packing and viscoelasticity of polyunsaturated omega-3 and omega-6 lipid bilayers as seen by (2)H NMR and X-ray diffraction. *Journal of the American Chemical Society, 127*(5), 1576–1588.

Ravacci, G. R., Brentani, M. M., Tortelli, T., Jr., Torrinhas, R. S. M. M., Saldanha, T., Torres, E. A. F. S., et al. (2013). Lipid raft disruption by docosahexaenoic acid induces apoptosis in transformed human mammary luminal epithelial cells harboring HER-2 overexpression. *Journal of Nutritional Biochemistry, 24*(3), 505–515.

Renner, L. D., & Weibel, D. B. (2011). Cardiolipin microdomains localize to negatively curved regions of *Escherichia coli* membranes. *Proceedings of the National Academy of Sciences of the United States of America, 108*(15), 6264–6269.

Rockett, B. D., Franklin, A., Harris, M., Teague, H., Rockett, A., & Shaikh, S. R. (2011). Membrane raft organization is more sensitive to disruption by (n-3) PUFA than nonraft organization in EL4 and B cells. *Journal of Nutrition, 141*(6), 1041–1048.

Rockett, B. D., Melton, M., Harris, M., Bridges, L. C., & Shaikh, S. R. (2013). Fish oil disrupts MHC class II lateral organization on the B-cell side of the immunological synapse independent of B-T cell adhesion. *Journal of Nutritional Biochemistry, 24*(11), 1810–1816.

Rockett, B. D., Teague, H., Harris, M., Melton, M., Williams, J., Wassall, S. R., et al. (2012). Fish oil increases raft size and membrane order of B cells accompanied by differential effects on function. *Journal of Lipid Research, 53*(4), 674–685.

Sabbatini, M., Apicella, L., Cataldi, M., Maresca, I., Nastasi, A., Vitale, S., et al. (2013). Effects of a diet rich in N-3 polyunsaturated fatty acids on systemic inflammation in renal transplant recipients. *Journal of the American College of Nutrition, 32*(6), 375–383.

Schley, P. D., Brindley, D. N., & Field, C. J. (2007). (n-3) PUFA alter raft lipid composition and decrease epidermal growth factor receptor levels in lipid rafts of human breast cancer cells. *Journal of Nutrition, 137*(3), 548–553.

Schwab, U., Lauritzen, L., Tholstrup, T., Haldorssoni, T., Riserus, U., Uusitupa, M., et al. (2014). Effect of the amount and type of dietary fat on cardiometabolic risk factors and risk of developing type 2 diabetes, cardiovascular diseases, and cancer: a systematic review. *Food and Nutrition Research, 58*.

Serhan, C. N. (2014). Pro-resolving lipid mediators are leads for resolution physiology. *Nature, 510*(7503), 92–101.

Shaikh, S. R. (2012). Biophysical and biochemical mechanisms by which dietary N-3 polyunsaturated fatty acids from fish oil disrupt membrane lipid rafts. *Journal of Nutritional Biochemistry, 23*(2), 101–105.

Shaikh, S. R., & Brown, D. A. (2013). Models of plasma membrane organization can be applied to mitochondrial membranes to target human health and disease with polyunsaturated fatty acids. *Prostaglandins Leukotrienes and Essential Fatty Acids, 88*(1), 21–25.

Shaikh, S. R., Brzustowicz, M. R., Gustafson, N., Stillwell, W., & Wassall, S. R. (2002). Monounsaturated PE does not phase-separate from the lipid raft molecules sphingomyelin and cholesterol: role for polyunsaturation? *Biochemistry, 41*(34), 10593–10602.

Shaikh, S. R., Cherezov, V., Caffrey, M., Soni, S. P., LoCascio, D., Stillwell, W., et al. (2006). Molecular organization of cholesterol in unsaturated phosphatidylethanolamines: X-ray diffraction and solid state 2H NMR reveal differences with phosphatidylcholines. *Journal of the American Chemical Society, 128*(16), 5375–5383.

Shaikh, S. R., Dumaual, A. C., Castillo, A., LoCascio, D., Siddiqui, R. A., Stillwell, W., et al. (2004). Oleic and docosahexaenoic acid differentially phase separate from lipid raft molecules: a comparative NMR, DSC, AFM, and detergent extraction study. *Biophysical Journal, 87*(3), 1752–1766.

Shaikh, S. R., Kinnun, J. J., Leng, X., Williams, J. A., & Wassall, S. R. (2015). How polyunsaturated fatty acids modify molecular organization in membranes: insight from NMR studies of model systems. *Biochimica et Biophysica Acta, 1848*, 211–219.

Shaikh, S. R., Rockett, B. D., Salameh, M., & Carraway, K. (2009). Docosahexaenoic acid modifies the clustering and size of lipid rafts and the lateral organization and surface expression of MHC class I of EL4 cells. *Journal of Nutrition, 139*(9), 1632–1639.

Shaikh, S. R., & Teague, H. (2012). N-3 fatty acids and membrane microdomains: from model membranes to lymphocyte function. *Prostaglandins Leukotrienes and Essential Fatty Acids, 87*(6), 205–208.

Shewchuk, B. M. (2014). Prostaglandins and n-3 polyunsaturated fatty acids in the regulation of the hypothalamic-pituitary axis. *Prostaglandins Leukotrienes and Essential Fatty Acids, 91*, 277–287.

Simopoulos, A. P. (2008). The importance of the omega-6/omega-3 fatty acid ratio in cardiovascular disease and other chronic diseases. *Experimental Biology and Medicine, 233*(6), 674–688.

Soni, S. P., LoCascio, D. S., Liu, Y., Williams, J. A., Bittman, R., Stillwell, et al. (2008). Docosahexaenoic acid enhances segregation of lipids between raft and non-raft domains: ^{2}H NMR study. *Biophysical Journal, 95*, 203–214.

Soubias, O., & Gawrisch, K. (2007). Docosahexaenoyl chains isomerize on the subnanosecond time scale. *Journal of the American Chemical Society, 129*, 6678–6679.

Stillwell, W., & Wassall, S. R. (2003). Docosahexaenoic acid: membrane properties of a unique fatty acid. *Chemistry and Physics of Lipids, 126*(1), 1–27.

Stulnig, T. M., Huber, J., Leitinger, N., Imre, E.-M., Angelisová, P., Nowotny, P., et al. (2001). Polyunsaturated eicosapentaenoic acid displaces proteins from membrane rafts by altering raft lipid composition. *Journal of Biological Chemistry, 276*(40), 37335–37340.

Talukdar, S., Olefsky, J. M., & Osborn, O. (2011). Targeting GPR120 and other fatty acid-sensing GPCRs ameliorates insulin resistance and inflammatory diseases. *Trends in Pharmacological Sciences, 32*(9), 543–550.

Teague, H., Ross, R., Harris, M., Mitchell, D. C., & Shaikh, S. R. (2013a). DHA-fluorescent probe is sensitive to membrane order and reveals molecular adaptation of DHA in ordered lipid microdomains. *Journal of Nutritional Biochemistry, 24*, 188–195.

Teague, H., Fhaner, C. J., Harris, M., Duriancik, D. M., Reid, G. E., & Shaikh, S. R. (2013b). .n-3 PUFAs enhance the frequency of murine B-cell subsets and restore the impairment of antibody production to a T-independent antigen in obesity. *Journal of Lipid Research, 54*(11), 3130–3138.

Teague, H., Harris, M., Fenton, J., Lallemand, P., Shewchuk, B., & Shaikh, S. R. (2014). Eicosapentaenoic and docosahexaenoic acid ethyl esters differentially enhance B-cell activity in murine obesity. *Journal of Lipid Research, 55*(7), 1420–1433.

Valk, E. E. J., & Hornstra, G. (2000). Relationship between vitamin E requirement and polyunsaturated fatty acid intake in man: a review. *International Journal for Vitamin and Nutrition Research, 70*, 31–42.

Vidgren, H. M., Agren, J. J., Schwab, U., Rissanen, T., Hänninen, O., & Uusitupa, M. I. (1997). Incorporation of n-3 fatty acids into plasma lipid fractions, and erythrocyte membranes and platelets during dietary supplementation with fish, fish oil, and docosahexaenoic acid-rich oil among healthy young men. *Lipids, 32*(7), 697–705.

Wassall, S. R., & Stillwell, W. (2008). Docosahexaenoic acid domains: the ultimate non-raft membrane domain. *Chemistry and Physics of Lipids, 153*(1), 57–63.

Wassall, S. R., & Stillwell, W. (2009). Polyunsaturated fatty acid-cholesterol interactions: domain formation in membranes. *Biochimica et Biophysica Acta, 1788*, 24–32.

Williams, J. A., Batten, S. E., Harris, M., Rockett, B. D., Shaikh, S. R., Stillwell, W., et al. (2012). Docosahexaenoic and eicosapentaenoic acids segregate differently between raft and nonraft domains. *Biophysical Journal, 103*(2), 228–237.

Williams, J. A., Wassall, C. D., Kemple, M. D., & Wassall, S. R. (2013). An electron paramagnetic resonance method for measuring the affinity of a spin-labeled analog of cholesterol for phospholipids. *Journal of Membrane Biology, 246*, 689–696.

Wong, S. W., Kwon, M.-J., Choi, A. M. K., Kim, H.-P., Nakahira, K., & Hwang, D. H. (2009). Fatty acids modulate toll-like receptor 4 activation through regulation of receptor dimerization and recruitment into lipid rafts in a reactive oxygen species-dependent manner. *Journal of Biological Chemistry, 284*(40), 27384–27392.

Yog, R., Barhoumi, R., McMurray, D. N., & Chapkin, R. S. (2010). n-3 polyunsaturated fatty acids suppress mitochondrial translocation to the immunologic synapse and modulate calcium signaling in T cells. *Journal of Immunology, 184*(10), 5865–5873.

CHAPTER SEVEN

Raft-Like Membrane Domains in Pathogenic Microorganisms

Amir M. Farnoud[1], Alvaro M. Toledo[1], James B. Konopka[1], Maurizio Del Poeta[1] and Erwin London[2,*]
[1]Department of Molecular Genetics and Microbiology, Stony Brook University, Stony Brook, NY, USA
[2]Department of Biochemistry and Cell Biology, Stony Brook University, Stony Brook, NY, USA
*Corresponding author: E-mail: erwin.london@stonybrook.edu

Contents

Abstract

The lipid bilayer of the plasma membrane is thought to be compartmentalized by the presence of lipid-protein microdomains. In eukaryotic cells, microdomains composed of sterols and sphingolipids, commonly known as lipid rafts, are believed to exist, and reports on the presence of sterol- or protein-mediated microdomains in bacterial cell membranes are also appearing. Despite increasing attention, little is known about microdomains in the plasma membrane of pathogenic microorganisms. This review attempts to provide an overview of the current state of knowledge of lipid rafts in pathogenic fungi and bacteria. The current literature on characterization of microdomains in pathogens is reviewed, and their potential role in growth, pathogenesis, and drug resistance is discussed. Better insight into the structure and function of membrane

Current Topics in Membranes, Volume 75
ISSN 1063-5823
http://dx.doi.org/10.1016/bs.ctm.2015.03.005

microdomains in pathogenic microorganisms might lead to a better understanding of their pathogenesis and development of raft-mediated approaches for therapy.

1. INTRODUCTION

Through the 1970s, the dominant view of the physical state of the cell membrane was that the membrane is a "fluid-mosaic," in which the proteins are randomly distributed (Singer & Nicolson, 1972). In the late 1960s and early 1970s, in vitro studies with lipid monolayers and bilayers showed that long-chain, saturated lipids pack favorably with sterols to form a compact state (Chapman, Owens, Phillips, & Walker, 1969; Lee, 1977; Papahadjopoulos, Jacobson, Nir, & Isac, 1973). These observations gradually raised questions about the physical state of the plasma membrane leading to suggestions that natural membranes could have a heterogenous lipid bilayer (Karnovsky, Kleinfeld, Hoover, & Klausner, 1982; Linden & Fox, 1975; Thompson & Tillack, 1985). Since then, evidence has accumulated indicating that ordered sphingolipid/sterol-rich domains (i.e., lipid rafts) that coexist with disordered fluid phases exist in cell membranes, and play an important role in cellular functions and in lipid—protein and protein—protein interactions by selective inclusion or exclusion of proteins in specific domains (Ahmed, Brown, & London, 1997; Brown & London, 1997; Brown & Rose, 1992; Fastenberg, Shogomori, Xu, Brown, & London, 2003; Field, Holowka, & Baird, 1999; Harder, Scheiffele, Verkade, & Simons, 1998; Levental, Grzybek, & Simons, 2010; Lin & London, 2013; Schroeder, London, & Brown, 1994; Simons & Ikonen, 1997; Simons & Van Meer, 1988; Skibbens, Roth, & Matlin, 1989).

Recent years have seen an increasing interest in the studies of lipid rafts in the function of different cells. Several recent reviews have discussed the importance of lipid rafts in various cell functions such as signal transduction, endocytosis, ion homeostasis, and cell growth, among others (Hicks, Nalivaeva, & Turner, 2012; Mollinedo, 2012; Yaqoob & Shaikh, 2010). However, the majority of studies have been focused on mammalian cells or model microorganisms such as *Saccharomyces cerevisiae* (Bagnat, Keränen, Shevchenko, Shevchenko, & Simons, 2000; Bagnat & Simons, 2002a). With lipid rafts being suggested to play a role in many important cellular functions, an understanding of their functions in pathogenic microorganisms could help better define their role in cell function and pathogenicity, and potentially pave the way for new therapeutic strategies. To this aim, the current chapter focuses on the role of membrane microdomains in pathogenic bacteria and fungi, and provides an overview of microdomain composition, function, and role in pathogenicity.

2. WHAT ARE MEMBRANE DOMAINS MADE OF?

Lipid phase segregation is the central element in the formation of lipid microdomains. Thus, many pioneering studies focused on lipid phase behavior in an attempt to understand the physical state of lipid domains in the plasma membrane. Early studies with synthetic bilayers of phospholipids using spin-labels and freeze fracture microscopy revealed the presence of lateral phase separations in the bilayer with areas of reduced fluidity in contact with more fluid areas (Grant, Wu, & McConnell, 1974; Kleemann, Grant, & McConnell, 1974). The solid-like areas of reduced fluidity were named the gel phase (Lβ), in which the lipid acyl chains are tightly packed and there is a low rate of lateral diffusion. The name solid-ordered (So) is also sometimes used for this state. In contrast, the more fluid areas were named the liquid crystalline (Lα) phase (Shimshick & McConnell, 1973; Wu & McConnell, 1975). This state is now more commonly called the liquid-disordered (Ld) state. The phase behavior of lipids is temperature-dependent and the Lβ phase transitions into Ld phase when the temperature is increased. The temperature at which this transition occurs is known as "gel to liquid transition temperature (Tm)" and depends on the structure of the lipid. The dependence of Tm on the acyl chains of lipids is well-characterized. Lipids with longer saturated fatty acyl chains (e.g., most sphingolipids) have high Tm, whereas lipids with fatty acids having cis double bonds (e.g., most cellular phospholipids) have low Tm.

A number of studies using lipid monolayers and bilayers showed that the presence of cholesterol can significantly affect the phase behavior of lipids (Chapman et al., 1969; Ladbrooke, Williams, & Chapman, 1968; Recktenwald & McConnell, 1981; Vist & Davis, 1990). Notably, high concentrations of cholesterol lead to the emergence of a lipid phase that is characterized by the tight packing of the Lβ phase and at the same time a relatively high lateral diffusion close to the Ld phase, leading to the name "liquid ordered (Lo)" phase (Ipsen, Karlström, Mourtisen, Wennerström, & Zuckermann, 1987). The Lo phase was observed in lipid bilayers which contained cholesterol and high Tm phospholipids or sphingolipids, and was suggested to be due to the favorable packing of cholesterol with these lipids (Sankaram & Thompson, 1990), an observation that was confirmed by numerous later studies (Ahmed et al., 1997; Almeida, Vaz, & Thompson, 1993; Schroeder, Ahmed, Zhu, London, & Brown, 1998).

The discovery of the cholesterol-mediated Lo phase led to numerous studies focused on the phase behavior of ternary lipid mixtures (consisting of a sterol, a high Tm lipid, and a low Tm lipid) as a simplistic model of a eukaryotic cell membrane (Ahmed et al., 1997; de Almeida, Fedorov, & Prieto, 2003; Feigenson & Buboltz, 2001; Schroeder et al., 1994; Silvius, del Giudice, & Lafleur, 1996; Veatch & Keller, 2003; Yuan, Furlong, Burgos, & Johnston, 2002; Zhao et al., 2007). Most such studies have used a three-component phase diagram to represent the phase behavior of ternary lipid mixtures. A general three-component phase diagram is shown in Figure 1. In this diagram, each vertex represents a pure lipid component and each side represents a binary mixture. Points inside the triangle refer to ternary mixtures, in which the concentration of the high Tm lipid and cholesterol increases from left to right and bottom to top, respectively.

In discussing three-phase diagrams, it should be borne in mind that apart from lipid composition, phases in such diagrams are highly dependent on temperature (de Almeida et al., 2003). The diagram in Figure 1 presents a situation in which the temperature is lower than the phase transition temperature of the high Tm lipid and higher than the transition temperature of the low Tm lipid. Such conditions are likely true for most mammalian

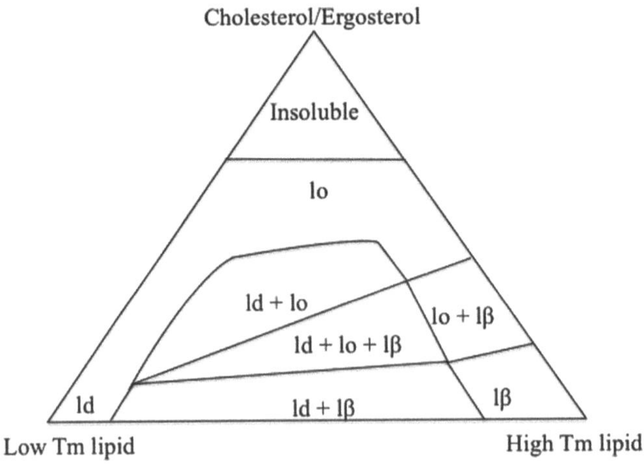

Figure 1 A hypothetical three-component phase diagram showing the interactions between a high Tm lipid, a low Tm lipid, and a sterol. Ld, liquid-disordered phase; lo, liquid-ordered phase (lipid raft); lβ, solid-ordered phase (gel).

cells at physiological temperature. Sphingomyelin, a saturated sphingolipid abundant in mammalian cells has a transition temperature of 35–45 °C (Calhoun & Shipley, 1979), whereas unsaturated phospholipids abundant in mammalian cells have Tm values of less than 0 °C (Koynova & Caffrey, 1998). The first and most important conclusion from a three-phase diagram is that multiple lipid phases may exist at the same time in a mixture of lipids (Figure 1). In conditions of interest for biological membranes where high Tm lipids and cholesterol are lower than 50% of the lipid content (Lange, Swaisgood, Ramos, & Steck, 1989), these coexisting phases include the Ld phase in the presence of Lo or So phases. This observation indicates that ordered and disordered phases can coexist in biologically relevant lipid ratios. In such conditions, an increase in cholesterol concentrations translates to an increase in lipid order and often favors the presence of more ordered phases. Note that the cholesterol solubility limit in binary mixtures with phospholipids has been reported to be between 51% and 66% with higher concentrations leading to crystallization (Huang, Buboltz, & Feigenson, 1999). Such conditions are usually physiologically irrelevant and are shown as "insoluble" in Figure 1.

Evidence from phase diagrams suggests that if ordered lipid phases were to exist in biological membranes, they would be rich in sterols and high Tm lipids as each of these compounds promote ordered phase formation. The outer leaflet of the plasma membrane of mammalian cells is rich in high Tm sphingolipids and cholesterol (Forstner, Tanaka, & Isselbacher, 1968; Ray, Skipski, Barclay, Essner, & Archibald, 1969), and thus would be the most likely site for such domains. The observation that glycosphingolipids have very high Tm suggested that they could form ordered domains in membranes (Thompson & Tillack, 1985). Interestingly, glycophosphatidylinositol (GPI)-anchored proteins were shown to be specifically localized on the apical surface of polarized epithelial cells (Lisanti, Sargiacomo, Graeve, Saltiel, & Rodriguez-Boulan, 1988), which are rich in glycosphingolipids, suggesting preferential segregation/sorting of GPI-anchored proteins in glycosphingolipid domains (Simons & Van Meer, 1988).

Other pioneering studies supporting the hypothesis that lipid domains might exist in the plasma membrane utilized an interesting characteristic of ordered lipid phases: detergent insolubility (explained in detail in Section 3.2.1). Lipid mixtures that are in the Lo phase do not solubilize in cold, nonionic detergents (depending to some degree on lipid-detergent ratio) and can be isolated from cells (Brown & Rose, 1992; London & Brown, 2000). Isolation of the detergent insoluble fractions from mammalian cell membranes revealed that GPI-anchored proteins were associated with detergent insoluble lipids, which were largely composed of sphingolipids and cholesterol (Brown & Rose, 1992). In vitro studies demonstrated that detergent insolubility of GPI-anchored proteins depends on the surrounding lipid medium. Purified GPI-anchored protein, placental alkaline phosphatase, showed high detergent solubility when incorporated in liposomes composed of low Tm phospholipids and cholesterol, but low solubility in liposomes composed of high Tm sphingolipids and cholesterol (Schroeder et al., 1994). This observation implied that detergent insolubility of cellular GPI-anchored proteins is due to their association with ordered lipid domains. In addition, it was observed that detergent-insolubility of lipid vesicles was associated with the formation of an Lo-like state. Further studies showed that detergent insolubility in model membranes composed of sphingomyelin, unsaturated phosphatidylcholine (PC), and cholesterol was only present under conditions when Lo domains could be detected spectroscopically (Ahmed et al., 1997). Taken together these studies suggested that sphingolipid- and cholesterol-rich domains form lateral platforms at the plasma membrane, which are able to preferentially segregate certain proteins. Sphingolipid- and cholesterol-rich membrane platforms were named lipid rafts (Simons & Ikonen, 1997). Although the details remain controversial, it is now widely accepted that lipid rafts exist in the plasma membrane of mammalian cells, and play a role in various cell functions mainly by affecting protein distribution within the membrane (Babina, Donatello, Nabi, & Hopkins, 2011; Brown, 2006; George & Wu, 2012; Grassme et al., 2003; Simons & Toomre, 2000; Suzuki, 2012; Triantafilou, Miyake, Golenbock, & Triantafilou, 2002).

In contrast to mammalian cells, membrane microdomains are highly understudied in bacteria and fungi. This is partly due to the smaller cell size of these microorganisms, differences in lipid compositions compared to mammalian cells, and diversity in morphology and growth conditions. The remainder of this chapter focuses on membrane microdomains in pathogenic fungi and bacteria with a view to provide insight on methods of study, microdomain composition, and potential roles in pathogenesis. It should be noted that studies on nonpathogenic microorganisms, most notably the yeast *Saccharomyces cerevisiae* (Bagnat, Chang, & Simons, 2001; Bagnat et al., 2000; Bagnat & Simons, 2002b; Klose et al., 2010), have been invaluable in understanding the role of rafts in cellular functions. However, such studies are not the focus of this chapter and the interested reader is referred to the pertinent literature for more information on those topics.

3. MEMBRANE DOMAINS IN PATHOGENIC FUNGI

3.1 Raft-forming lipids in fungi

Due to their different lipid composition, the sterols and saturated lipids that can form the Lo phase in a fungal plasma membrane are different than those in mammalian cells. The primary sterol in fungal cells is ergosterol, not cholesterol. Also, the sphingolipids in fungi are generally glycosphingolipids (i.e., sphingolipids connected to sugar molecules), not sphingomyelin. Glycosphingolipids in fungi include glucosylceramide (GlcCer), inositolphosphorylceramide (IPC), mannosyl-inositolphosphoylceramide (MIPC), and mannosyl-diinositolphosphorylceramide (MIP2C) (Bagnat et al., 2000; Rittershaus et al., 2006) (inositol is not a sugar, but does have a carbohydrate-like structure). These sphingolipids differ significantly from the glycosphingolipids of mammalian cells.

Sterols and saturated sphingolipids in fungi are better raft formers compared to their mammalian counterparts based on in vitro studies. Using fluorescence quenching experiments, Xu and colleagues (Xu et al., 2001) showed that mixtures of ergosterol with various saturated lipids form more stable lipid domains compared to mixtures of cholesterol with the same lipids. This was likely due to the presence of an additional double bond in the B ring, which led to better sterol-saturated lipid packing. Glycosphingolipids are also better ordered domain formers than nonglycosylated sphingolipids, i.e., they form gel states that melt at higher temperatures than sphingomyelins. This is likely due to their long acyl chains and the structure of their headgroups, which allows for extensive hydrogen bonding due to the presence of many hydroxyl groups (Redfern & Gericke, 2004). IPC has been shown to have higher Tm compared to acyl-chain matched sphingomyelin consistent with a role of the headgroup structure in promoting raft formation (Björkbom et al., 2010). In addition, IPC isolated from the yeast *S. cerevisiae*, has been shown to form ordered domains in mixtures with ergosterol and phosphatidylinositol (Klose et al., 2010). The strong raft-promoting properties of ergosterol (Xu et al., 2001) and fungal glycosphingolipids do not necessarily mean that all fungal cells contain lipid rafts, as the degree of unsaturation in unsaturated phospholipids and sterol-sphingolipid-unsaturated lipid ratios play a significant role in raft formation, and is different for various cell types. However, in vitro observations reveal that fungal cells possess the compositional requirements for the formation of lipid rafts.

3.2 Methods to study membrane domains in fungi

Except for the model yeast *S. cerevisiae*, a limited number of methods have been used to study membrane domains in pathogenic fungi. Isolation of detergent resistant membranes (DRMs), staining, and extraction/substitution of sterols are the main methods used in fungal cells, which are explained in detail below. Several other methods can potentially be used for domain studies in pathogenic fungi; however, this section is focused on methods that have been used to detect membrane domains in pathogenic fungi and other methods that might be applicable are not discussed.

3.2.1 Isolation of detergent resistant membranes

The most widely used method for characterization of lipid rafts in cells is the isolation of DRMs. Due to their tight packing, lipids in the gel (Dennis, 1974) and liquid-ordered phase (Schroeder et al., 1994) do not solubilize in ice-cold nonionic detergents such as Triton X-100 (TX-100). Exposure to TX-100 followed by density centrifugation will result in floatation of insoluble lipids known as DRMs (Brown & Rose, 1992; London & Brown, 2000). DRMs have been generally used as a surrogate for lipid rafts and their lipid and protein composition has been characterized to provide information on raft composition. A number of proteins such as Plasma membrane ATPase 1 (Pma1p) and Glycosylphosphatidylinositol-anchored surface protein 1 (Gas1p), have been shown to be concentrated in the DRMs of *S. cerevisiae* (Bagnat et al., 2000) and have been typically used as a marker for DRMs in other fungi.

DRM isolation is a powerful and widely used tool for biochemical studies of cell membranes. However, DRMs are not equal to lipid rafts and should not be identified as such. DRM isolation is traditionally performed by exposure of cell lysates to ice-cold detergents (Lingwood & Simons, 2007). Since low temperatures increase lipid order, this procedure might result in the expansion of preexisting lipid rafts and overestimation of the amount of rafts in a cell membrane. Enlargement of rafts can also lead to the recruitment of proteins that otherwise would not be raft-localized (Brown, 2006; London & Brown, 2000). Furthermore, different detergents have been shown to result in DRMs that have different protein compositions suggesting that DRMs should not be identified with preexisting lipid rafts in the cell membrane (Schuck, Honsho, Ekroos, Shevchenko, & Simons, 2003). Thus, the information provided by DRM isolation should be interpreted with caution and validated by another method whenever possible.

3.2.2 Staining and light microscopy

The polyene antibiotic filipin has been the molecule of choice for sterol visualization in fungi. Filipin is a sterol-binding molecule, which has a structure similar to amphotericin B (Loura, Castanho, Fedorov, & Prieto, 2001). Filipin is fluorescent and allows for visualization of sterols using fluorescence microscopy. Since sterols are a crucial component of lipid rafts, filipin fluorescence can be used to detect sterol-rich areas in the cell, which might correspond to areas in which lipid rafts are present. A problem in working with filipin is that it is very sensitive to photobleaching. Also, filipin can disrupt membrane architecture at high concentrations or long incubation times and care should be taken in using filipin for imaging (Smejkal & Hoff, 1994; Takeshita, Higashitsuji, Konzack, & Fischer, 2008).

Sterol rich domains (SRDs) observed in cells using light microscopy should not necessarily be equated with lipid raft domains. Sterol rich regions can be as large as a few microns, whereas rafts are believed to be most frequently nano-sized (London, 2005; Pike, 2009); thus, SRDs could likely represent a region containing a high concentration of lipid rafts. However, given the metabolic activity of cells, sterol rich regions could represent membrane regions to which a high level of sterols have been delivered, and not sphingolipid enriched liquid-ordered domains that are at equilibrium. It should be noted that in addition to filipin, by staining nonraft regions, the dye *FAST* diI (with affinity toward Ld phases) has also been used for detection of large lipid rafts in *S. cerevisiae* (Toulmay & Prinz, 2013). However, to the best of our knowledge, its use has been limited to nonpathogenic fungi.

3.2.3 Sterol depletion/substitution

The use of chemical compounds to deplete sterols has been a widely used method to study lipid rafts. Since lipid rafts in eukaryotes are made of sterols and sphingolipids, depletion of sterols should often lead to raft disruption allowing for studies of their role in cellular processes. The cyclic oligosaccharide, methyl-beta cyclodextrin (MβCD), has been widely used to extract cholesterol from mammalian and in some cases fungal cells (Siafakas, Wright, Sorrell, & Djordjevic, 2006; Zidovetzki & Levitan, 2007). MβCD is water soluble, but has a hydrophobic core, which can efficiently bind sterols. Much evidence confirms that cholesterol depletion by MβCD alters lipid packing in eukaryotic cells. For example, cholesterol depletion in human aortic endothelial cells has been shown to displace the raft-localized lipid monosialotetrahexosyl ganglioside (GM1) from the plasma membrane to the intercellular space, possibly due to raft disruption (Byfield, Tikku, Rothblat, Gooch, & Levitan, 2006). In addition, studies with a fluorescent probe based on Nile Red (NR12S) (Kucherak et al., 2010) have shown that cholesterol extraction with MβCD leads to a reduction in the membrane order of the outer leaflet of the plasma membrane in HeLa cells (Darwich, Klymchenko, Kucherak, Richert, & Mély, 2012).

Cyclodextrins are sterol donor—acceptor molecules and have also been used to extract one sterol and replace it with another (LaRocca et al., 2013; Tagliari et al., 2012). Sterol substitution experiments have the added benefit that they allow for studies of sterol structure without drastically changing the sterol mass balance in the cell, thus reducing the undesired effects that might be due to mechanisms other than raft disruption. Cyclodextrins have been used to extract ergosterol from fungal cells and also to perform sterol substitution studies (Siafakas et al., 2006; Tagliari et al., 2012). However, their use has been limited and there has been a report suggesting that cyclodextrins are not effective in ergosterol extraction from *S. cerevisiae* (Bagnat et al., 2000). This might be due to the presence of a cell wall that hinders the interaction of MβCD with the plasma membrane.

3.2.4 Inhibiting sterol/sphingolipid biosynthesis

In addition to MβCD, inhibitors of sterol and sphingolipid synthesis have also been used to study raft function in cells. Azoles are a class of drugs that block the 14-α-demethylation of lanosterol in the ergosterol biosynthesis pathway (Ghannoum & Rice, 1999; Martin & Konopka, 2004) and have been used to study the role of sterols in raft formation in fungi. Compounds that inhibit sphingolipid biosynthesis have also been used to disrupt fungal lipid rafts as a method to study their function in cells. Primary among such compounds are myriocin, which inhibits serine palmitoyltransferase at the first committed step of sphingolipid biosynthesis (Lattif et al., 2011; Martin & Konopka, 2004; Miyake, Kozutsumi, Nakamura, Fujita, & Kawasaki, 1995) and aureobasidin A, which inhibits IPC synthase and thus the synthesis of glycosphingolipids (Lattif et al., 2011; Takesako et al., 1993).

3.3 Observation of membrane domains in pathogenic fungi

Many of the concepts later used in studies of pathogenic fungi were first discovered in the nonpathogenic yeast *S. cerevisiae*. The first observation of lipid rafts in fungal cells was reported in *S. cerevisiae* (Bagnat et al., 2000). In this work, DRMs were isolated from *S. cerevisiae* and analyzed for their lipids and proteins. Glycosphingolipids and ergosterol were reported as the main raft-forming lipids, while the GPI-anchored protein Gas1p and the membrane bound proton pump protein Pma1p were found to be two of the most abundant raft proteins. The findings of this study were important as these lipids and proteins have been generally used to identify lipid rafts in other fungi. There is a growing body of research further analyzing the role and composition of lipid rafts in *S. cerevisiae* (Aresta-Branco et al., 2011; Klose et al., 2010; Toulmay & Prinz, 2013). However, in this section we have focused on reports regarding lipid rafts in pathogenic fungi, namely *Candida albicans*, *Cryptococcus neoformans*, *Aspergillus nidulans*, and *Histoplasma capsulatum*.

Candida albicans. *C. albicans* is a dimorphic opportunistic pathogen capable of growing both by budding (yeast) and hyphal (mold) formation. *C. albicans* is generally benign, but can cause mucosal or even bloodstream infections under certain conditions, and immunocompromised individuals are particularly susceptible. *C. albicans* is also able to form biofilms on medical implants, causing infections which might require the removal of the implant (Ramage, Martínez, & Lopez-Ribot, 2006).

The first evidence for the existence of lipid rafts in *C. albicans* was generated by Martin and Konopka (Martin & Konopka, 2004), who assessed the role of sterols in morphogenesis by filipin staining *C. albicans* cells during budding and hyphal growth. It was shown that sterols are homogenously distributed in the plasma membrane during budding growth, but are significantly concentrated at hyphal tips during hyphal growth. DRMs were isolated from budding cells and hyphae and showed similar protein composition indicating that differences in staining are due to localization of SRDs at hyphal tips. Disrupting sphingolipid biosynthesis by myriocin and sterol biosynthesis by ketoconazole led to defects in hyphal formation demonstrating the importance of sphingolipids and SRDs for hyphal growth of *C. albicans*.

Lipid microdomains in *C. albicans* seem to play a role in another change of morphology: changing from planktonic form to biofilms. A study by Lattif and colleagues (Lattif et al., 2011) showed that filipin stained *C. albicans* biofilms much more intensely than planktonic forms, suggesting the presence of more lipid rafts. This proposal is supported by the presence of significantly larger amounts of glycosphingolipids in biofilms than in planktonic forms. In addition, the ability of *C. albicans* to form biofilms was significantly reduced after treatment with myriocin and aureobacidin A, suggesting that sphingolipids play an important role in biofilm formation, likely by regulating the amount of lipid rafts.

The protein composition of DRMs in *C. albicans* has also been the focus of various studies. Proteome analysis has revealed the presence of 29 proteins in the DRM of *C. albicans* (Insenser, Nombela, Molero, & Gil, 2006). Interestingly, many of the identified proteins were those known to be located in the mitochondria, the Golgi apparatus, or the endoplasmic reticulum suggesting that lipid rafts may also be present in intracellular membranes, as has been suggested by others (Mollinedo et al., 2011). The presence of lipid rafts in intracellular membranes of *Candida* spp. is also supported by another study reporting a number of mitochondrial proteins as part of the proteome of DRMs in *Candida guilliermondii* (An, Chen, Li, Qin, & Tian, 2014). It should be noted that the possibility that detergent insolubility of mitochondria arises from properties other than the presence of Lo domains has not been ruled out.

Proteomic studies performed on the DRMs isolated from budding and hyphal *C. albicans* cells revealed that the plasma membrane proteome differs based on morphology (Alvarez & Konopka, 2007). Several notable proteins have been identified in *C. albicans* DRM. A novel protein, Ngt1, was discovered and shown to be a transporter of *N*-acetylglucosamine, a known inducer of hyphae in *C. albicans* (Simonetti, Strippoli, & Cassone, 1974). The presence of Ngt1 in DRMs further supports the importance of lipid rafts in regulating morphological changes. Other important proteins suggested to be part of the DRM proteome in *C. albicans* include CaCdr1P, an efflux pump protein and a drug transporter (Pasrija, Panwar, & Prasad, 2008), and Rta2p, which acts to transfer long-chain sphingoid bases from the cytofacial to the exofacial side of the plasma membrane (Wang et al., 2012). Both of these proteins were highly concentrated in the DRM fractions and localized at the plasma membrane of *C. albicans*. Detergent resistance and plasma membrane localization of these proteins was abolished in response to mutations or drugs that affect ergosterol and sphingolipid biosynthesis, suggesting that raft-association is required for the proper localization of these proteins (Pasrija et al., 2008; Wang et al., 2012).

C. neoformans. The budding yeast *C. neoformans* is a facultative intracellular pathogen and the leading cause of fungal meningitis. *C. neoformans* is able to grow at both acidic and neutral conditions allowing it to survive and proliferate inside immune cells and in the bloodstream. Studies of lipid rafts in *C. neoformans* have been focused on virulence factors. Virulence factors have been defined as "mechanisms that allow the fungus to efficiently divide and survive inside the host,…resist host immune attack, and…damage the host" (Coelho, Bocca, & Casadevall, 2014). Similar to *C. albicans*, the first observation of membrane domains in *C. neoformans* was provided by filipin staining, revealing the presence of SRDs in the bud tips of *C. neoformans* (Nichols, Fraser, & Heitman, 2004). The first isolation of DRMs in *C. neoformans* was performed by Siafakas and colleagues (Siafakas et al., 2006). This analysis revealed the presence of phosphatidylethanolamine (PE), PC, phosphatidylserine (PS), ergosterol, and ceramide monohexosides (such as GlcCer) as the main lipids of *C. neoformans* DRMs. Interestingly, IPC, MIPC, and MIP2C were not reported as part of the DRM lipidome in *C. neoformans*. This might be due to the techniques used in the study (these lipids have been generally visualized using radiolabelling (Rittershaus et al., 2006)) or lower amounts of these lipids in *C. neoformans* compared to other fungi. Two virulence factors of *C. neoformans* were found in the DRM fractions: Phospholipase B1 (which acts to degrade the phospholipids of the host cell (Chrisman, Albuquerque, Guimaraes, Nieves, & Casadevall, 2011)) and Cu/Zn superoxide dismutase (which protects the fungus against oxidative stress (Cox et al., 2003)).

Studies have also focused on the role of lipid rafts in the ability of *C. neoformans* to survive at different levels of ambient pH. A *C. neoformans* mutant that lacks the gene sphingolipid C9 methyltransferase (*Smt1*) produces GlcCer that is not methylated. This strain loses its ability to grow at neutral pH and shows attenuated virulence (Singh et al., 2012). This mutant strain also showed increased membrane fluidity as evidenced by a slight reduction in the fluorescent anisotropy of live cells. In addition, atomic force microscopy analysis of reconstituted lipid extracts showed a reduction in the height of the ordered lipid phases suggesting a less ordered packing of the vertical fatty acyl chains. The role of GlcCer methylation in membrane fluidity is unclear; however, it is possible that methylation is required for efficient packing of GlcCer with sterols and other sphingolipids (Del Poeta, Nimrichter, Rodrigues, & Luberto, 2014). Another possibility is that a mutation in the sphingolipid pathway might result in complex changes in membrane composition by affecting other pathways and the synthesis of sterols and/or unsaturated phospholipids.

The connection between membrane fluidity and growth remains unclear. A recent study has shown that similar to *S. cerevisiae* and *C. albicans* the protein Pma1p is concentrated in DRMs of *C. neoformans* and plays an important role in fungal growth and survival in acidic conditions (Farnoud, Mor, Singh, & Del Poeta, 2014). Further studies on characterization of lipid raft proteins might help elucidate the role of lipid rafts in growth at pH 7.4.

Aspergillus nidulans. *Aspergillus* is a fungus commonly found as mold in the environment. *Aspergillus* spp. are able to cause disease in immunocompromised patients with inhalation of conidia (spores) being the common route of infection. *Aspergillus* infections are generally limited to the lungs, but can disseminate to the bloodstream in severely immunocompromised patients. Most *Aspergillus* infections are caused by *Aspergillus fumigatus*; however, other strains, such as *A. nidulans* are also known to cause infections (Segal et al., 1998). Compared to *C. albicans* and *C. neoformans*, much less is known about the membrane microdomains in *Aspergillus*. However, in similarity to those fungi, filipin staining has revealed the presence of SRDs at the hyphal tips and the sites of septation in *A. nidulans* (Pearson, Xu, Sharpless, & Harris, 2004; Takeshita et al., 2008). SRDs localize the cell end marker protein, Tea1, and disruption of SRDs displaces this protein from the hyphal tips (Takeshita et al., 2008). Interestingly, the interactions between SRDs and proteins appear to be reciprocal. Flotillin is a lipid raft marker in mammalian cells (Dermine et al., 2001), which has a homologue in *Aspergillus*. *A. nidulans* mutants lacking this protein showed disrupted SRDs and reduced growth rate (Takeshita, Diallinas, & Fischer, 2012). However, the flotillin homologue did not localize in the SRDs, suggesting an indirect relationship between its localization and SRD formation.

Histoplasma capsulatum. *H. capsulatum* is a temperature-dependent dimorphic fungus, which is found as mold in the environment but grows as yeast in physiological temperatures. Inhalation of conidia can lead to severe lung infections in immunocompromised patients. To our knowledge, only one study has focused on characterization of membrane microdomains in *H. capsulatum* (Tagliari et al., 2012). In this study, DRMs were isolated from *H. capsulatum* revealing the presence of ergosterol, PE, PC, phosphatidylinositol (PI) and glycosphingolipids GlcCer, IPC, MIPC, and MIP2C as detergent insoluble lipids. Extraction of sterols using MβCD resulted in the displacement of Pma1p and several other proteins from the detergent resistant fraction. It should be noted that sterol extraction can act not only by abolishing ordered domains, but also by simply reducing the resistance of ordered domains to solubilization to detergent (London, 2005; Xu & London, 2000). However, in this study residual DRMs containing sphingolipids were detected after MβCD treatment. Interestingly, sterol depletion also significantly reduced the infectivity of alveolar macrophages by *H. capsulatum*. Both protein displacement and reduced infectivity persisted even after substituting ergosterol with cholesterol, indicating the importance of sterol chemical structure in proper protein localization and infectivity.

3.4 The role of membrane domains in fungal pathogenesis

A summary of studies of lipid rafts in pathogenic fungi is provided in Table 1. As shown in the Table, the literature on lipid rafts in pathogenic fungi remains fairly small and is divided between various microorganisms. Based on the limited literature, it is difficult to pinpoint one universal role for lipid rafts in all pathogenic fungi. However, one area in which lipid rafts appear to facilitate fungal infections is by regulating changes in fungal morphology. SRDs were observed in hyphal tips in *C. albicans* (Martin & Konopka, 2004) and *A. nidulans* (Pearson et al., 2004; Takeshita et al., 2008) suggesting that lipid rafts regulate hyphal growth. In addition, the involvement of lipid rafts in biofilm formation (Lattif et al., 2011) suggests that rafts might play a role in all forms of changes in morphology. Hyphal growth and biofilm formation are important determinants of *C. albicans* pathogenicity. Hyphae have been suggested to promote tissue penetration and facilitate invasive growth (Sudbery, Gow, & Berman, 2004). In fact, *C. albicans* mutants that are locked in the yeast form are avirulent in a mouse model (Lo et al., 1997). In addition, biofilms are significantly more resistant to antifungal drugs than planktonic cells (Ramage, VandeWalle, Bachmann, Wickes, & López-Ribot, 2002) and are able to colonize implant devices causing significant health problems.

Lipid rafts may also contribute to fungal pathogenesis by regulating drug resistance. CaCdr1p is an efflux pump protein involved in multidrug resistance in *C. albicans* (Shukla, Saini, Jha, Ambudkar, & Prasad, 2003). When expressed in *S. cerevisiae*, this protein was associated with DRMs and disruption of ergosterol or sphingolipid biosynthesis pathway displaced this protein from detergent resistant fractions and inhibited the ability of the fungus to resist methotrexate and fluconazole (Pasrija et al., 2008). It has also been suggested that lipid rafts might be involved in drug resistance of filamentous fungi by regulating membrane fluidity. The polymeric compound chitosan has been shown to have a fungicidal effect against the filamentous fungi by permeabilizing the fungal plasma membrane (Palma-Guerrero et al., 2009). Interestingly, a study of chitosan-resistant and chitosan-sensitive filamentous fungi revealed that a *Neurospora crassa* desaturase mutant with greatly reduced levels of polyunsaturated fatty acids, and thus increased membrane rigidity, was chitosan-resistant (Palma-Guerrero et al., 2010). These observations suggest that a rigid plasma membrane, mainly or entirely in the Lo state, might act as a barrier to membrane-permeabilizing drugs. Membrane rigidity has also been suggested to act as

Table 1 Published studies of lipid rafts in pathogenic fungi

Microorganism	Study	Methods	Proposed raft function
Candida albicans	Martin and Konopka (2004)	Filipin staining, DRM isolation, blocking sterol/ sphingolipid synthesis	Hyphal growth
	Alvarez and Konopka (2007)	DRM isolation, proteomics	Hyphal growth
	Lattif et al. (2011)	Filipin staining, blocking sphingolipid synthesis	Biofilm formation
	Inserner, Nombela, Molero, and Gil (2006)	DRM isolation, proteomics	Not applicable, proteomic study
	Wang et al. (2012)	Filipin staining, DRM isolation, blocking sterol/ sphingolipid synthesis, proteomics	Protein sorting
Cryptococcus neoformans	Nichols et al. (2004)	Filipin staining	Budding growth
	Siafakas et al. (2006)	DRM isolation, sterol extraction	Protein sorting
	Singh et al. (2012)	Fluorescent anisotropy, atomic force microscopy	Membrane integrity, growth at neutral pH
	Farnoud et al. (2014)	DRM isolation	Protein sorting, growth at acidic pH
Aspergillus nidulans	Pearson et al. (2004)	Filipin staining	Hyphal growth, protein sorting
	Takeshita et al. (2008)	Filipin staining	Hyphal growth, protein sorting
	Takeshita et al. (2012)	Filipin staining, DRM isolation	Hyphal growth, protein sorting
Histoplasma capsulatum	Tagliari et al. (2012)	DRM isolation, sterol substitution, lipidomics	Infectivity

a barrier to membrane permeabilization in response to osmotic pressure in the yeast *S. cerevisiae* (Dupont, Beney, Ferreira, & Gervais, 2011).

Lipid rafts also play a role in pathogenicity by concentrating virulence factors. Phospholipase B1 (Plb1) and Cu/Zn SOD, two important virulence factors of *C. neoformans*, were reported to be concentrated in DRMs (Siafakas et al., 2006). Plb1 acts to degrade the phospholipids of the host immune cells, possibly allowing phagocytosed *C. neoformans* to escape from the phagolysosomal membrane into the cytosol, where it can acquire nutrients (Chrisman et al., 2011). Plb1 mutants in *C. neoformans* show less nonlytic exocytosis (escape of cryptococcal cells from the macrophages after ingestion, with survival of both fungal and mammalian cells (Nicola, Robertson, Albuquerque, da Silveira Derengowski, & Casadevall, 2011)), likely due to reduced ability to survive the hostile environment of host phagolysosomes (Chayakulkeeree et al., 2011). Cu/Zn superoxide dismutase protects the fungus against oxidative stress inside the phagolysosomes by reducing superoxide anions to hydrogen peroxide (Cox et al., 2003). Lipid rafts may also play a role in pathogenicity of *C. neoformans* by concentrating the cryptococcal Pma1p. As noted above, Pma1p is a proton pump, which maintains the intracellular pH of fungal cells in the physiological range (Holyoak et al., 1996). Pma1p is raft-localized and its association with rafts is required for surface delivery and protein stability at the plasma membrane (Bagnat et al., 2001; Gong & Chang, 2001). *C. neoformans* mutants in which DRM-associated Pma1p was not functioning properly showed attenuated virulence, likely due to their inability to survive the acidic environment of the phagolysosome (Farnoud et al., 2014; Shea, Kechichian, Luberto, & Del Poeta, 2006). Finally, reduced phagocytosis of ergosterol-depleted *H. capsulatum* cells by macrophages suggests the fungal lipid rafts may also play a role in fungal entry into mammalian cells (Tagliari et al., 2012). It is possible that changes in membrane fluidity upon ergosterol depletion trigger the secretion of *Histoplasma* proteins with antiphagocytic function. The secretion of such proteins have been reported in *C. neoformans* (Luberto et al., 2003), but their dependence on membrane fluidity has not been studied.

It is likely that rafts play other roles in pathogenic fungi or that they play different roles at different times. For example, it is likely that fungal raft proteins may have multiple conformations or varying protein—protein interactions, and thus different functions in different host environments. In addition, the principles of raft organization have not been fully explored, and may involve more than just lipid—lipid interaction. Some specialized fungal membrane domains (e.g., eisosomes) may be organized by plasma membrane proteins (Douglas & Konopka, 2014). Thus, more studies and better characterization of raft lipids and proteins are needed to elucidate all raft functions in pathogenic fungi.

4. MEMBRANE DOMAINS IN PATHOGENIC BACTERIA

4.1 Domain-forming lipids in bacteria

Bacterial microdomains are significantly different from those in eukaryotes. Bacterial membranes are enriched in phospholipids, lipopolysaccharides, and various lipoproteins (Huijbregts, de Kroon, & de Kruijff, 2000). However, most bacteria do not have sphingolipids (Parsons & Rock, 2013) and only a few bacteria contain sterols, with sterol-containing bacteria including *Helicobacter*, *Mycoplasma*, *Ehrlichia*, *Anaplasma*, *Brachyspira*, and *Borrelia* (Haque, Hirai, Yokota, & Oguma, 1995; Hirai et al., 1995; Lin & Rikihisa, 2003; Livermore, Bey, & Johnson, 1978; Smith, 1971; Trott, Alt, Zuerner, Wannemuehler, & Stanton, 2001). Due to the important role of sterols in promoting lipid rafts in eukaryotic cells, much of the studies on lipid rafts in bacteria has been focused on sterol-containing species, and in particular *Borrelia burgdorferi*. However, the presence of microdomains has been shown even in sterol-lacking bacteria, as will be described below. It is worth mentioning that some nonpathogenic bacteria synthesize hopanoids, which have a chemical structure similar to that of cholesterol (Rohmer, Bouvier, & Ourisson, 1979) and studies have already started addressing the question whether these bacteria contain membrane microdomains (Sáenz, Sezgin, Schwille, & Simons, 2012). Such studies will not be discussed in the following section, which is focused on pathogenic bacteria. In addition, many of the methods to study lipid rafts in bacteria are similar to those in fungal cells and thus are not discussed in detail.

4.2 Observation of membrane domains in pathogenic bacteria

Borrelia burgdorferi. *B. burgdorferi* is an extracellular tick-borne spirochete that causes Lyme disease. The most prominent manifestation of Lyme disease is a skin rash with central clearing at the site of inoculation known as erythema migrans. Other organs and tissues including the heart, joints, and the nervous system can be affected due to the ability of bacterium to disseminate through the bloodstream (Benach et al., 1983). The membrane of *B. burgdorferi* contains phosphatidylcholine, phosphatidylglycerol, and lipoproteins (Belisle, Brandt, Radolf, & Norgard, 1994; Brooks, Vuppala, Jett, & Akins, 2006; Jones, Bourell, Norgard, & Radolf, 1995; Kenedy, Lenhart, & Akins, 2012; Radolf et al., 1995). In addition, it has free cholesterol and two cholesterol glycolipids, acylated cholesteryl galactoside (ACGal), and cholesteryl galactoside (CGal) as well as a noncholesterol glycolipid, monogalactosyl diacylglycerol (MGalD) (Ben-Menachem, Kubler-Kielb, Coxon, Yergey, & Schneerson, 2003; Schroder et al., 2003; Stubs et al., 2009, 2011; Wheeler et al., 1993). Interestingly, *Borrelia* does not have the biosynthetic ability to synthesize cholesterol and has to acquire it from the mammalian host. *B. burgdorferi* recruits the cholesterol needed to form the glycolipids from its host cells. The mechanisms by which *B. burgdorferi* interacts with the host cell has not been elucidated yet, but it has been hypothesized that such interaction could be mediated through lipid raft—lipid raft interactions (Crowley et al., 2013). In any case, in this recruitment process there is an exchange of cholesterol, antigens, and lipids between the host cell and the spirochete (Crowley et al., 2013). The acquisition of cholesterol in *B. burgdorferi* is time dependent (Crowley et al., 2013) and cholesterol is glycosylated in the spirochete by undetermined bacterial enzymes (Östberg, Berg, Comstedt, Wieslander, & Bergström, 2007).

The presence of sterols in *B. burgdorferi* has led to questions whether *Borrelia* contains lipid rafts similar to those observed in eukaryotic cells. To answer these questions, LaRocca and colleagues (LaRocca et al., 2010) performed ultrastructure analysis on *B. burgdorferi*. Previous studies established the cross-reaction between *Borrelia* glycolipids and host gangliosides by showing that *B. burgdorferi* elicits an antibody response in infected rats that targeted the host gangliosides (Garcia-Monco, Seidman, & Benach, 1995). The cross-reaction between GM1 and *Borrelia* cholesterol glycolipids allowed for the use of an asialo-GM1 antibody to study the localization of cholesterol glycolipids with transmission electron microscopy (TEM). Using a particle-conjugated asialo-GM1 antibody and another particle-conjugated CB2 antibody, which binds to the outer surface protein B (OspB), it was shown that the surface protein and cholesteryl glycolipids form microdomains in *B. burgdorferi*. Moreover, the size and the number of these clusters were temperature-dependent and were present in both culture- and mouse-derived spirochetes (LaRocca et al., 2010). DRMs similar to those of eukaryotic cells could be isolated from the bacterium and revealed the presence of cholesteryl glycolipids and proteins, in the floating (detergent insoluble) fractions, further corroborating the presence of lipid rafts containing specific proteins. (It should be noted that steryl glycolipids are also found in the DRM fraction from plant cells (Lefebvre et al., 2007; Simon-Plas, Perraki, Bayer, Gerbeau-Pissot, & Mongrand, 2011).) Treatment with MβCD, which led to extraction of cholesterol and cholesterol-glycolipids, resulted in the relocalization of proteins in the detergent soluble fractions (LaRocca et al., 2010). Also, both lipoproteins detected in the insoluble fraction were shown to co-localize with *Borrelia* glycolipids in the membrane of the spirochete by TEM.

Persuasive proof for the existence of lipid rafts in *B. burgdorferi* was found when treatment with MβCD was followed by supplementing the bacterium with a set of different sterols (LaRocca et al., 2013). In vitro experiments with synthetic lipid bilayers have demonstrated that the ability of sterols to promote raft formation is dependent on their chemical structure (Megha, Bakht, & London, 2006; Wang, Megha, & London, 2004; Xu et al., 2001; Xu & London, 2000). Substituting cholesterol with strongly raft-promoting sterols (e.g., ergosterol) or intermediate raft-promoting sterols (e.g., lanosterol) did not affect the presence of sterol glycolipids domains in the ultrastructure of *B. burgdorferi* as shown by TEM and Förster resonance electron transfer (FRET) in live cells. However, raft-disrupting sterols (e.g., coprostanol) led to disruption of these structures. Interestingly, upon prolonged incubation, the bacterium lost its membrane integrity after sterol depletion or in the presence of raft-disrupting sterols, as evidenced by the visualization of exposed periplasmic flagella, not visible in normal spirochetes, and by the release of cytosolic proteins into the external environment (LaRocca et al., 2013). These observations indicated that sterol-mediated membrane microdomains exist in *Borrelia* and regulate its membrane integrity.

Although lipid rafts contain a subset of specific proteins, little is known about the contribution of proteins to the stability of the microdomains. A recent study showed that outer surface proteins play a role in membrane domain stability in *B. burgdorferi* although their role is not as crucial as that of sterols. A double-mutant lacking raft-associated proteins, OspA and OspB, showed significantly reduced domain stability as detected by FRET. Membrane domain stability was recovered in the double-mutant when it was complemented with OspA, but not when it was complemented with another surface lipoprotein, OspC, that is not associated with the microdomains, suggesting that only those proteins that localize into the microdomain regulate membrane domain stability (Toledo et al., 2014). The evidence that bacterial surface proteins play an important role in the stability of microdomains is of significance not only for *Borrelia* but also for bacteria that lack sterols. Although, only a few such bacteria have been studied for their ability to form microdomains (see below), the possibility of domain promotion by surface proteins suggests that many more bacteria might contain membrane microdomains, which should be a focus of future studies.

Bacillus subtilis. *B. subtilis* is a rod-shaped gram-positive bacterium commonly found in soil. *B. subtilis* is generally used in probiotics; however, infections in immune compromised individuals have been reported both with *B. subtilis* and with other strains (Nelms, Larson, & Barnes-Josiah, 1997; Oggioni, Pozzi, Valensin, Galieni, & Bigazzi, 1998). *B. subtilis* cells lack cholesterol; however, evidence is accumulating that membrane proteins play a role in microdomain formation in their membranes. Flotillin is a protein frequently reported in the DRM of eukaryotic cells (Langhorst, Reuter, & Stuermer, 2005) and is commonly used as a raft-marker protein (Macdonald & Pike, 2005). *B. subtilis* contains two flotillin-like proteins, floT and floA (Mielich-Süss, Schneider, & Lopez, 2013). To examine the role of flotillin in microdomain formation in *B. subtilis*, Donovan and Bramkamp (Donovan & Bramkamp, 2009) used a genetically modified strain producing fluorescently tagged floT. Observation of this strain under fluorescence microscopy revealed the presence of several flotillin domains across the membranes. These domains were dynamic and changed their size and location with time. FloT was only found in the membrane and floT-enriched DRMs were isolated from *B. subtilis* providing further evidence for the involvement of floT in the formation of ordered membrane domains.

Analogously to lipid rafts in budding yeast, membrane microdomains in *B. subtilis* appear to play a role in morphogenetic processes, such as sporulation, biofilm formation, and cell division. A genetically modified *B. subtilis* strain lacking floT showed defective sporulation (Donovan & Bramkamp, 2009). In addition, floA and floT were observed in the septum of dividing cells indicating that they interact with septum-localized proteins (Yepes et al., 2012). Overexpression of flotillins promoted biofilm formation and increased cell division leading to a reduction in cell size and loss of the rod shape (Mielich-Süss et al., 2013).

Even in sterol-lacking bacteria, sterol-like molecules are required for organization of microdomains. The *yisP* gene in *B. subtilis* shows similarity with squalene synthases in fungi. Since squalene is a molecule involved in ergosterol biosynthesis in fungi (Ghannoum et al., 1994), this finding implied that sterol-like molecules might be present in *B. subtilis* (López & Kolter, 2010). Incubation of bacteria with zaragozic acid, a molecule known to inhibit squalene synthases, inhibited biofilm formation by *B. subtilis*, changed the protein profile of DRMs and abrogated flotillin microdomains suggesting that flotillin microdomains depend on a sterol-like molecule for stability, and are analogous to eukaryotic lipid rafts (López & Kolter, 2010). These observations were confirmed by Laurdan spectroscopy experiments (Bach & Bramkamp, 2013). The emission spectrum of the fluorescent dye Laurdan is dependent on the polarity of the surrounding environment and shows a difference in disordered versus ordered lipid phases due to the presence of less water molecules in ordered phases. Staining wild-type *B. subtilis* cells with Laurdan revealed the presence of dispersed areas of high lipid order due to the presence of membrane microdomains. However, in *B. subtilis* *ΔyisP* a coalescence of these ordered areas into a single larger domain was observed indicating a change in membrane heterogeneity. Similar phenomena were observed in mutants lacking flotillin-like proteins suggesting that these proteins and sterol-like molecules are both involved in maintaining domain integrity (Bach & Bramkamp, 2013).

Staphylococcus aureus. *S. aureus* is a gram-positive bacterium generally found in hair and noses of humans and animals. *S. aureus* commonly causes skin infections; however, it can also infect the bloodstream and various organs. Infections with *S. aureus* can be hard to treat due to the emergence of methicillin-resistant *S. aureus* (MRSA) (Enright et al., 2002). A sequence homologue of flotillin exists in *S. aureus* and the possibility for the presence of membrane microdomains containing flotillin-like proteins was examined by fusion of a fluorescent tag to this protein (López & Kolter, 2010). The fused fluorescent protein was localized to a single domain in the bacterial membrane suggesting the presence of flotillin-induced microdomains. The localization of this protein at the membrane was abolished by zaragozic acid, suggesting the importance of sterol-like molecules in domain formation.

Isolation of DRMs from *S. aureus* followed by protein sequencing showed the presence of 18 proteins in the DRMs including elastin-binding protein EbpS, which is involved in biofilm formation (López & Kolter, 2010). Treating the bacterium with clomazone, statin, and zaragozic acid (inhibitors of various steps of sterol biosynthesis) abrogated biofilm formation, and in the cases of clomazone and statin, led to cell death, suggesting that sterol-like molecules play a role in biofilm formation and survival of *S. aureus*. Interestingly, *S. aureus*, contains a lipid, staphyloxanthin, that is structurally similar to cholesteryl glycolipids. In staphyloxanthin a polyisoprene replaces cholesterol (which itself is derived biosynthetically from isoprene units). Whether this lipid participates in raft formation is not clear.

Based on the current evidence from the literature, microdomains differ significantly in fungal cells, sterol-containing, and sterol-lacking bacteria (Figure 2). While in fungal cells, sphingolipids and ergosterol pack to form microdomains (Figure 2(A)), in sterol-containing bacteria either cholesterol glycolipids play the role of both sterols and sphingolipids (Figure 2(B)) or other lipids substitute for sphingolipids. In these microorganisms, membrane proteins stabilize microdomains but are not necessary for microdomain formation. Finally, in sterol-lacking bacteria, microdomain formation appears to be mediated by the packing of flotillin-like proteins

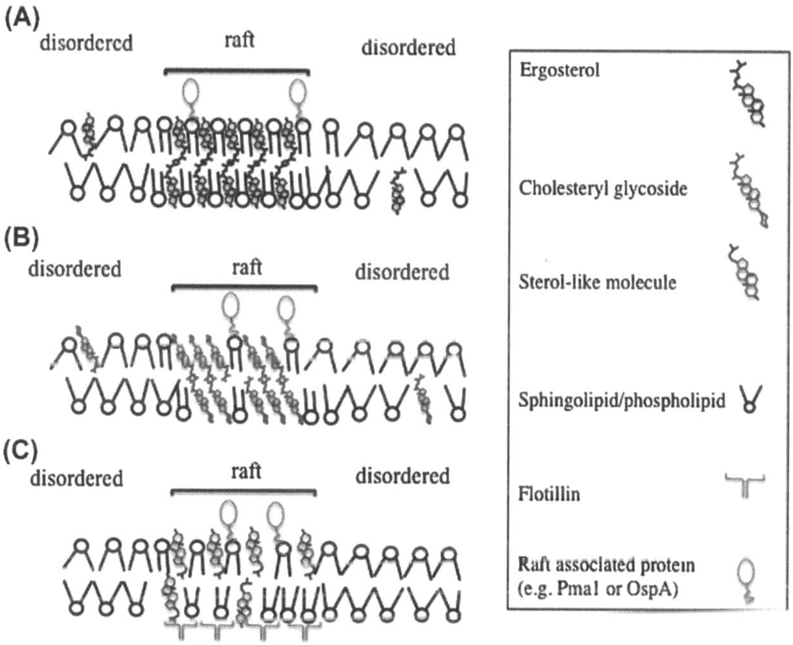

Figure 2 A hypothetical schematic of lipid rafts in (A) fungal, (B) sterol-containing and (C) sterol-lacking bacteria: (A) in fungal cells, ergosterol and glycosphingolipids pack to form membrane microdomains, (B) in sterol-containing bacteria, cholesteryl glycosides play the role of both sterols and sphingolipids (much of the cholesteryl glycosides are acylated, which is not shown in the schematic), and (C) in sterol-lacking bacteria, flotillin and sterol-like molecules mediate raft formation. It should be emphasized that the lipid composition of lipid rafts in bacteria is only poorly characterized. Certain membrane lipids and proteins are not included for simplicity. (See color plate)

and yet unknown sterol-like lipids (Figure 2(C)). Although membrane microdomains have been identified only in *B. burgdorferi*, *B. subtilis*, and *S. aureus*, it seems likely that sterol- or protein-mediated membrane microdomains will soon be discovered in other bacteria.

4.3 Role of membrane domains in bacterial pathogenesis

Prokaryotic membrane microdomains appear to play a similar role to eukaryotic microdomains with regard to cell division and morphogenesis. Similarly to SRDs in fungi, flotillin-like proteins are concentrated in the septa in dividing *B. subtilis* cells. In both cases, the microdomains likely act as a hub for concentrating proteins required for septum formation. Microdomains also play a significant role in biofilm formation in bacteria, and *B. subtilis* mutants deficient in flotillin-like proteins (and thus heterogenous microdomains) do not form biofilms. The inability of bacteria to form biofilms upon microdomain disruptions is strikingly similar to the inability of fungi to form biofilms upon lipid raft disruption (Lattif et al., 2011). This phenomenon is likely caused by disruption of microdomain-mediated, protein–protein interaction. FtsH is a protease essential for the differentiation of biofilm matrix producers. This protein is shown to directly interact with flotillin-like proteins and does not localize to DRM fractions in the absence of flotillin-like proteins suggesting a link between protein–protein interaction, microdomains, and biofilm formation (Yepes et al., 2012). This is of biomedical importance because biofilms have superior resistance to antibiotics and host immune response, and so significantly contribute to the persistence of bacterial infections.

Protein function and secretion machinery also depend on the function of membrane microdomains. Analysis of the protein profile of detergent resistant fractions in *B. subtilis* revealed that flotillin-like proteins interact with proteins that have a role in diverse cell functions, such as cell wall and energy metabolism and protein secretion, and mutants deficient in flotillin-like proteins also show reduced protein secretion (Bach & Bramkamp, 2013). Extracellular secretion is an important part of bacterial pathogenesis and allows bacteria to release toxins and virulence factors to the host (Lee & Schneewind, 2001). For example *S. aureus* produces extracellular toxins able to form pores in a eukaryotic plasma membrane (Dinges, Orwin, & Schlievert, 2000). Extracellular release is also utilized by *B. burgdorferi*, in which raft components such as cholesteryl glycolipids are released into the medium via outer membrane vesicles (Crowley et al., 2013). If delivered to the host cell, cholesteryl glycolipids might not be recognized by the host immune response due to their structural similarity to host lipids, or cause the host cells to become the target of immune response due to the presence of lipid antigens (Crowley et al., 2013).

Lipid rafts might also play the role of an environmental sensor for the pathogen. It has been shown that the size of microdomains in *B. burgdorferi* increases with decreasing temperature (LaRocca et al., 2010). In another study, it has been shown that adaptation of *Bacillus cereus* to low temperatures depends on maintaining a certain membrane fluidity (de Sarrau et al., 2013), which is consistent with the idea of lipid rafts acting as environmental sensors. *Borrelia* is exposed to a wide range of temperatures as it has to survive both inside ticks and in mammalian hosts. Temperature-dependent changes in microdomain size and membrane fluidity might act as a sensor of the extracellular environment, so that, for example, a reduction in microdomain size can signal the entrance to the mammalian host. Changes in microdomains could also affect the raft-localized lipoproteins with further implications in signal transduction.

From a therapeutic point of view, bacterial lipid rafts may also be a drug target. The bactericidal CB2 antibody against *Borrelia* depends on sterols for its efficacy and sterol depletion significantly reduces its bactericidal activity (LaRocca et al., 2010). Also, zaragozic acid, which inhibits the activity of squalene synthase, has been shown to prevent biofilm synthesis in *B. subtilis* and *S. aureus*. In fact, inhibitors of squalene synthase have been shown to inhibit the virulence of *S. aureus* (Liu et al., 2008). Thus, drugs that affect membrane structure via altering lipid raft formation might be an effective therapy against bacterial infections, and in particular, resistant biofilms.

5. UNEXPLORED AREAS AND FUTURE DIRECTIONS

In the past two decades, there has been increasing interest in the presence of membrane microdomains in living cells and the literature in this field is rapidly expanding. The fundamental knowledge gained on lipid–lipid interactions has been significantly improved by in vitro studies, while the presence of membrane microdomains in living cells has been demonstrated by a number of in vivo studies. Despite this progress, there are currently many outstanding questions and areas that need to be further explored to provide an understanding on the role of membrane domains in pathogenesis. The number of studies performed in pathogenic microorganisms is significantly less than those performed in model microorganisms such as

S. cerevisiae, and pathogenic bacteria in particular are still highly understudied. While many outstanding questions remain regarding the role of membrane domains in pathogenic microorganisms, we believe the following are some of the major unexplored areas that need to be addressed.

Do all bacteria have membrane domains? While the presence of lipid rafts in eukaryotic cells is now widely accepted, it is unclear whether all bacteria include microdomains. Of the several bacterial species that contain sterols only *Borrelia* has been studied in detail and it is not clear if microdomains exist in other sterol-containing pathogenic bacteria. Also it is unclear whether sterol-like molecules mediate domain formation in sterol-lacking bacteria. It has been suggested that sterol-like molecules mediate membrane domain in *B. subtilis* (Bach & Bramkamp, 2013; López & Kolter, 2010), while other sterol-like molecules such as hopanoids have been suggested to induce raft formation in bacteria (Sáenz et al., 2012). The answers to these questions can have important implications for therapeutic strategies. For example, if sterol-like molecules contribute to membrane integrity in bacterial pathogens, then inhibitors of their synthesis might be very potent drugs against a variety of infections, as noted above (Liu et al., 2008).

More methods are needed to study microdomains. The paucity of appropriate methods has been a road-block in studies of lipid rafts. This is perhaps epitomized by the fact that the conventional method of DRM extraction has many limitations (London, 2005; London & Brown, 2000). Biophysical methods such as FRET, fluorescent anisotropy, and Laurdan spectroscopy have provided a significant amount of information about lipid rafts in live bacterial cells; however, such methods have been rarely used in fungal cells. These methods depend on effective delivery of fluorescent probes into the cell membrane, a requirement that is hampered by the presence of cell wall and in some cases a capsule in fungal cells. Nanotechnology-based applications involving degradation of the capsule or cell wall, or the use of small fluorescent particles may make it possible to overcome this problem. In addition, the use of new microscopy techniques and lipidomics can significantly improve the current state of knowledge on molecular diffusion in rafts and the molecular species of raft-forming molecules, respectively.

How do lipid rafts change during the course of growth and pathogenesis? Although existing studies have provided information on the proteins associated with lipid rafts, the changes in the protein profile of lipid rafts during various stages of pathogenesis is still unknown. For example, it has been proposed that different proteins are expressed in *B. burgdorferi* during various stages of infection (Tilly, Bestor, Dulebohn, & Rosa, 2009; Tilly, Bestor, & Rosa, 2013). However, it is unclear whether these proteins perform similar functions in maintaining lipid rafts and how these changes in protein composition occur. Do they reflect changes in expression, or release to the extracellular environment? It is also not known if lipid rafts differ in commensal and noncommensal microorganisms. It is possible that the amount of lipid rafts increases during infections, presenting a platform for the concentration of virulence factors. Studies comparing pathogenic and nonpathogenic strains, especially in microorganisms that switch from commensalism to pathogenicity, might help to investigate this possibility. A related issue is that the role of rafts in various stages of bacterial or fungal growth remains unclear. *B. subtilis* cells show few microdomains during exponential growth, but the number of microdomains increases as the bacterium reaches the stationary phase (Donovan & Bramkamp, 2009). A similar appearance of vacuolar microdomains at various stages of growth has been noted in *S. cerevisiae* (Toulmay & Prinz, 2013). Finally, the role of microdomains in maintaining cell viability during various stages of growth and inside the host requires further studies. Clearly, the structure and biological function of lipid rafts in pathogenic microorganisms is a field that is only in its infancy, and is likely to be a major area of future research.

ACKNOWLEDGMENTS

This work was supported by NIH grants GM 099892 (to E.L.), RO1 AI47837 (to J.B.K.), and AI56168, AI71142, and AI100631 (to M.D.P.). M.D.P. is a Burroughs Wellcome Investigator in Infectious Diseases.

REFERENCES

Ahmed, S. N., Brown, D. A., & London, E. (1997). On the origin of sphingolipid/cholesterol-rich detergent-insoluble cell membranes: physiological concentrations of cholesterol and sphingolipid induce formation of a detergent-insoluble, liquid-ordered lipid phase in model membranes. *Biochemistry, 36*, 10944–10953.

de Almeida, R. F., Fedorov, A., & Prieto, M. (2003). Sphingomyelin/phosphatidylcholine/cholesterol phase diagram: boundaries and composition of lipid rafts. *Biophysical Journal, 85*, 2406–2416.

Almeida, P., Vaz, W., & Thompson, T. (1993). Percolation and diffusion in three-component lipid bilayers: effect of cholesterol on an equimolar mixture of two phosphatidylcholines. *Biophysical Journal, 64*, 399–412.

Alvarez, F. J., & Konopka, J. B. (2007). Identification of an *N*-acetylglucosamine transporter that mediates hyphal induction in *Candida albicans*. *Molecular Biology of the Cell, 18*, 965–975.

An, B., Chen, Y., Li, B., Qin, G., & Tian, S. (2014). Ca^{2+}—CaM regulating viability of *Candida guilliermondii* under oxidative stress by acting on detergent resistant membrane proteins. *Journal of Proteomics, 109*, 38—49.

Aresta-Branco, F., Cordeiro, A. M., Marinho, H. S., Cyrne, L., Antunes, F., & de Almeida, R. F. (2011). Gel domains in the plasma membrane of *Saccharomyces cerevisiae* Highly ordered, ergosterol-free, and sphingolipid-enriched lipid rafts. *Journal of Biological Chemistry, 286*, 5043—5054.

Babina, I. S., Donatello, S., Nabi, I. R., & Hopkins, A. M. (2011). Lipid rafts as master regulators of breast cancer cell function. In M. Gunduz, & E. Gunduz (Eds.), *Breast Cancer - Carcinogenesis, Cell Growth and Signalling Pathways* (pp. 401—428). Riejka, Croatia: InTech.

Bach, J. N., & Bramkamp, M. (2013). Flotillins functionally organize the bacterial membrane. *Molecular Microbiology, 88*, 1205—1217.

Bagnat, M., & Simons, K. (2002a). Cell surface polarization during yeast mating. *Proceedings of the National Academy of Sciences, 99*, 14183—14188.

Bagnat, M., & Simons, K. (2002b). Lipid rafts in protein sorting and cell polarity in budding yeast *Saccharomyces cerevisiae*. *Biological Chemistry, 383*, 1475—1480.

Bagnat, M., Keränen, S., Shevchenko, A., Shevchenko, A., & Simons, K. (2000). Lipid rafts function in biosynthetic delivery of proteins to the cell surface in yeast. *Proceedings of the National Academy of Sciences, 97*, 3254—3259.

Bagnat, M., Chang, A., & Simons, K. (2001). Plasma membrane proton ATPase Pma1p requires raft association for surface delivery in yeast. *Molecular Biology of the Cell, 12*, 4129—4138.

Belisle, J. T., Brandt, M. E., Radolf, J. D., & Norgard, M. V. (1994). Fatty acids of *Treponema pallidum* and *Borrelia burgdorferi* lipoproteins. *Journal of Bacteriology, 176*, 2151—2157.

Benach, J. L., Bosler, E. M., Hanrahan, J. P., Coleman, J. L., Habicht, G. S., Bast, T. F., & Burgdorfer, W. (1983). Spirochetes isolated from the blood of two patients with Lyme disease. *New England Journal of Medicine, 308*, 740—742.

Ben-Menachem, G., Kubler-Kielb, J., Coxon, B., Yergey, A., & Schneerson, R. (2003). A newly discovered cholesteryl galactoside from *Borrelia burgdorferi*. *Proceedings of the National Academy of Sciences, 100*, 7913—7918.

Björkbom, A., Ohvo-Rekilä, H., Kankaanpää, P., Nyholm, T. K., Westerlund, B., & Slotte, J. P. (2010). Characterization of membrane properties of inositol phosphorylceramide. *Biochimica et Biophysica Acta (BBA)—Biomembranes, 1798*, 453—460.

Brooks, C. S., Vuppala, S. R., Jett, A. M., & Akins, D. R. (2006). Identification of *Borrelia burgdorferi* outer surface proteins. *Infection and Immunity, 74*, 296—304.

Brown, D. A., & London, E. (1997). Structure of detergent-resistant membrane domains: does phase separation occur in biological membranes? *Biochemical and Biophysical Research Communications, 240*, 1—7.

Brown, D. A., & Rose, J. K. (1992). Sorting of GPI-anchored proteins to glycolipid-enriched membrane subdomains during transport to the apical cell surface. *Cell, 68*, 533—544.

Brown, D. A. (2006). Lipid rafts, detergent-resistant membranes, and raft targeting signals. *Physiology, 21*, 430—439.

Byfield, F. J., Tikku, S., Rothblat, G. H., Gooch, K. J., & Levitan, I. (2006). OxLDL increases endothelial stiffness, force generation, and network formation. *Journal of Lipid Research, 47*, 715—723.

Calhoun, W. I., & Shipley, G. G. (1979). Sphingomyelin-lecithin bilayers and their interaction with cholesterol. *Biochemistry, 18*, 1717—1722.

Chapman, D., Owens, N., Phillips, M., & Walker, D. (1969). Mixed monolayers of phospholipids and cholesterol. *Biochimica et Biophysica Acta (BBA)—Biomembranes, 183*, 458—465.

Chayakulkeeree, M., Johnston, S. A., Oei, J. B., Lev, S., Williamson, P. R., Wilson, C. F., & Meyer, W. (2011). SEC14 is a specific requirement for secretion of phospholipase B1 and pathogenicity of *Cryptococcus neoformans*. *Molecular Microbiology, 80*, 1088—1101.

Chrisman, C. J., Albuquerque, P., Guimaraes, A. J., Nieves, E., & Casadevall, A. (2011). Phospholipids trigger *Cryptococcus neoformans* capsular enlargement during interactions with amoebae and macrophages. *PLoS Pathogens, 7*, e1002047.

Coelho, C., Bocca, A. L., & Casadevall, A. (2014). The tools for virulence of *Cryptococcus neoformans*. *Advances in Applied Microbiology, 87*, 1–41.

Cox, G. M., Harrison, T. S., McDade, H. C., Taborda, C. P., Heinrich, G., Casadevall, A., et al. (2003). Superoxide dismutase influences the virulence of *Cryptococcus neoformans* by affecting growth within macrophages. *Infection and Immunity, 71*, 173–180.

Crowley, J. T., Toledo, A. M., LaRocca, T. J., Coleman, J. L., London, E., & Benach, J. L. (2013). Lipid exchange between Borrelia burgdorferi and host cells. *PLoS Pathogens, 9*, e1003109.

Darwich, Z., Klymchenko, A. S., Kucherak, O. A., Richert, L., & Mély, Y. (2012). Detection of apoptosis through the lipid order of the outer plasma membrane leaflet. *Biochimica et Biophysica Acta (BBA)—Biomembranes, 1818*, 3048–3054.

Del Poeta, M., Nimrichter, L., Rodrigues, M. L., & Luberto, C. (2014). Synthesis and biological properties of fungal glucosylceramide. *PLoS Pathogens, 10*, e1003832.

Dennis, E. A. (1974). Formation and characterization of mixed micelles of the nonionic surfactant Triton X-100 with egg, dipalmitoyl, and dimyristoyl phosphatidylcholines. *Archives of Biochemistry and Biophysics, 165*, 764–773.

Dermine, J. F., Duclos, S., Garin, J., St-Louis, F., Rea, S., Parton, R. G., et al. (2001). Flotillin-1-enriched lipid raft domains accumulate on maturing phagosomes. *Journal of Biological Chemistry, 276*, 18507–18512.

Dinges, M. M., Orwin, P. M., & Schlievert, P. M. (2000). Exotoxins of *Staphylococcus aureus*. *Clinical Microbiology Reviews, 13*, 16–34.

Donovan, C., & Bramkamp, M. (2009). Characterization and subcellular localization of a bacterial flotillin homologue. *Microbiology, 155*, 1786–1799.

Douglas, L. M., & Konopka, J. B. (2014). Fungal membrane organization: the eisosome concept. *Annual Review of Microbiology, 68*, 377–393.

Dupont, S., Beney, L., Ferreira, T., & Gervais, P. (2011). Nature of sterols affects plasma membrane behavior and yeast survival during dehydration. *Biochimica et Biophysica Acta (BBA)—Biomembranes, 1808*, 1520–1528.

Enright, M. C., Robinson, D. A., Randle, G., Feil, E. J., Grundmann, H., & Spratt, B. G. (2002). The evolutionary history of methicillin-resistant *Staphylococcus aureus* (MRSA). *Proceedings of the National Academy of Sciences, 99*, 7687–7692.

Farnoud, A. M., Mor, V., Singh, A., & Del Poeta, M. (2014). Inositol phosphosphingolipid phospholipase C1 regulates plasma membrane ATPase (Pma1) stability in *Cryptococcus neoformans*. *FEBS Letters, 588*, 3932–3938.

Fastenberg, M. E., Shogomori, H., Xu, X., Brown, D. A., & London, E. (2003). Exclusion of a transmembrane-type peptide from ordered-lipid domains (rafts) detected by fluorescence quenching: extension of quenching analysis to account for the effects of domain size and domain boundaries. *Biochemistry, 42*, 12376–12390.

Feigenson, G. W., & Buboltz, J. T. (2001). Ternary phase diagram of dipalmitoyl-PC/dilauroyl-PC/cholesterol: Nanoscopic domain formation driven by cholesterol. *Biophysical Journal, 80*, 2775–2788.

Field, K. A., Holowka, D., & Baird, B. (1999). Structural aspects of the association of FcεRI with detergent-resistant membranes. *Journal of Biological Chemistry, 274*, 1753–1758.

Forstner, G., Tanaka, K., & Isselbacher, K. (1968). Lipid composition of the isolated rat intestinal microvillus membrane. *Biochemical Journal, 109*, 51–59.

Garcia-Monco, J. C., Seidman, R. J., & Benach, J. L. (1995). Experimental immunization with *Borrelia burgdorferi* induces development of antibodies to gangliosides. *Infection and Immunity, 63*, 4130–4137.

George, K. S., & Wu, S. (2012). Lipid raft: a floating island of death or survival. *Toxicology and Applied Pharmacology, 259*, 311–319.

Ghannoum, M. A., & Rice, L. B. (1999). Antifungal agents: mode of action, mechanisms of resistance, and correlation of these mechanisms with bacterial resistance. *Clinical Microbiology Reviews, 12*, 501–517.

Ghannoum, M. A., Spellberg, B. J., Ibrahim, A. S., Ritchie, J. A., Currie, B., Spitzer, E. D., & Casadevall, A. (1994). Sterol composition of *Cryptococcus neoformans* in the presence and absence of fluconazole. *Antimicrobial Agents and Chemotherapy, 38*, 2029–2033.

Gong, X., & Chang, A. (2001). A mutant plasma membrane ATPase, Pma1-10, is defective in stability at the yeast cell surface. *Proceedings of the National Academy of Sciences, 98*, 9104–9109.

Grant, C. W., Wu, S. H.-W., & McConnell, H. M. (1974). Lateral phase separations in binary lipid mixtures: correlation between spin label and freeze-fracture electron microscopic studies. *Biochimica et Biophysica Acta (BBA)—Biomembranes, 363*, 151–158.

Grassme, H., Jendrossek, V., Riehle, A., Von Kürthy, G., Berger, J., Schwarz, H., & Gulbins, E. (2003). Host defense against *Pseudomonas aeruginosa* requires ceramide-rich membrane rafts. *Nature Medicine, 9*, 322–330.

Haque, M., Hirai, Y., Yokota, K., & Oguma, K. (1995). Steryl glycosides: a characteristic feature of the *Helicobacter* spp.? *Journal of Bacteriology, 177*, 5334–5337.

Harder, T., Scheiffele, P., Verkade, P., & Simons, K. (1998). Lipid domain structure of the plasma membrane revealed by patching of membrane components. *The Journal of Cell Biology, 141*, 929–942.

Hicks, D. A., Nalivaeva, N. N., & Turner, A. J. (2012). Lipid rafts and Alzheimer's disease: protein-lipid interactions and perturbation of signaling. *Frontiers in Physiology, 3*, 189.

Hirai, Y., Haque, M., Yoshida, T., Yokota, K., Yasuda, T., & Oguma, K. (1995). Unique cholesteryl glucosides in *Helicobacter pylori*: composition and structural analysis. *Journal of Bacteriology, 177*, 5327–5333.

Holyoak, C., Stratford, M., McMullin, Z., Cole, M., Crimmins, K., Brown, A., et al. (1996). Activity of the plasma membrane H(+)-ATPase and optimal glycolytic flux are required for rapid adaptation and growth of *Saccharomyces cerevisiae* in the presence of the weak-acid preservative sorbic acid. *Applied and Environmental Microbiology, 62*, 3158–3164.

Huang, J., Buboltz, J. T., & Feigenson, G. W. (1999). Maximum solubility of cholesterol in phosphatidylcholine and phosphatidylethanolamine bilayers. *Biochimica et Biophysica Acta (BBA)—Biomembranes, 1417*, 89–100.

Huijbregts, R. P., de Kroon, A. I., & de Kruijff, B. (2000). Topology and transport of membrane lipids in bacteria. *Biochimica et Biophysica Acta (BBA)—Reviews on Biomembranes, 1469*, 43–61.

Insenser, M., Nombela, C., Molero, G., & Gil, C. (2006). Proteomic analysis of detergent-resistant membranes from *Candida albicans*. *Proteomics, 6*, S74–S81.

Ipsen, J. H., Karlström, G., Mourtisen, O., Wennerström, H., & Zuckermann, M. (1987). Phase equilibria in the phosphatidylcholine-cholesterol system. *Biochimica et Biophysica Acta (BBA)—Biomembranes, 905*, 162–172.

Jones, J. D., Bourell, K. W., Norgard, M. V., & Radolf, J. D. (1995). Membrane topology of *Borrelia burgdorferi* and *Treponema pallidum* lipoproteins. *Infection and Immunity, 63*, 2424–2434.

Karnovsky, M. J., Kleinfeld, A. M., Hoover, R. L., & Klausner, R. D. (1982). The concept of lipid domains in membranes. *The Journal of Cell Biology, 94*, 1–6.

Kenedy, M. R., Lenhart, T. R., & Akins, D. R. (2012). The role of *Borrelia burgdorferi* outer surface proteins. *FEMS Immunology and Medical Microbiology, 66*, 1–19.

Kleemann, W., Grant, C., & McConnell, H. (1974). Lipid phase separations and protein distribution in membranes. *Journal of Supramolecular Structure, 2*, 609–616.

Klose, C., Ejsing, C. S., García-Sáez, A. J., Kaiser, H.-J., Sampaio, J. L., Surma, M. A., & Simons, K. (2010). Yeast lipids can phase-separate into micrometer-scale membrane domains. *Journal of Biological Chemistry, 285*, 30224–30232.

Koynova, R., & Caffrey, M. (1998). Phases and phase transitions of the phosphatidylcholines. *Biochimica et Biophysica Acta (BBA)—Reviews on Biomembranes, 1376*, 91–145.

Kucherak, O. A., Oncul, S., Darwich, Z., Yushchenko, D. A., Arntz, Y., Didier, P., & Klymchenko, A. S. (2010). Switchable nile red-based probe for cholesterol and lipid order at the outer leaflet of biomembranes. *Journal of the American Chemical Society, 132*, 4907–4916.

Ladbrooke, B. D., Williams, R. M., & Chapman, D. (1968). Studies on lecithin-cholesterol-water interactions by differential scanning calorimetry and X-ray diffraction. *Biochimica et Biophysica Acta—Biomembranes, 150*, 333–340.

Lange, Y., Swaisgood, M., Ramos, B., & Steck, T. (1989). Plasma membranes contain half the phospholipid and 90% of the cholesterol and sphingomyelin in cultured human fibroblasts. *Journal of Biological Chemistry, 264*, 3786–3793.

Langhorst, M. F., Reuter, A., & Stuermer, C. (2005). Scaffolding microdomains and beyond: the function of reggie/flotillin proteins. *Cellular and Molecular Life Sciences CMLS, 62*, 2228–2240.

LaRocca, T. J., Crowley, J. T., Cusack, B. J., Pathak, P., Benach, J., London, E., & Benach, J. L. (2010). Cholesterol lipids of *Borrelia burgdorferi* form lipid rafts and are required for the bactericidal activity of a complement-independent antibody. *Cell Host & Microbe, 8*, 331–342.

LaRocca, T. J., Pathak, P., Chiantia, S., Toledo, A., Silvius, J. R., Benach, J. L., et al. (2013). Proving lipid rafts exist: membrane domains in the prokaryote *Borrelia burgdorferi* have the same properties as eukaryotic lipid rafts. *PLoS Pathogens, 9*, e1003353.

Lattif, A. A., Mukherjee, P. K., Chandra, J., Roth, M. R., Welti, R., Rouabhia, M., et al. (2011). Lipidomics of *Candida albicans* biofilms reveals phase-dependent production of phospholipid molecular classes and role for lipid rafts in biofilm formation. *Microbiology, 157*, 3232–3242.

Lee, V. T., & Schneewind, O. (2001). Protein secretion and the pathogenesis of bacterial infections. *Genes & Development, 15*, 1725–1752.

Lee, A. (1977). Lipid phase transitions and phase diagrams II. Mixtures involving lipids. *Biochimica et Biophysica Acta (BBA)—Reviews on Biomembranes, 472*, 285–344.

Lefebvre, B., Furt, F., Hartmann, M.-A., Michaelson, L. V., Carde, J.-P., Sargueil-Boiron, F., & Bessoule, J. J. (2007). Characterization of lipid rafts from *Medicago truncatula* root plasma membranes: a proteomic study reveals the presence of a raft-associated redox system. *Plant Physiology, 144*, 402–418.

Levental, I., Grzybek, M., & Simons, K. (2010). Greasing their way: lipid modifications determine protein association with membrane rafts. *Biochemistry, 49*, 6305–6316.

Lin, Q., & London, E. (2013). Transmembrane protein (Perfringolysin O) association with ordered membrane domains (rafts) depends upon the raft-associating properties of protein-bound sterol. *Biophysical Journal, 105*, 2733–2742.

Lin, M., & Rikihisa, Y. (2003). *Ehrlichia chaffeensis* and *Anaplasma phagocytophilum* lack genes for lipid A biosynthesis and incorporate cholesterol for their survival. *Infection and Immunity, 71*, 5324–5331.

Linden, C. D., & Fox, C. F. (1975). Membrane physical state and function. *Accounts of Chemical Research, 8*, 321–327.

Lingwood, D., & Simons, K. (2007). Detergent resistance as a tool in membrane research. *Nature Protocols, 2*, 2159–2165.

Lisanti, M. P., Sargiacomo, M., Graeve, L., Saltiel, A. R., & Rodriguez-Boulan, E. (1988). Polarized apical distribution of glycosyl-phosphatidylinositol-anchored proteins in a renal epithelial cell line. *Proceedings of the National Academy of Sciences, 85*, 9557–9561.

Liu, C., Liu, G. Y., Song, Y., Yin, F., Hensler, M. E., Jeng, W., & Oldfield, E. (2008). A cholesterol biosynthesis inhibitor blocks *Staphylococcus aureus* virulence. *Science, 319*, 1391–1394.

Livermore, B., Bey, R., & Johnson, R. (1978). Lipid metabolism of *Borrelia hermsi*. *Infection and Immunity, 20*, 215–220.

Lo, H., Köhler, J. R., DiDomenico, B., Loebenberg, D., Cacciapuoti, A., & Fink, G. R. (1997). Nonfilamentous *C. albicans* mutants are avirulent. *Cell, 90*, 939–949.

London, E., & Brown, D. A. (2000). Insolubility of lipids in triton X-100: physical origin and relationship to sphingolipid/cholesterol membrane domains (rafts). *Biochimica et Biophysica Acta (BBA)—Biomembranes, 1508*, 182–195.

London, E. (2005). How principles of domain formation in model membranes may explain ambiguities concerning lipid raft formation in cells. *Biochimica et Biophysica Acta (BBA)—Molecular Cell Research, 1746*, 203–220.

López, D., & Kolter, R. (2010). Functional microdomains in bacterial membranes. *Genes & Development, 24*, 1893–1902.

Loura, L., Castanho, M. A., Fedorov, A., & Prieto, M. (2001). A photophysical study of the polyene antibiotic filipin: self-aggregation and filipin–ergosterol interaction. *Biochimica et Biophysica Acta (BBA)—Biomembranes, 1510*, 125–135.

Luberto, C., Martinez-Mariño, B., Taraskiewicz, D., Bolaños, B., Chitano, P., Toffaletti, D. L., & Balish, E. (2003). Identification of App1 as a regulator of phagocytosis and virulence of *Cryptococcus neoformans*. *Journal of Clinical Investigation, 112*, 1080–1094.

Macdonald, J. L., & Pike, L. J. (2005). A simplified method for the preparation of detergent-free lipid rafts. *Journal of Lipid Research, 46*, 1061–1067.

Martin, S. W., & Konopka, J. B. (2004). Lipid raft polarization contributes to hyphal growth in *Candida albicans*. *Eukaryotic Cell, 3*, 675–684.

Megha, Bakht, O., & London, E. (2006). Cholesterol precursors stabilize ordinary and ceramide-rich ordered lipid domains (lipid rafts) to different degrees Implications for the Bloch hypothesis and sterol biosynthesis disorders. *Journal of Biological Chemistry, 281*, 21903–21913.

Mielich-Süss, B., Schneider, J., & Lopez, D. (2013). Overproduction of flotillin influences cell differentiation and shape in *Bacillus subtilis*. *mBio, 4*, e00719–00713.

Miyake, Y., Kozutsumi, Y., Nakamura, S., Fujita, T., & Kawasaki, T. (1995). Serine palmitoyltransferase is the primary target of a sphingosine-like immunosuppressant, ISP-1/myriocin. *Biochemical and Biophysical Research Communications, 211*, 396–403.

Mollinedo, F., Fernández, M., Hornillos, V., Delgado, J., Amat-Guerri, F., Acuña, A. U., & Ceña, V. (2011). Involvement of lipid rafts in the localization and dysfunction effect of the antitumor ether phospholipid edelfosine in mitochondria. *Cell Death & Disease, 2*, e158.

Mollinedo, F. (2012). Lipid raft involvement in yeast cell growth and death. *Frontiers in Oncology, 2*, 140.

Nelms, P. K., Larson, O., & Barnes-Josiah, D. (1997). Time to *B. cereus* about hot chocolate. *Public Health Reports, 112*, 240–244.

Nichols, C. B., Fraser, J. A., & Heitman, J. (2004). PAK kinases Ste20 and Pak1 govern cell polarity at different stages of mating in *Cryptococcus neoformans*. *Molecular Biology of the Cell, 15*, 4476–4489.

Nicola, A. M., Robertson, E. J., Albuquerque, P., da Silveira Derengowski, L., & Casadevall, A. (2011). Nonlytic exocytosis of *Cryptococcus neoformans* from macrophages occurs in vivo and is influenced by phagosomal pH. *mBio, 2*, e00167–00111.

Oggioni, M. R., Pozzi, G., Valensin, P. E., Galieni, P., & Bigazzi, C. (1998). Recurrent septicemia in an immunocompromised patient due to probiotic strains of *Bacillus subtilis*. *Journal of Clinical Microbiology, 36*, 325–326.

Östberg, Y., Berg, S., Comstedt, P., Wieslander, Å., & Bergström, S. (2007). Functional analysis of a lipid galactosyltransferase synthesizing the major envelope lipid in the Lyme disease spirochete *Borrelia burgdorferi*. *FEMS Microbiology Letters, 272*, 22–29.

Palma-Guerrero, J., Huang, I.-C., Jansson, H.-B., Salinas, J., Lopez-Llorca, L., & Read, N. (2009). Chitosan permeabilizes the plasma membrane and kills cells of *Neurospora crassa* in an energy dependent manner. *Fungal Genetics and Biology, 46*, 585–594.
Palma-Guerrero, J., Lopez-Jimenez, J., Pérez-Berná, A., Huang, I., Jansson, H., Salinas, J., & Lopez-Llorca, L. (2010). Membrane fluidity determines sensitivity of filamentous fungi to chitosan. *Molecular Microbiology, 75*, 1021–1032.
Papahadjopoulos, D., Jacobson, K., Nir, S., & Isac, I. (1973). Phase transitions in phospholipid vesicles fluorescence polarization and permeability measurements concerning the effect of temperature and cholesterol. *Biochimica et Biophysica Acta (BBA)—Biomembranes, 311*, 330–348.
Parsons, J. B., & Rock, C. O. (2013). Bacterial lipids: metabolism and membrane homeostasis. *Progress in Lipid Research, 52*, 249–276.
Pasrija, R., Panwar, S. L., & Prasad, R. (2008). Multidrug transporters CaCdr1p and CaMdr1p of *Candida albicans* display different lipid specificities: both ergosterol and sphingolipids are essential for targeting of CaCdr1p to membrane rafts. *Antimicrobial Agents and Chemotherapy, 52*, 694–704.
Pearson, C. L., Xu, K., Sharpless, K. E., & Harris, S. D. (2004). MesA, a novel fungal protein required for the stabilization of polarity axes in *Aspergillus nidulans*. *Molecular Biology of the Cell, 15*, 3658–3672.
Pike, L. J. (2009). The challenge of lipid rafts. *Journal of Lipid Research, 50*, S323–S328.
Radolf, J. D., Goldberg, M. S., Bourell, K., Baker, S. I., Jones, J. D., & Norgard, M. V. (1995). Characterization of outer membranes isolated from *Borrelia burgdorferi*, the Lyme disease spirochete. *Infection and Immunity, 63*, 2154–2163.
Ramage, G., VandeWalle, K., Bachmann, S. P., Wickes, B. L., & López-Ribot, J. L. (2002). In vitro pharmacodynamic properties of three antifungal agents against preformed *Candida albicans* biofilms determined by time-kill studies. *Antimicrobial Agents and Chemotherapy, 46*, 3634–3636.
Ramage, G., Martínez, J. P., & Lopez-Ribot, J. L. (2006). *Candida* biofilms on implanted biomaterials: a clinically significant problem. *FEMS Yeast Research, 6*, 979–986.
Ray, T. K., Skipski, V. P., Barclay, M., Essner, E., & Archibald, F. M. (1969). Lipid composition of rat liver plasma membranes. *Journal of Biological Chemistry, 244*, 5528–5536.
Recktenwald, D. J., & McConnell, H. M. (1981). Phase equilibriums in binary mixtures of phosphatidylcholine and cholesterol. *Biochemistry, 20*, 4505–4510.
Redfern, D. A., & Gericke, A. (2004). Domain formation in phosphatidylinositol monophosphate/phosphatidylcholine mixed vesicles. *Biophysical Journal, 86*, 2980–2992.
Rittershaus, P. C., Kechichian, T. B., Allegood, J. C., Merrill, A. H., Hennig, M., Luberto, C., et al. (2006). Glucosylceramide synthase is an essential regulator of pathogenicity of *Cryptococcus neoformans*. *The Journal of Clinical Investigation, 116*, 1651–1659.
Rohmer, M., Bouvier, P., & Ourisson, G. (1979). Molecular evolution of biomembranes: structural equivalents and phylogenetic precursors of sterols. *Proceedings of the National Academy of Sciences, 76*, 847–851.
Sáenz, J. P., Sezgin, E., Schwille, P., & Simons, K. (2012). Functional convergence of hopanoids and sterols in membrane ordering. *Proceedings of the National Academy of Sciences, 109*, 14236–14240.
Sankaram, M. B., & Thompson, T. E. (1990). Interaction of cholesterol with various glycerophospholipids and sphingomyelin. *Biochemistry, 29*, 10670–10675.
de Sarrau, B., Clavel, T., Zwickel, N., Despres, J., Dupont, S., Beney, L., & Nguyen-the, C. (2013). Unsaturated fatty acids from food and in the growth medium improve growth of *Bacillus cereus* under cold and anaerobic conditions. *Food Microbiology, 36*, 113–122.
Schroder, N. W., Schombel, U., Heine, H., Gobel, U. B., Zahringer, U., & Schumann, R. R. (2003). Acylated cholesteryl galactoside as a novel immunogenic motif in Borrelia burgdorferi sensu stricto. *Journal of Biological Chemistry, 278*, 33645–33653.

Schroeder, R., London, E., & Brown, D. (1994). Interactions between saturated acyl chains confer detergent resistance on lipids and glycosylphosphatidylinositol (GPI)-anchored proteins: GPI-anchored proteins in liposomes and cells show similar behavior. *Proceedings of the National Academy of Sciences, 91*, 12130–12134.
Schroeder, R. J., Ahmed, S. N., Zhu, Y., London, E., & Brown, D. A. (1998). Cholesterol and sphingolipid enhance the Triton X-100 insolubility of glycosylphosphatidylinositol-anchored proteins by promoting the formation of detergent-insoluble ordered membrane domains. *Journal of Biological Chemistry, 273*, 1150–1157.
Schuck, S., Honsho, M., Ekroos, K., Shevchenko, A., & Simons, K. (2003). Resistance of cell membranes to different detergents. *Proceedings of the National Academy of Sciences, 100*, 5795–5800.
Segal, B. H., DeCarlo, E. S., Kwon-Chung, K., Malech, H. L., Gallin, J. I., & Holland, S. M. (1998). *Aspergillus nidulans* infection in chronic granulomatous disease. *Medicine, 77*, 345–354.
Shea, J. M., Kechichian, T. B., Luberto, C., & Del Poeta, M. (2006). The cryptococcal enzyme inositol phosphosphingolipid-phospholipase C confers resistance to the antifungal effects of macrophages and promotes fungal dissemination to the central nervous system. *Infection and Immunity, 74*, 5977–5988.
Shimshick, E. J., & McConnell, H. M. (1973). Lateral phase separation in phospholipid membranes. *Biochemistry, 12*, 2351–2360.
Shukla, S., Saini, P., Jha, S., Ambudkar, S. V., & Prasad, R. (2003). Functional characterization of *Candida albicans* ABC transporter Cdr1p. *Eukaryotic Cell, 2*, 1361–1375.
Siafakas, A. R., Wright, L. C., Sorrell, T. C., & Djordjevic, J. T. (2006). Lipid rafts in *Cryptococcus neoformans* concentrate the virulence determinants phospholipase B1 and Cu/Zn superoxide dismutase. *Eukaryotic Cell, 5*, 488–498.
Silvius, J. R., del Giudice, D., & Lafleur, M. (1996). Cholesterol at different bilayer concentrations can promote or antagonize lateral segregation of phospholipids of differing acyl chain length. *Biochemistry, 35*, 15198–15208.
Simonetti, N., Strippoli, V., & Cassone, A. (1974). Yeast-mycelial conversion induced by *N*-acetyl-D-glucosamine in *Candida albicans*. *Nature, 250*, 344–346.
Simon-Plas, F., Perraki, A., Bayer, E., Gerbeau-Pissot, P., & Mongrand, S. (2011). An update on plant membrane rafts. *Current Opinion in Plant Biology, 14*, 642–649.
Simons, K., & Ikonen, E. (1997). Functional rafts in cell membranes. *Nature, 387*, 569–572.
Simons, K., & Toomre, D. (2000). Lipid rafts and signal transduction. *Nature Reviews Molecular Cell Biology, 1*, 31–39.
Simons, K., & Van Meer, G. (1988). Lipid sorting in epithelial cells. *Biochemistry, 27*, 6197–6202.
Singer, S. J., & Nicolson, G. L. (1972). The fluid mosaic model of the structure of cell membranes. *Science, 175*, 720–731.
Singh, A., Wang, H., Silva, L. C., Na, C., Prieto, M., Futerman, A. H., & Del Poeta, M. (2012). Methylation of glycosylated sphingolipid modulates membrane lipid topography and pathogenicity of *Cryptococcus neoformans*. *Cellular Microbiology, 14*, 500–516.
Skibbens, J. E., Roth, M. G., & Matlin, K. S. (1989). Differential extractability of influenza virus hemagglutinin during intracellular transport in polarized epithelial cells and nonpolar fibroblasts. *The Journal of Cell Biology, 108*, 821–832.
Smejkal, G. B., & Hoff, H. F. (1994). Filipin staining of lipoproteins in polyacrylamide gels: sensitivity and photobleaching of the fluorophore and its use in a double staining method. *Electrophoresis, 15*, 922–925.
Smith, P. F. (1971). Biosynthesis of cholesteryl glucoside by *Mycoplasma gallinarum*. *Journal of Bacteriology, 108*, 986–991.
Stubs, G., Fingerle, V., Wilske, B., Gobel, U. B., Zahringer, U., Schumann, R. R., et al. (2009). Acylated cholesteryl galactosides are specific antigens of borrelia causing lyme

disease and frequently induce antibodies in late stages of disease. *Journal of Biological Chemistry, 284*, 13326–13334.

Stubs, G., Fingerle, V., Zahringer, U., Schumann, R. R., Rademann, J., & Schroder, N. W. (2011). Acylated cholesteryl galactosides are ubiquitous glycolipid antigens among Borrelia burgdorferi sensu lato. *FEMS Immunology and Medical Microbiology, 63*, 140–143.

Sudbery, P., Gow, N., & Berman, J. (2004). The distinct morphogenic states of *Candida albicans. Trends in Microbiology, 12*, 317–324.

Suzuki, K. G. (2012). Lipid rafts generate digital-like signal transduction in cell plasma membranes. *Biotechnology Journal, 7*, 753–761.

Tagliari, L., Toledo, M. S., Lacerda, T. G., Suzuki, E., Straus, A. H., & Takahashi, H. K. (2012). Membrane microdomain components of *Histoplasma capsulatum* yeast forms, and their role in alveolar macrophage infectivity. *Biochimica et Biophysica Acta (BBA)–Biomembranes, 1818*, 458–466.

Takesako, K., Kuroda, H., Inoue, T., Haruna, F., Yoshikawa, Y., Kato, I., & Yamaguchi, H. (1993). Biological properties of aureobasidin A, a cyclic depsipeptide antifungal antibiotic. *The Journal of Antibiotics, 46*, 1414–1420.

Takeshita, N., Higashitsuji, Y., Konzack, S., & Fischer, R. (2008). Apical sterol-rich membranes are essential for localizing cell end markers that determine growth directionality in the filamentous fungus *Aspergillus nidulans. Molecular Biology of the Cell, 19*, 339–351.

Takeshita, N., Diallinas, G., & Fischer, R. (2012). The role of flotillin FloA and stomatin StoA in the maintenance of apical sterol-rich membrane domains and polarity in the filamentous fungus *Aspergillus nidulans. Molecular Microbiology, 83*, 1136–1152.

Thompson, T., & Tillack, T. W. (1985). Organization of glycosphingolipids in bilayers and plasma membranes of mammalian cells. *Annual Review of Biophysics and Biophysical Chemistry, 14*, 361–386.

Tilly, K., Bestor, A., Dulebohn, D. P., & Rosa, P. A. (2009). OspC-independent infection and dissemination by host-adapted *Borrelia burgdorferi. Infection and Immunity, 77*, 2672–2682.

Tilly, K., Bestor, A., & Rosa, P. A. (2013). Lipoprotein succession in *Borrelia burgdorferi*: similar but distinct roles for OspC and VlsE at different stages of mammalian infection. *Molecular Microbiology, 89*, 216–227.

Toledo, A., Crowley, J. T., Coleman, J. L., LaRocca, T. J., Chiantia, S., London, E., et al. (2014). Selective association of outer surface lipoproteins with the lipid rafts of *Borrelia burgdorferi. mBio, 5*, e00899–00814.

Toulmay, A., & Prinz, W. A. (2013). Direct imaging reveals stable, micrometer-scale lipid domains that segregate proteins in live cells. *The Journal of Cell Biology, 202*, 35–44.

Triantafilou, M., Miyake, K., Golenbock, D. T., & Triantafilou, K. (2002). Mediators of innate immune recognition of bacteria concentrate in lipid rafts and facilitate lipopolysaccharide-induced cell activation. *Journal of Cell Science, 115*, 2603–2611.

Trott, D. J., Alt, D. P., Zuerner, R. L., Wannemuehler, M. J., & Stanton, T. B. (2001). The search for *Brachyspira* outer membrane proteins that interact with the host. *Animal Health Research Reviews, 2*, 19–30.

Veatch, S. L., & Keller, S. L. (2003). Separation of liquid phases in giant vesicles of ternary mixtures of phospholipids and cholesterol. *Biophysical Journal, 85*, 3074–3083.

Vist, M. R., & Davis, J. H. (1990). Phase equilibria of cholesterol/dipalmitoylphosphatidylcholine mixtures: deuterium nuclear magnetic resonance and differential scanning calorimetry. *Biochemistry, 29*, 451.

Wang, J., Megha, & London, E. (2004). Relationship between sterol/steroid structure and participation in ordered lipid domains (lipid rafts): implications for lipid raft structure and function. *Biochemistry, 43*, 1010–1018.

Wang, L., Jia, Y., Tang, R.-J., Xu, Z., Cao, Y.-B., Jia, X.-M., et al. (2012). Proteomic analysis of Rta2p-dependent raft-association of detergent-resistant membranes in *Candida albicans*. *PLoS One, 7*, e37768.

Wheeler, C. M., Garcia Monco, J. C., Benach, J. L., Golightly, M. G., Habicht, G. S., & Steere, A. C. (1993). Nonprotein antigens of Borrelia burgdorferi. *Journal of Infectious Diseases, 167*, 665–674.

Wu, S. H., & McConnell, H. M. (1975). Phase separations in phospholipid membranes. *Biochemistry, 14*, 847–854.

Xu, X., & London, E. (2000). The effect of sterol structure on membrane lipid domains reveals how cholesterol can induce lipid domain formation. *Biochemistry, 39*, 843–849.

Xu, X., Bittman, R., Duportail, G., Heissler, D., Vilcheze, C., & London, E. (2001). Effect of the structure of natural sterols and sphingolipids on the formation of ordered sphingolipid/sterol domains (rafts). Comparison of cholesterol to plant, fungal, and disease-associated sterols and comparison of sphingomyelin, cerebrosides, and ceramide. *Journal of Biological Chemistry, 276*, 33540–33546.

Yaqoob, P., & Shaikh, S. R. (2010). The nutritional and clinical significance of lipid rafts. *Current Opinion in Clinical Nutrition & Metabolic Care, 13*, 156–166.

Yepes, A., Schneider, J., Mielich, B., Koch, G., García-Betancur, J. C., Ramamurthi, K. S., & López, D. (2012). The biofilm formation defect of a *Bacillus subtilis* flotillin-defective mutant involves the protease FtsH. *Molecular Microbiology, 86*, 457–471.

Yuan, C., Furlong, J., Burgos, P., & Johnston, L. J. (2002). The size of lipid rafts: an atomic force microscopy study of ganglioside GM1 domains in sphingomyelin/DOPC/cholesterol membranes. *Biophysical Journal, 82*, 2526–2535.

Zhao, J., Wu, J., Heberle, F. A., Mills, T. T., Klawitter, P., Huang, G., et al. (2007). Phase studies of model biomembranes: Complex behavior of DSPC/DOPC/Cholesterol. *Biochimica et Biophysica Acta (BBA)-Biomembranes, 1768*, 2764–2776.

Zidovetzki, R., & Levitan, I. (2007). Use of cyclodextrins to manipulate plasma membrane cholesterol content: evidence, misconceptions and control strategies. *Biochimica et Biophysica Acta (BBA)—Biomembranes, 1768*, 1311–1324.

CHAPTER EIGHT

Trafficking and Membrane Organization of GPI-Anchored Proteins in Health and Diseases

Simona Paladino[1,2,a], **Stéphanie Lebreton**[3,a] and **Chiara Zurzolo**[1,3,*]

[1]Dipartimento di Medicina Molecolare e Biotecnologie Mediche, Università Federico II, Napoli, Italy
[2]CEINGE Biotecnologie Avanzate, Napoli, Italy
[3]Unité de Trafic Membranaire et Pathogénèse, Institut Pasteur, Paris, France
*Corresponding author: E-mail: zurzolo@pasteur.fr

Contents

Abstract

Glycosylphosphatidylinositol (GPI)-anchored proteins (GPI-APs) are a class of lipid-anchored proteins attached to the membranes by a glycolipid anchor that is added, as posttranslation modification, in the endoplasmic reticulum. GPI-APs are expressed at the cell surface of eukaryotes where they play diverse vital functions. Like all plasma membrane proteins, GPI-APs must be correctly sorted along the different steps of the secretory pathway to their final destination. The presence of both a glycolipid anchor and a protein portion confers special trafficking features to GPI-APs. Here, we discuss the recent

[a] These authors contributed equally to this work

Current Topics in Membranes, Volume 75
ISSN 1063-5823
http://dx.doi.org/10.1016/bs.ctm.2015.03.006

advances in the field of GPI-AP trafficking, focusing on the mechanisms regulating their biosynthetic pathway and plasma membrane organization. We also discuss how alterations of these mechanisms can result in different diseases. Finally, we will examine the strict relationship between the trafficking and function of GPI-APs in epithelial cells.

Abbreviations

DRMs Detergent-resistant membranes
ER Endoplasmic reticulum
FCS Fluorescence correlation spectroscopy
FRAP Fluorescence recovery after photobleaching
FRET Fluorescence resonance energy transfer
GH Growth hormone
GPI-APs GPI-anchored proteins
HMW High molecular weight
MBCD Methyl-ß-cyclodextrin
N&B Number and brightness
SPT Single particle tracking
TOCCSL Thinning out clusters while conserving stoichiometry of labeling

1. INTRODUCTION

Glycosylphosphatidylinositol (GPI)-anchored proteins (GPI-APs) are a class of lipid-anchored proteins expressed from lower (yeast and protozoa) to higher eukaryotes. So far, more than 150 GPI-APs with diverse functions, including surface antigens, receptors, hydrolytic enzymes, cell adhesion molecules, and complement regulatory components, have been identified in mammalian cells (Chatterjee & Mayor, 2001; Ferguson, Kinoshita, & Hart, 2009; Kinoshita, Fujita, & Maeda, 2008). They are attached to the luminal/external leaflet of the membrane bilayer via the lipid portion of a conserved posttranslation glycolipid modification (GPI anchor). The highly conserved core of the GPI structure consists of a phospholipid moiety (phosphatidylinositol (PI)), a glycan backbone composed of a single glucosamine residue (which is linked to inositol by glycosidic bond), and three mannoses in turn linked to an ethanolamine phosphate engaged in an amide bond with the protein part (Ikezawa, 2002; McConville & Menon, 2000; Orlean & Menon, 2007).

GPI biosynthesis requires roughly 20 distinct gene products and occurs in the endoplasmic reticulum (ER) through at least 10 steps, involving the sequential addition of monosaccharides to PI (Kinoshita et al., 2008; Orlean & Menon, 2007). The mature GPI is then attached to ER-translocated proteins carrying a C-terminal GPI-attachment signal sequence. Variations of the structure include substituent on each of the mannose residues (e.g., phosphoethanolamine group, monosaccharide, etc.), differences in the lipid moiety (diacylglycerol, alkylacylglycerol, etc.), and the presence of acylation of the inositol. Some of these variations result from the GPI anchor modifications, which occur either in ER or in the Golgi of mammalian cells (Fujita & Jigami, 2008; Fujita & Kinoshita, 2010).

In addition to serving in membrane attachment, GPI anchors are crucial for protein functions. Loss of GPI anchoring is embryonically lethal (Nozaki et al., 1999), and defects in different GPI anchor biosynthetic steps cause several types of inheritable and acquired diseases (Almeida, Layton, & Karadimitris, 2009). The presence of a GPI anchor could confer unique characteristics to these proteins, resulting in exclusive functions. Indeed, often GPI-anchored and transmembrane forms of the same proteins play different or even opposite roles, as in the case of adhesion molecules NCAM and ApCAM or for CEA (Martin & Kandel, 1996; Screaton, DeMarte, Draber, & Stanners, 2000). This could depend on the affinity of GPI-APs for lipid rafts (membrane domains enriched in sphingolipid and cholesterol, see (Box 1)) and, therefore, on their specific submembrane organization.

GPI-APs are almost exclusively localized on the cell surface, confined to the outer leaflet of the bilayer, where they must exert their physiological functions. Thus, it is crucial to understand how the trafficking of GPI-APs is regulated, as well as to identify the molecular machinery required for their transport through the secretory pathway to their final destination. The presence of both the glycolipid and protein portions (and their subsequent modifications) confers special trafficking features to these proteins. Therefore, pathways involved in lipid and protein metabolisms may influence the intracellular trafficking of GPI-APs.

Here, we will summarize the current knowledge on the biosynthetic and exocytic pathways followed by GPI-APs to reach the cell surface. We will also examine their plasma membrane organization and discuss the mechanisms regulating it. Finally, on the basis of our recent findings, we will discuss the strict relationship that exists between GPI-AP sorting, trafficking, and function in polarized epithelial cells.

2. MECHANISMS OF INTRACELLULAR TRAFFICKING

Once having entered the secretory pathway, proteins are delivered to their correct destinations via the sequential action of several sorting signals

BOX 1 Lipid Rafts

In 1997, Kai Simons and Elina Ikonen reported a new concept regarding plasma membrane organization. They described the existence of membrane microdomains enriched in cholesterol and sphingolipids named rafts, in which cholesterol and sphingolipids have a critical role in the segregation of different types of proteins such as GPI-APs, certain transmembrane proteins, and acylated proteins (Simons & Ikonen, 1997). The raft concept has enabled better understanding of how protein function could be spatially and temporally regulated within the plasma membrane, by assigning new roles to lipids in addition to their function in bilayer assembly. Raft domains have been especially well studied in fibroblasts and immune cells, where they have been shown to be involved in the regulation of many cellular processes such as protein sorting, endocytosis, virus and bacteria infection, and cell signaling (Harder & Sangani, 2009; Parton & Richards, 2003; Pike, 2003; Simons & Gerl, 2010; Simons & Toomre, 2000).

It is important to note that as the size of raft domains is below the classical optical resolution of fluorescence microscopy, the study of raft domains is tightly linked to methodologies available to analyze their dynamics and visualize these membrane microdomains. Indeed, their existence has been questioned because of lack of detection using conventional fluorescence microscopy approaches. Another controversial issue is whether lipid microdomains preexist in a form similar to that observed in model membranes during phase separation of liquid-ordered and liquid-disordered domains, and/or whether they are created by mutual interactions with proteins. Recent evidence supports this latter hypothesis, assigning to proteins an important role (Kusumi et al., 2011; Owen et al., 2012a; Suzuki et al., 2012).

Thanks to the improvement of many biophysical approaches in 2015, including SPT, FRET, N&B, FCS, FRAP, and super-resolution microscopy, there is no doubt that the plasma membrane is organized in microdomains enriched in cholesterol and sphingolipids segregating different lipids and proteins and regulating their functions.

and multiple sorting events (Rodriguez-Boulan, Kreitzer, & Musch, 2005; Rodriguez-Boulan & Macara, 2014; Rodriguez-Boulan, Musch, & Le Bivic, 2004; Weisz & Rodriguez-Boulan, 2009). Proteins contain specific sorting signals which can be recognized by cellular machineries able to decode that information, and that can be used at each intermediate step of their journey: for forward transport, for retention in a specific compartment, or for retrieval to the compartment of origin. As for other secretory proteins, GPI-AP trafficking also appears to be regulated, and important discoveries have emerged in the last years shedding light on the molecular mechanisms involved.

2.1 ER-Golgi transport

Synthesis of GPI-APs and attachment of the GPI anchor occurs in the ER. A cleavable, hydrophobic amino-terminal signal sequence targets the protein to the ER, while a hydrophobic signal sequence at the C-terminus (named GPI attachment signal) directs GPI anchoring (Kinoshita et al., 2008; Orlean & Menon, 2007). The GPI attachment signal is recognized in the ER, cleaved (at the level of the omega site), and replaced by a preassembled GPI moiety, thanks to the action of a multimeric enzyme, the GPI-transamidase, in a cotranslational process (Kinoshita et al., 2008; Orlean & Menon, 2007). GPI anchoring is indispensable for further transport to the Golgi apparatus (Doering & Schekman, 1996; Field, Moran, Li, Keller, & Caras, 1994), as well as for the acquisition of the proper structure, which requires posttranslational modifications of the glycolipid and protein portions (Tanaka, Maeda, Tashima, & Kinoshita, 2004; Watanabe & Riezman, 2004).

Like other secretory cargos, both in yeast and in mammalian cells, GPI-APs utilize COPII machinery for their export from the ER (Bonnon, Wendeler, Paccaud, & Hauri, 2010; Doering & Schekman, 1996; Stephens & Pepperkok, 2004; Sutterlin, Doering, Schimmoller, Schroder, & Riezman, 1997). In yeast, GPI-APs are segregated from other plasma membrane proteins at the level of the ER (Muniz, Morsomme, & Riezman, 2001; Muniz & Zurzolo, 2014), and incorporated into different populations of COPII-coated vesicles upon concentration at specific ER export sites (ERES) (Castillon, Watanabe, Taylor, Schwabe, & Riezman, 2009; Muniz et al., 2001). This process is mediated by the small GTPase Ypt1p, and by specific tethering factors and SNARE proteins (Morsomme, Prescianotto-Baschong, & Riezman, 2003; Morsomme & Riezman, 2002; Muniz & Zurzolo, 2014). In contrast, in mammalian cells GPI-APs are concentrated at the same ERES (Stephens & Pepperkok, 2004; Rivier et al., 2010) and are transported to the Golgi in the same vesicles together with other secretory proteins (Muniz & Zurzolo, 2014; Rivier et al., 2010; Stephens & Pepperkok, 2004). Nevertheless, the expression of GTP-restricted mutant of Sar1 (Sar1H79G) in HeLa and CHO cells selectively inhibits the ER–Golgi transport of GPI-APs, which become unable to be incorporated into ERES and accumulate within the ER lumen (Rivier et al., 2010; Stephens & Pepperkok, 2004). Thus, although GPI-APs may use the same carriers of other plasma membrane proteins, these data indicate that the mechanism of recruitment into ERES is specific. Consistently, recent findings highlighted that ER export of GPI-APs is mediated by specific molecular requirements. Several studies revealed a role for p24 proteins, a family of type I transmembrane proteins (conserved from yeast to mammals) that contain binding motifs for COPI and COPII coats and cycle between ER and Golgi compartments (Strating, Hafmans, & Martens, 2009; Strating & Martens, 2009; Strating, van Bakel, Leunissen, & Martens, 2009). This protein family can be divided into four subfamilies: α, β, γ, δ (also classified as p25, p24, p28, and p23, respectively) (Strating, van Bakel, et al., 2009, Strating, Hafmans, et al., 2009). The traffic from the ER to the Golgi of GPI-APs, but not other cargo molecules, is slowed down upon knockdown of p23 and p24 (Bonnon et al., 2010; Fujita et al., 2011; Takida, Maeda, & Kinoshita, 2008) and their recruitment to ERES is drastically affected (Fujita et al., 2011). Moreover, it has been found that p23/p24 proteins physically interact with GPI-APs suggesting that p24 family acts as cargo receptor for GPI-APs by forming a large complex (Kinoshita, Maeda, & Fujita, 2013). Furthermore, a recent study showing that among the five γ isoforms, p24γ2 impairs the transport of a GPI-AP reporter, providing evidence that p24γ2 could play a crucial role in the cargo recognition (Theiler et al., 2014). The alpha-helical region of p24γ2 (containing coil-coiled domain in all p24 proteins) is crucial for GPI-AP transport (Theiler et al., 2014); however, further studies are necessary to determine the minimal amino acid sequence and/or specific amino acids critical for the binding. Moreover, it will be interesting to understand whether and how the other γ isoforms, which increase upon p24γ2 silencing, are involved. In addition, the precise stoichiometry and dynamics of the p24 complex is unknown. Some of the p24 proteins could have a role in favoring the partitioning of GPI-APs in specific ER membrane domains, since their association with GPI-APs appears to occur in detergent-resistant membranes (DRMs) and seems to be dependent on cholesterol (Bonnon et al., 2010).

Similar to the effect of p24/p23 knockdown, the ER-Golgi transport of CD59 and other GPI-AP reporters has been found to be delayed in CHO cells defective of post-GPI-attachment to protein 1 (PGAP1) or post-GPI-attachment to protein 5 (PGAP5), two enzymes mediating structural remodeling of the GPI anchor (inositol deacylation and removal of EtNP on second mannose, respectively) in the ER (Fujita et al., 2009, 2011; Tanaka et al., 2004). In both these mutants, GPI-APs are not recruited into ERES and do not coimmunoprecipitate with p23/p24 proteins (Fujita et al., 2011), indicating that the structural remodeling of the GPI anchor is crucial for the sorting of GPI-APs to the prebudding complexes at ERES. This could be explained by the fact that p24γ2 recognizes the glycan moiety of the lipid anchor or a specific conformation of the entire GPI-AP acquired as result of the GPI remodeling. It will be interesting to test these two hypotheses; however, because this interaction requires the coil-coiled domains of p24 (Theiler et al., 2014), which are generally implicated in protein–protein interactions, the second possibility may be more plausible. Moreover, while PGAP1 is distributed throughout the ER (Tanaka et al., 2004), PGAP5 is mainly localized into ERES (Fujita et al., 2009). This could therefore guarantee more efficient transport and/or assure another step of quality control.

Overall, these data indicate that GPI-APs are actively transported out the ER through a p24-dependent mechanism and that remodeling of the GPI anchor is required for ERES concentration. Following this step, Sec24C/D (components of inner COPII coat) have been found to be specifically required for GPI-AP transport (Bonnon et al., 2010), as they are recruited in Sar1-dependent manner and associate with p24 complex, thus ensuring the incorporation of GPI-APs in the nascent COPII vesicle. Interestingly, the association of p24 proteins with GPI-APs is pH dependent (Fujita et al., 2011). Therefore, one could imagine that within the neutral pH of the ER, p24 proteins could efficiently bind GPI-APs and promote their recruitment into COPII-coated prebudding complexes, while they would dissociate once reaching the acidic *cis*-Golgi compartments.

An intriguing question remains: Why are different mechanisms used for GPI-APs and other secretory proteins to enter in the same vesicle? Excluding the possibility that GPI-APs and other secretory cargos are detected in the same carriers because of the limits of optical resolution, it is possible that the use of specific mechanisms ensures a finer regulation by allowing switching on/off one specific secretory pathway in response to specific physiological requirements. This, for example, could be important upon ER stress, because the unfolded protein response (UPR) activates the expression of genes encoding key enzymes in lipid biosynthesis (Cox, Chapman, & Walter, 1997; Harding, Calfon, Urano, Novoa, & Ron, 2002) in addition to genes encoding ER-resident chaperones, members of the ER-associated degradation machinery, and other components of the secretory system. Thus, it could be hypothesized that upon UPR activation the trafficking of GPI-APs is guaranteed. Further studies will be necessary to test this hypothesis.

2.2 Golgi to plasma membrane transport

2.2.1 Mechanism of intracellular sorting in the Golgi complex

After leaving the ER, all plasma membrane proteins are transported through the Golgi complex where they are subjected to posttranslational modifications before being sorted to their final destination. In polarized epithelial cells, which must maintain the asymmetric composition of apical and basolateral plasma membrane domains required to carry out specialized vectorial functions (secretion, re-absorption, hormone response, etc.), the *trans*-Golgi network (TGN) is considered to be the main sorting station (Cao, Surma, & Simons, 2012; Rodriguez-Boulan et al., 2005; Rodriguez-Boulan & Macara, 2014; Rodriguez-Boulan & Musch, 2005). Indeed, at the level of the TGN apical and basolateral proteins are segregated into distinct vesicles and undertake different routes to the surface upon the recognition of specific apical or basolateral sorting signals (Cao et al., 2012; Mellman & Nelson, 2008; Mostov, Verges, & Altschuler, 2000; Rodriguez-Boulan et al., 2005; Rodriguez-Boulan & Macara, 2014). The TGN also appears to be a site of sorting for GPI-APs. Indeed, GPI-APs progressively segregate from other cargo molecules through the Golgi stack, and they have been shown to be completely segregated from transmembrane proteins in the TGN from where they exit in distinct vesicles (Keller, Toomre, Diaz, White, & Simons, 2001; Rustom et al., 2002).

Emerging data showing that both newly synthesized apical and basolateral proteins traverse endocytic compartments *en route* to the respective cell surface (Ang et al., 2004; Cresawn et al., 2007; Cramm-Behrens, Dienst, & Jacob, 2008; Farr, Hull, Mellman, & Caplan, 2009; Folsch, Mattila, & Weisz, 2009; Lock & Stow, 2005) broaden this view and suggest that protein sorting can also occur in post-TGN compartments (Folsch et al., 2009). This could imply either that proteins have to be sorted twice (i.e., in the Golgi and in post-TGN compartments) or that sorting occurs only in post-TGN compartments depending either on the protein or on the cell type. Studies based on monitoring the real-time transport of various GPI-APs to the surface in fully polarized live MDCK cells grown on filters clearly showed that, after exiting from the Golgi, GPI proteins gradually accumulate on the apical surface (Hua, Sheff, Toomre, & Mellman, 2006; Paladino, Pocard, Catino, & Zurzolo, 2006). Moreover, it was observed that their surface arrival was largely completed within 20 min and that the rate-limiting step for GPI-AP transport was the exit from the Golgi (Hua et al., 2006), supporting the hypothesis that apical sorting of GPI-APs occurs at the TGN. On the other hand, it has been reported that the GPI-anchored form of endolyn transits through early endosomes before reaching the apical surface (Cresawn et al., 2007). Thus, the existence of an endosomal sorting station for GPI-APs is conceivable and the use of this pathway could be a protein-specific feature (Cresawn et al., 2007). Further evidence assessing the use of this pathway for native GPI-APs needs to be gathered.

A peculiarity of GPI-APs is that they are mainly sorted to the apical surface of the majority of epithelia (Brown, Crise, & Rose, 1989; Lisanti, Caras, Davitz, & Rodriguez-Boulan, 1989). GPI-APs have been found to partition into sphingolipid- and cholesterol-rich microdomains (or rafts, see Box 1), which were proposed to act as apical sorting platforms because of their capacity to segregate specific classes of lipids and proteins enriched at the apical membranes (Simons & Ikonen, 1997). This original hypothesis was supported by the findings that GPI-APs acquire detergent resistance during their passage through the Golgi apparatus (Brown & Rose, 1992; Lipardi, Nitsch, & Zurzolo, 2000; Paladino et al., 2004) and that cholesterol and/or sphingolipid depletion impairs their apical sorting (Ehehalt et al., 2008; Lipardi et al., 2000; Paladino et al., 2004). However, there are some exceptions: (1) FRT cells sort the majority of endogenous DRM-associated GPI-APs to the basolateral surface (Lipardi et al., 2000; Zurzolo, Lisanti, Caras, Nitsch, & Rodriguez-Boulan, 1993); and (2) in MDCK cells, some GPI-APs are basolaterally sorted although DRM-associated (Benting, Rietveld, & Simons, 1999; Paladino et al., 2004; Sarnataro et al., 2002). Thus, the association of GPI-APs with detergent-resistant domains or lipid rafts is a general property of GPI-APs acquired as a consequence of fatty acid remodeling of the GPI anchor in the Golgi (Fujita & Kinoshita, 2012; Maeda & Kinoshita, 2011; Maeda et al., 2007), and is not sufficient for their apical sorting.

On the other hand we have shown that protein oligomerization is a key step to determine the apical sorting of GPI-APs (Paladino et al., 2004; Paladino & Zurzolo, 2009). Indeed, only apical GPI-APs are able to form high molecular weight (HMW) complexes during their transport to the plasma membrane, and disruption of this GPI-AP oligomerization capacity impairs their apical sorting (Paladino et al., 2004). This mechanism appears to be functional in epithelial cells of different origin (FRT and Caco-2, other than MDCK cells) (Paladino, Sarnataro, Tivodar, & Zurzolo, 2007), indicating that oligomerization might be the prime mechanism determining GPI-AP apical sorting in different epithelia.

Formation of HMW complexes of apical GPI-APs occurs in the Golgi concomitantly with their acquisition of detergent resistance (Paladino et al., 2004, 2007) suggesting that oligomerization and raft association cooperate in promoting apical sorting of GPI-APs. Therefore, two requirements appear to be necessary for apical sorting of GPI-APs (Figure 1(A)). The first one is partitioning into rafts. Apical and basolateral GPI-APs partition into rafts because of the natural chemical affinity of the GPI anchor for these lipid microdomains (Benting, Rietveld, Simons, et al., 1999; Lipardi et al., 2000; Paladino et al., 2004), and this might allow for their segregation from Golgi resident and other secretory proteins. The second one is clustering and stabilization into rafts. Only GPI-APs that are stabilized into rafts by protein oligomerization are able to be included in a raft-enriched apical vesicle (Figure 1(A)). Oligomerization might promote this process by increasing the affinity of GPI-APs for rafts, which in turn could result in a longer residency time in rafts, as previously also suggested for the case of GPI-AP sorting in early endosomes of GPI-APs (Fivaz et al., 2002; Howes, Mayor, & Parton, 2010). Alternatively, oligomerization could drive the coalescence of small rafts into a larger functional raft with subsequent formation of the apical vesicle. Rafts clustering could provide a mechanism for formation of the apical vesicles by increasing the local line energy at the boundary between raft and nonraft phases, thus enabling the membrane curvature necessary to drive vesicle formation (Huttner & Zimmerberg, 2001; Liu, Kaksonen, Drubin, & Oster, 2006; McMahon & Gallop, 2005; Tian & Baumgart, 2009). Indeed several studies showed that curvature can be modulated by changes in lipid composition as well as by the oligomerization of scaffolding proteins (McMahon & Gallop, 2005). Moreover, clustering of individual small rafts should reinforce the segregation of different lipid phases and consequently of membrane components associated with them, as proposed to be critical in other membrane functions such as signaling, protein internalization, and pathogen entry (Parton & Richards, 2003; Schuck & Simons, 2004; Surma, Klose, & Simons, 2012). Whether oligomerization per se or the interaction of GPI-AP oligomers with a putative receptor drives the coalescence of rafts into an apical vesicle is still unknown.

2.2.2 Mechanism of oligomerization

Lipid–lipid, glycan–lipid, protein–lipid, and protein–protein interactions could drive GPI-AP oligomerization and these interactions may involve the ectodomain, the GPI anchor, or both. The precise mechanisms by which apical GPI-APs oligomerize and how GPI-AP clustering is needed

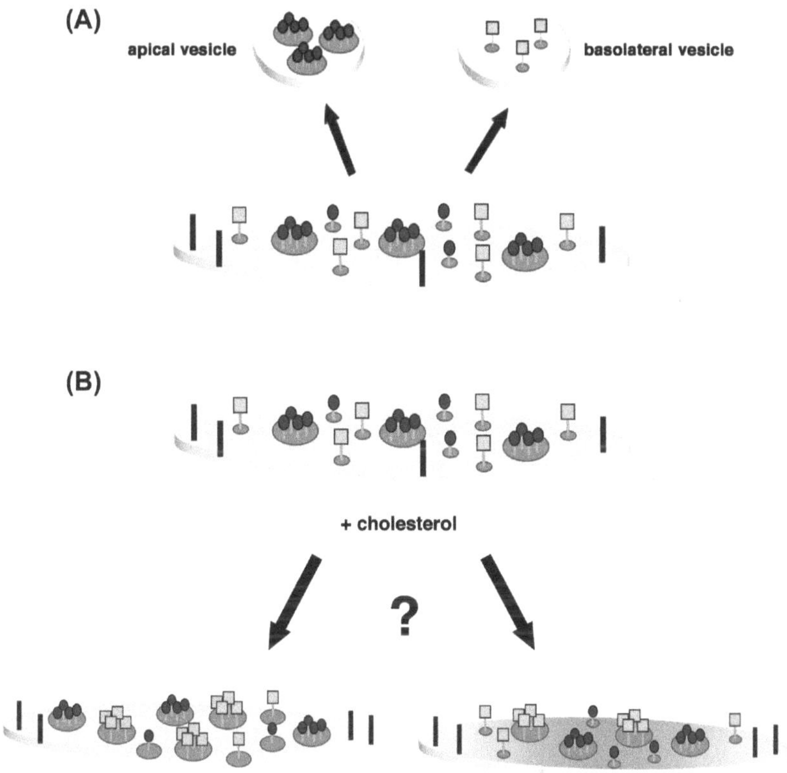

Figure 1 ***TGN sorting of GPI-APs in polarized epithelial cells.*** (A) At the TGN, the GPI anchor, both apical (red) and basolateral (cyan) GPI-APs, partition into lipid rafts and segregate from nonraft-transmembrane proteins (blue). Only apical GPI-APs are stabilized into rafts by protein oligomerization/clustering and are able to be included in a raft-enriched apical vesicle. Thus, two requirements are necessary for GPI-AP apical sorting. GPI-AP homoclustering might promote raft coalescence that reinforces the segregation between different lipid phases and consequently favors the formation of apical vesicles. Alternatively, a putative apical receptor (not shown here) could drive the raft coalescence by interacting directly with GPI-APs (by recognition of some modifications on the protein or more likely on the GPI anchor) or binding specific raft lipids. (B) Cholesterol has a crucial role in GPI-AP sorting in the TGN. In addition to cholesterol being necessary for oligomer formation, the addition of exogenous cholesterol is sufficient to lead to the oligomerization of basolateral GPI-APs and to redirect them to apical surface. Increased levels of cholesterol could stabilize the interaction of basolateral GPI-AP with rafts by favoring their oligomerization (left). Alternatively, cholesterol could change the properties of the Golgi lipid environment, thus enabling GPI-AP oligomerization and apical sorting (right). (See color plate)

for apical sorting are still not totally clear. However, as we explain below, in the last few years we have added many pieces to this puzzle.

GPI-AP oligomers, once formed, are insensitive to cholesterol depletion and are resistant to conditions in which DRMs are disrupted (e.g., SDS extraction), suggesting that they are maintained by protein—protein interactions (Paladino et al., 2004). Nonetheless, the fact that oligomerization occurs concomitantly with raft-association and cholesterol depletion impairs the oligomer formation in the Golgi apparatus (Paladino et al., 2004) suggests that rafts (or cholesterol-enriched membranes) may constitute a favorable environment for the formation of HMW complexes. It is therefore possible that in addition to the protein ectodomain, the GPI anchor also has a role in favoring clustering of apical GPI-APs. Several studies showed that differences in the lipid anchor (such as the length of acyl and alkyl chains, or the remodeling of glycan portion) as well as fatty acid remodeling resulting in the replacement of an unsaturated chain in *sn*-2 position with a long saturated one could mediate a different affinity for lipid rafts (Benting, Rietveld, Ansorge, & Simons, 1999; Fujita & Jigami, 2008; Jaensch, Correa, & Watanabe, 2014; Legler et al., 2005; Maeda et al., 2007; Tashima et al., 2006). Thus, a GPI lipid remodeling-dependent mechanism similar to the one that drives the GPI-AP concentration into ERES in yeast (Castillon et al., 2009; Muniz & Zurzolo, 2014) could contribute to Golgi sorting of GPI-APs in mammalian cells. In agreement with this hypothesis, it has been recently shown that different GPI-attachment signals (derived from an apical or basolateral native GPI-AP, respectively FR and PrP) affect the ability of the same protein ectodomain (GFP) to oligomerize and to be apically sorted (Paladino et al., 2008). Indeed, while GFP-FR is able to oligomerize and is apically sorted, GFP-PrP does not cluster and is direct to the basolateral domain (Paladino et al., 2008). According to their oligomeric state, at the level of the Golgi membranes GFP-PrP diffuses faster compared to GFP-FR (Lebreton, Paladino, & Zurzolo, 2008). Interestingly, exogenous cholesterol addition is sufficient to induce the basolateral GFP-PrP to oligomerize and to be redirected to the apical surface (Paladino et al., 2008). Cholesterol could stabilize the interaction of the basolateral GPI-AP with rafts (e.g., rigidifying the raft membranes or inducing the coalescence of small rafts) and therefore allowing its oligomerization and apical sorting (Figure 1(B), left). Alternatively, cholesterol could change the properties of the Golgi lipid environment thus enabling GPI-AP oligomerization and apical sorting (Figure 1(B), right). All these data suggest that a specific lipid environment is required for oligomerization and consequently for apical sorting of GPI-APs. It will be interesting to understand whether different GPI-attachment signals allow the attachment of a different anchor or how they could influence the remodeling of the anchor itself.

2.2.3 Protein ectodomain

Remarkably, recent findings highlighted that the lipid remodeling could influence the oligomerization of GPI-APs (Seong, Wang, Kinoshita, & Maeda, 2013), supporting the hypothesis that GPI anchor could play a role in GPI-AP clustering and sorting. However, the intrinsic capability of the ectodomain to oligomerize (so-called "permissive ectodomain") is required in order that GPI-APs are apically sorted. Indeed, cholesterol addition does not affect the sorting of the GFP-FR mutant carrying a double cys (S49/71) mutation which compromises its ability to oligomerize (Paladino et al., 2008). On the basis of all these findings, we have proposed that in order to be apically sorted, a GPI-AP should (1) partition in a favorable environment that allows oligomerization to occur and (2) have an intrinsic capability to form HMW complexes (Paladino et al., 2008; Paladino & Zurzolo, 2009). This might occur, thanks to the integrated action of the GPI anchor and the protein ectodomain.

Noncovalent interactions, mediated by posttranslational modifications could be responsible for the formation of oligomers. Glycosylation determinants within protein ectodomains could facilitate weak interactions and could therefore be suitable candidates. However, there are contrasting data regarding the role of *N*- and *O*-glycosylation in the sorting of GPI-APs. The addition of *N*-glycans to the GPI-anchored form of growth hormone confers its apical sorting (Benting, Rietveld, Ansorge, et al., 1999). While *N*-glycans promote apical targeting of the native GPI-AP dipeptidase (Pang, Urquhart, & Hooper, 2004), they are not required for the apical sorting of the GPI form of endolyn (Potter, Ihrke, Bruns, Weixel, & Weisz, 2004). Moreover, mutagenesis of putative *N*- and *O*-glycosylation sites does not affect the oligomerization and apical sorting of a native (PLAP) and a chimeric GPI-AP (NTR-PLAP) in MDCK cells (Catino, Paladino, Tivodar, Pocard, & Zurzolo, 2008), whereas inhibition of *N*-glycosylation by using tunicamycin does (Catino et al., 2008). This indicates that *N*-glycans of the protein ectodomain are not directly involved in GPI-AP sorting, but a putative *N*-glycosylated receptor could be implicated. Interestingly, in FRT cells the same mutant of PLAP lacking both *N*-glycosylation sites does not oligomerize and is missorted (Imjeti et al., 2011). This different behavior could be explained by the fact that the Golgi of FRT cells contains more cholesterol than MDCK cells and, thus, implies that at least two oligomerization mechanisms exist: one cholesterol-dependent and one cholesterol-independent and dependent on glycosylation (Imjeti et al., 2011).

2.2.4 Putative receptor or additional factors

In addition to oligomerization interaction with a putative receptor(s), additional factors could also be implicated in the formation of apical raft-enriched vesicle at the Golgi rim. For example, VIP17/MAL, a lipid raft-associated protein, has been shown to be involved in apical transport in MDCK cells (Cheong, Zacchetti, Schneeberger, & Simons, 1999). However, it does not seem to be specific for GPI-APs, as it is also necessary for apical transport of transmembrane proteins independently of their association with rafts (Martin-Belmonte et al., 2000).

Similarly, caveolins, which are involved in the internalization of cross-linked GPI-APs (Kirkham & Parton, 2005; Parton & Richards, 2003), do not appear to be good candidates because caveolin-1 knockdown does not affect sorting and kinetics of the apical transport of raft-associated transmembrane and GPI-linked proteins (Manninen et al., 2005). Other possible candidates could be some annexin isoforms (such as annexin II and annexin XIII), which are implicated in the apical transport of raft transmembrane proteins (Fiedler, Lafont, Parton, & Simons, 1995; Jacob et al., 2004; Lafont, Lecat, Verkade, & Simons, 1998; Lecat et al., 2000), although their mechanistic role is still unclear.

Other factors could be proteins belonging to the galectin family. While galectin-3 is clearly implicated in the transport of apical transmembrane proteins in different epithelial cells by sugar recognition (Delacour et al., 2006, 2008), galectin-4, which binds sulfatides carrying long-chain hydroxylated fatty acids of glycosphingolipids and is specifically enriched in lipid rafts could be a good candidate. However, knockdown of galectin-4 impairs the apical trafficking of both transmembrane and GPI-anchored raft-associated proteins (Delacour et al., 2005). This suggests that galectin-4 is not a specific sorting receptor for GPI-APs, but could favor the formation of apical lipid-raft vesicles by cross-linking different raft domains. It will be interesting to explore whether transmembrane and GPI-APs share the same TGN exit sites and/or whether they travel in the same carriers. A recent study showing that galectin-9, which binds the Forssmann glycolipid, regulates the apical trafficking in MDCK cells highlights a possible role in promoting raft clustering (Mishra, Grzybek, Niki, Hirashima, & Simons, 2010).

Another interesting candidate could be FAPP2 (four-phosphate-adaptor protein 2), which localizes to the TGN through binding the phosphoinositide PtdIns(4)P via a pleckstrin homology domain, since its knockdown resulted in delayed delivery and/or intracellular accumulation of apical raft-associated proteins (either transmembrane and GPI-APs) in polarized MDCK cells, whereas the basolateral transport was unaffected (Vieira, Verkade, Manninen, & Simons, 2005). Because FAPP2 is involved in the production of GSLs in the Golgi by transferring GlcCer (D'Angelo et al., 2007), it could promote their enrichment in specific Golgi subdomains in turn favoring the formation of apical raft carriers. A recent finding showing that in vitro FAPP2 has the capability to form tubules from membrane sheets by inducing positive curvature (Cao et al., 2009) supports this hypothesis. We can imagine that FAPP2 and galectins might work together as the two faces of the same machinery acting at the TGN: the first one modifying the lipid environment, and the second acting as bending factor, together promoting apical raft vesicle formation.

Although these are interesting hypotheses, further studies are needed to better clarify how these possible candidates work, as well as to identify novel factors.

3. GPI-AP PLASMA MEMBRANE ORGANIZATION

Our view of the plasma membrane organization has drastically changed from the fluid mosaic model of membranes described in the last century (Singer & Nicolson, 1972). Indeed, increasing evidence indicates that biological membranes are organized into compositionally and functionally distinct membrane domains, among which rafts are described as microdomains enriched in sphingolipids and cholesterol (Brown, 2006; Edidin, 2003; Kusumi, Suzuki, Kasai, Ritchie, & Fujiwara, 2011; Pike, 2006; Simons & Gerl, 2010; Simons & Sampaio, 2011; Simons & Vaz, 2004). As a consequence, proteins do not freely diffuse within plasma membrane, but are constrained depending on their affinity for these lipid microdomains. GPI-APs have been used for decades as raft markers and their plasma membrane organization has been the object of several studies (Brameshuber et al., 2010; Goswami et al., 2008; Sharma et al., 2004; Suzuki et al., 2012; Paladino et al., 2014). Thanks to the recent improvement of biophysical techniques allowing for studies of single molecules in vivo, our knowledge of GPI-AP organization has advanced considerably in the last 10 years.

The first evidence showing that GPI-AP are clustered at the plasma membrane was given by Friedrichson and Kurzchalia which showed that in living cells, two chimeric GPI-APs (GH-DAF and GFP-FR) could be cross-linked on the cell surface in clusters containing at least 15 molecules independently of the level of protein expression (Friedrichson & Kurzchalia, 1998). Similarly, two native GPI-APs (PLAP and folate receptor) are clustered at the surface of different cell lines (BHK, Jurkat, 3T3-L1) when they are cross-linked with specific antibodies (co-patching studies) (Harder, Scheiffele, Verkade, & Simons, 1998; Mayor, Rothberg, & Maxfield, 1994; Meder, Moreno, Verkade, Vaz, & Simons, 2006), thus indicating that GPI-APs are confined in restricted domains of the plasma membrane. Importantly, the clustering requires the GPI anchor since the transmembrane form of GH was not found in these cross-linkable complexes (Friedrichson & Kurzchalia, 1998). This organization is specific for raft proteins because it has been shown that GPI-APs co-patch with other GPI-APs or raft-associated transmembrane proteins (Harder et al., 1998; Meder et al., 2006). Although co-patching and cross-linking experiments provided interesting data, they did not demonstrate that in absence of these perturbations GPI-APs are clustered.

Pioneering experiments using fluorescence resonance energy transfer (FRET) (measured by photobleaching of the acceptor) in fixed cells revealed significant energy transfer between donor- and acceptor-labeled antibodies against the GPI-AP 5′ nucleotidase at the apical membrane of MDCK cells (Kenworthy & Edidin, 1998). However, the energy transfer was dependent on the surface density of molecules, indicating that in those experimental conditions FRET was due to the crowding of molecules (Kenworthy & Edidin, 1998).

The clustered organization of GPI-APs was also shown in native conditions for the first time in living cells by Mayor and colleagues using FRET anisotropy (Sharma et al., 2004; Varma & Mayor, 1998). This technique detects the proximity between similar fluorophores at 1- to 10-nm scale. These studies showed that 20–40% of GPI-APs are organized in nanoclusters that accommodate at most four molecules of different species (Sharma et al., 2004; Varma & Mayor, 1998) and exclude transmembrane proteins. Importantly, this nanocluster organization is independent of the surface density of GPI-APs at the cell surface and is dependent on cholesterol (Sharma et al., 2004; Varma & Mayor, 1998). In contrast, sphingolipid depletion had no effect on the homo-FRET of GPI-APs, therefore suggesting that sphingolipids do not have a direct role (Sharma et al., 2004).

Most of these studies were carried at the plasma membrane of nonpolarized fibroblastic cells. However, as described above, one specific characteristic of GPI-APs is to be selectively sorted to the apical plasma membrane in polarized epithelial cells. Interestingly, recent evidence indicates that the organization of GPI-AP in fully polarized epithelial cells is different from fibroblasts (Figure 2). Indeed, at the apical membrane of polarized MDCK cells, GPI-APs are organized in small clusters of single GPI-AP species (named homo-clusters), which are independent of cholesterol. These complexes, but not monomeric GPI-APs, coalesce into larger cholesterol-dependent clusters formed by multiple GPI-APs species termed heteroclusters (Paladino et al., 2014). Notably, GPI-AP heteroclustering is

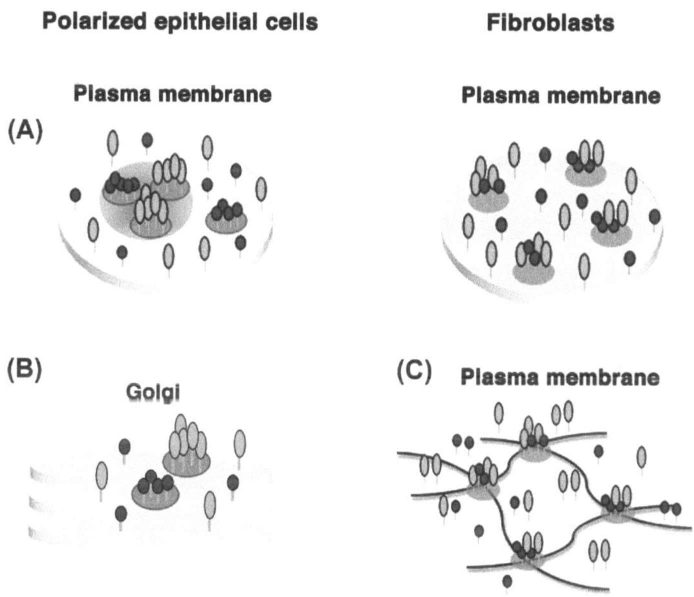

Figure 2 ***Plasma membrane organization of GPI-anchored proteins in polarized epithelial cells and in fibroblasts.*** The plasma membrane organization of GPI-APs is different in epithelial cells and fibroblasts. (A) While in fibroblasts, GPI-AP cholesterol-dependent nanoclusters accommodate three to five different proteins (red and green), in polarized epithelial cells apical GPI-APs are organized into cholesterol independent homoclusters of the same GPI-AP species (red and green aggregates), and cholesterol-dependent heteroclusters are derived from coalescence of homoclusters. (B) Homoclusters form in the Golgi and are a prerequisite for heterocluster formation. (C) Cortical actin regulates the formation/maintenance of GPI-AP clusters at the plasma membrane of fibroblasts, while it is unknown whether it plays a role in epithelial cells. (See color plate)

independent of the protein concentration, suggesting that these heteroclusters are actively formed (Paladino et al., 2014).

Interestingly, this heteroclustered organization of GPI-APs is achieved only when they are properly sorted to the apical plasma membrane in fully polarized cells, thus correlating Golgi GPI-AP sorting with their surface organization. Indeed, as described above, prior to their apical sorting GPI-APs form homoclusters in the Golgi in a cholesterol-dependent manner (Paladino et al., 2004); once having reached the apical surface these homoclusters coalesce in heteroclusters. Homoclusters are required for heteroclustering, as monomeric or dimeric GPI-APs (found in nonpolarized cells) do not form heteroclusters (Paladino et al., 2014). This results in a fundamentally different organization of GPI-APs in epithelial cells compared to fibroblasts, where GPI-AP clusters form at the plasma membrane from the coalescence of different GPI-APs and do not rely on pre-homoclustering in the Golgi apparatus (Paladino et al., 2014). The interdependence of the sorting of GPI-APs in the Golgi with their plasma membrane organization is critical for epithelial cells, since this will allow GPI-APs to exert their specific functions only at the apical surface. Hence, we discovered that this mechanism allows fine-tuning of GPI-AP activity and maintains the proteins in a suboptimal state at the cell surface until cells attain a polarized phenotype and the proteins are properly polarized. Consistently, in nonpolarized MDCK cells, GPI-APs leave the Golgi in monomers/dimers, remain unclustered at the cell surface, and display lower protein activity (Paladino et al., 2014). Together with the fact that cholesterol addition is sufficient to induce homo- and heterocluster formation in nonpolarized MDCK cells, cholesterol appears to be the major driver of GPI-AP clustering in epithelial cells (Lebreton et al., 2008; Paladino et al., 2008, 2014).

While cholesterol appears to be critical for GPI-AP clustering at the plasma membrane of both epithelial and fibroblastic cells, recent findings showed that in fibroblasts GPI-AP nanoclusters also depend on cortical actin (Goswami et al., 2008). Indeed it was shown that both cholesterol and actin perturbation affect the nanocluster organization of GPI-APs in fibroblasts and that actomyosin contractility is also required for their formation (Goswami et al., 2008). Interestingly, it has been shown that cholesterol depletion impacts on the organization of cortical actin (Chubinskiy-Nadezhdin, Efremova, Khaitlina, & Morachevskaya, 2013; Kwik et al., 2003; Sun et al., 2007). However, the mechanistic link between cholesterol and actin in GPI-AP organization is unknown, and possible molecules involved have yet to be identified. Furthermore, it will be interesting to investigate the role of actin cytoskeleton in GPI-AP organization in polarized cells as rearrangement of the actin cytoskeleton occurs during epithelial cell polarization.

3.1 Dynamics of GPI-AP clusters

Another important open question regarding GPI-APs organization is how dynamic GPI-APs clusters are. This property of GPI-APs has mainly been evaluated by analyzing the diffusional properties of GPI-APs with different imaging approaches. FRAP studies showed that GPI-APs exhibit higher apparent diffusion coefficient and higher mobile fraction compared to transmembrane proteins at the surface of different cell lines. Furthermore, upon MBCD treatment the apparent diffusion coefficient of both raft- and nonraft-associated proteins was affected (Kenworthy et al., 2004). Interestingly, GPI-APs behave differently in nonpolarized (Ptk2) and polarized (MDCK, Caco2 cells) epithelial cells as shown by an exhaustive study by Meder and colleagues which compared the diffusion properties of GPI-APs with raft- or nonraft-associated transmembrane proteins in these cells (Meder et al., 2006). While in nonpolarized Ptk2 cells all the proteins studied had similar diffusion characteristics, at the apical surface of polarized cells the same proteins exhibited different diffusion properties and could be divided into two different groups. The raft-associated proteins (both GPI-APs and transmembrane) diffused rapidly and recovered to nearly 100%, whereas the nonraft-associated proteins displayed slower diffusion with only partial recovery, indicating that the membrane environment of these proteins is different (Meder et al., 2006). Moreover, other FRAP studies reported that in polarized MDCK cells apical and basolateral GPI-APs diffuse much faster than apical and basolateral transmembrane nonraft-associated proteins (Lebreton et al., 2008).

While in fibroblast cells, cholesterol depletion affects the diffusion of raft (GPI-AP and transmembrane) and nonraft-proteins (Kenworthy et al., 2004), it alters only the diffusion of GPI-APs in polarized epithelial cells (Lebreton et al., 2008). Moreover, in polarized cells GPI-AP diffusion is independent of actin pertubation, which instead affects raft- and nonraft-associated transmembrane proteins (Lebreton et al., 2008). In agreement with these studies, Lenne et al. (2006), using FCS, arrived at the same conclusion by performing an analysis in fibroblasts (COS-7 cells), thus suggesting that GPI-AP dynamics are not regulated by actin. More experiments are needed to understand these findings in light of the proposed role of actin in the organization and formation of GPI-AP clusters (Goswami et al., 2008; Gowrishankar et al., 2009). On the other hand, in COS-7 cells the diffusion of GPI-APs has been found dependent on sphingomyelinase (Lenne et al., 2006). Considering that the clusters are unaffected by sphingolipid depletion as above described (Sharma et al., 2004), one could speculate that sphingolipids are not essential to maintain the clusters, but are necessary to regulate their lateral organization. Further studies are necessary to test this hypothesis.

Single particle tracking (SPT) revealed that GPI-APs are mobile and move in a transient confinement zone of 30–250 nm, exhibiting alternating periods of apparently simple Brownian diffusion and temporary immobilization (Fujiwara, Ritchie, Murakoshi, Jacobson, & Kusumi, 2002; Schutz, Kada, Pastushenko, & Schindler, 2000; Sheets, Lee, Simson, & Jacobson, 1997; Suzuki, Fujiwara, Edidin, & Kusumi, 2007; Suzuki, Fujiwara, Sanematsu, et al., 2007). Along the same lines, it has recently been shown that from transient homodimers (200 ms) in resting cells, GPI-APs form stable clusters (up to four molecules) upon ligand binding (Suzuki et al., 2012), thus suggesting that transient immobilization functions as a platform for transducing extracellular signals (Suzuki, Fujiwara, Edidin, et al., 2007; Suzuki, Fujiwara, Sanematsu, et al., 2007; Suzuki et al., 2012).

All of these data indicate that GPI-APs diffuse in the membrane although a fraction of the molecules are present in small clusters. Nevertheless, the scenario depicted by other techniques is less clear: while by homo-FRET anisotropy, Goswami reported that GPI-AP nanoclusters are immobile (Goswami et al., 2008), Brameshuber and colleagues, using TOCCSL, propose that in the plasma membrane of living CHO and Jurkat cells small, mobile, and long-lived nanoplatforms are present (Brameshuber et al., 2010). Whether these discrepancies are derived from differences in the techniques used, the identity of the clusters observed, or the timescales of the different types of experiments needs to be further investigated.

3.2 Size of GPI-AP clusters

By immunoelectron microscopy GPI-APs have been detected in clusters as well as in caveolae after being cross-linked with antibodies (Harder et al., 1998; Howell, Reuter-Carlson, Devaney, Luzio, & Fuller, 1987; Mayor et al., 1994). Moreover, in co-patching studies it has been inferred that the cluster size of PLAP in MDCK cells is much smaller than in fibroblast BHK cells (Meder et al., 2006), although mechanism underlying this difference is unknown.

To overcome the limitations of electron microscopy (e.g., efficiency in protein labeling, possible artifacts due to fixation procedures), super-resolution fluorescence imaging methods such as stochastic optical reconstruction microscopy (STORM), photo activated localization microscopy (PALM), stimulated emission depletion (STED), and near-field scanning optical microscope (NSOM) have been applied to investigate GPI-AP organization in the plasma membrane (Galbraith & Galbraith, 2011; Owen, Magenau, Williamson, & Gaus, 2013, Owen et al., 2010; Sengupta, Jovanovic-Talisman, & Lippincott-Schwartz, 2013; Sengupta & Lippincott-Schwartz, 2012; Sengupta, Van Engelenburg, & Lippincott-Schwartz, 2012).

So far, different super-resolution methods revealed that at the cell surface of fibroblasts, immune and nonpolarized epithelial cells, about 20–40% of GPI-APs are organized in clusters with a size below 70 nm and containing three to five proteins (Eggeling et al., 2009; Sengupta et al., 2011; van Zanten et al., 2009). Furthermore, super-resolution has also been applied to study the organization and dynamics of lipids in cell membranes (Mueller et al., 2010; Owen et al., 2013; Owen et al., 2012b). For instance, STED-FCS has been shown to be a sensitive tool to study lipid dynamics and lipid–lipid and lipid–cytoskeleton interactions (Mueller et al., 2010). By using PALM, a recent publication showed that cholesterol- and sphingomyelin-enriched domains occupy different regions on the plasma membrane of HeLa cells (Mizuno et al., 2011). Overall, the development of these methods will provide a better understanding of features of plasma membrane organization and dynamics that are critical for biological protein activities.

4. ROLE OF INTRACELLULAR TRAFFICKING IN GPI-AP FUNCTION

GPI-APs play a wide variety of physiological roles including nutrient uptake, signaling, enzymatic catalysis, cell surface protection, cell adhesion and migration, and immune recognition (Chatterjee & Mayor, 2001; Kinoshita et al., 2008; Lakhan, Sabharanjak, & De, 2009). Hence, it is not surprising that alterations of the intracellular trafficking of GPI-APs and/or their plasma membrane organization can result in the development of different diseases. Because the presence of the GPI anchor is critical for protein functions and correct processing of GPI anchors is fundamental for protein targeting, it is likely that defects in GPI anchor biosynthesis/remodeling can cause inheritable and acquired disease.

While a defective PIG-A, responsible for the first step in GPI biosynthesis, causes early arrest of embryonic development (ninth day of gestation of transgenic mice) (Nozaki et al., 1999), its somatic mutation in hematopoietic cells leads to paroxysmal nocturnal hemoglobinuria (PNH) disease (Brodsky,

2014; Karadimitris & Luzzatto, 2001), a rare human disease characterized by intravascular hemolysis, thrombosis, and bone marrow failure. Red blood cells of PNH patients do not express GPI-APs, which include complement regulatory proteins such as CD55 and CD59. As a consequence, they become susceptible to complement-mediated lysis. Nevertheless, the role of PIG-A in development as well as in bone marrow failure is still unclear. Interestingly, a recent study showed that PIG-A-null iPSCs (induced pluripotent stem cells) were unable to generate hematopoietic cells and to differentiate mesodermal cells, thus suggesting that the expression of certain GPI-APs are required for the development of mesoderm (Yuan et al., 2013).

Moreover, emerging data have highlighted that other genes involved in GPI anchor biosynthesis (PIG-L, PIG-M, PIG-V, PIG-N, PIG-O) have been associated with the onset of different diseases (Almeida et al., 2009, 2006; Hansen et al., 2013; Krawitz et al., 2012; Maydan et al., 2011; Nakamura et al., 2014; Ng et al., 2012). Hyperphosphatasia with mental retardation syndrome (HPMRS), an autosomal-recessive form of intellectual disability characterized by facial dysmorphism, seizures, brachytelephalangy, and hyperphosphatasia, was recently shown to be caused by mutations in PIG-V or in PIG-O (Krawitz et al., 2012, 2010). While it is clear that deficiency in GPI-anchor synthesis is crucial for the development of this disease, the underlying molecular pathophysiological mechanism is not yet understood. Interestingly, a recent paper reported that missense mutations in PIG-T cause multiple malformations and intellectual disability (Kvarnung et al., 2013), thus suggesting that GPI synthesis is critical for the correct development of the nervous system. Consistently, mutations in PIG-L and PIG-N are associated with syndromes characterized prevalently by intellectual disability, hypotonia epilepsy, and even progressive cerebella atrophy (Maydan et al., 2011; Ng et al., 2012; Ohba et al., 2013).

Recent findings showed that mutations in genes involved in the remodeling of the GPI anchor which also lead to a wide phenotypic spectrum ranging from syndromic disorders with various malformations to nonspecific forms of intellectual disability (Hansen et al., 2013; Howard et al., 2014; Krawitz et al., 2013; Murakami et al., 2014). Interestingly, different mutations in PGAP-2 genes cause HPMRS (Krawitz et al., 2013) or an autosomal-recessive intellectual disability (Hansen et al., 2013). Although unknown, the pathogenic mechanism of these disorders could be associated to defects in the proper membrane organization of GPI-APs since it has been shown that in mutant CHO cells lacking PGAP2 or PGAP3, the enzymes responsible in the substitution of unsaturated chain with saturated one,

GPI-APs are unable to associate with lipid rafts (Maeda et al., 2007). A homozygous three base pair deletion in the GPI remodeling gene PGAP1 has been identified for the first time in a family with encephalopathy and nonspecific intellectual disability (Murakami et al., 2014). Interestingly, the expression levels of GPI-APs on B lymphoblastoid cells derived from an affected person were normal, but the GPI-APs were PIPL-C resistant, thus suggesting that the correct structure of GPI anchor could be critical for neuronal development. However, considering that the ER-Golgi transport rate was reduced fourfold in PGAP-1 defective CHO cells compared to wild-type cells, another possibility to consider for the pathogenesis of the diseases is that this inefficient trafficking could affect neuronal differentiation in the early phases of development.

Much evidence over the past decades suggests that different GPI-APs are involved in a range of cancers (Gamage & Hendrickson, 2013; Kalli et al., 2008). Interestingly, recent data highlight that some subunits of GPI-transamidase, the multienzymatic complex responsible for the GPI attachment to protein, can act as oncogenes. For instance, PIG-U has been found overexpressed in bladder cancer and the induction of its overexpression in cell cultures leads to an increase in cell growth (Guo et al., 2004). Significant overexpression was observed in colon, thyroid, lung, and pancreatic cancers (Nagpal et al., 2008) as well as for PIG-K (the catalytic subunit) in ovarian and uterine cancers. Interestingly, PIG-K has been found to be downregulated in other cancers like bladder, colon, and hepatocellular carcinoma (Nagpal et al., 2008). It will be important to understand how the overexpression or downregulation of these proteins correlates with tumorigenesis. This could be dependent in turn on the specific GPI-APs that become overexpressed or downregulated in the different tissues.

In the case of GPI-AP surface receptors involved in the regulation of proliferation and/or differentiation pathways, it is likely that alteration of their intracellular trafficking as well as their suborganization at the plasma membrane could lead to changes in these signaling pathways and therefore to cancer development. Moreover, the altered localization of GPI-APs could imply loss or gain of function as consequence of their partitioning into a different environment. This seems to be the case for the apical GPI-AP CEA that has been found basolaterally localized in colon carcinoma cells (Ehehalt et al., 2008). This altered localization appears to promote invasion by interfering with normal intercellular adhesion (Ilantzis, DeMarte, Screaton, & Stanners, 2002).

Another interesting case is the folate receptor alpha (FRα), an apical GPI-AP that has a restricted expression in normal tissues but that is overexpressed in several human malignancies (breast, ovarian, lung) (Parker et al., 2005; Shia et al., 2008). FRα is one of the four isoforms that allow internalization of folate that is critical for all dividing cells. It was shown that FR bearing a transmembrane domain does not efficiently uptake the ligand and is not regulated in a similar manner as the GPI-anchored form (Ritter, Fajardo, Matsue, Anderson, & Lacey, 1995; Sabharanjak & Mayor, 2004). In MA104 cells, it has been reported that clustering of GPI-anchored folate receptor is critical for uptake of the ligand (Smart, Mineo, & Anderson, 1996), therefore establishing a direct link between the plasma membrane organization of GPI-APs and their physiological functions. Thus, it will be interesting to characterize the functional organization of FRα (e.g., correlation between plasma membrane organization and function) in ovarian cancer cells, where this receptor is a target for therapeutic approaches (Kalli et al., 2008). Further studies are necessary to investigate these different aspects and gain new insight into the molecular basis of cancer.

5. CONCLUSIONS AND FUTURE PROSPECTS

In the last decade, especially thanks to the advent of new imaging techniques, novel findings have accumulated on the trafficking of GPI-APs and on the molecular mechanisms regulating this process, leading to the emergence of new concepts regarding the cell biology of GPI-proteins.

First, it is clear that in mammals the trafficking of GPI-APs is strictly regulated at all steps along the secretory pathway. Evidence indicating that GPI-APs are actively exported from the ER through a selective mechanism based on recognition of specific signals has changed previous ideas that these proteins exit the ER by passive bulk flow. However, it remains unclear why different mechanisms of ER recruitment/export exist when GPI-APs and other secretory proteins use the same post-ER carriers. It will be therefore important to understand the physiological relevance of this process.

It is now clear that in polarized epithelial cells, apical sorting of GPI-APs occurs in the Golgi apparatus and is dependent on two factors: association with rafts and protein oligomerization. While it is clear that this depends on cholesterol, the next challenge is to understand the exact molecular mechanism regulating GPI-AP oligomerization, as well as whether additional factors are implicated in the formation of the raft-enriched apical vesicle. Furthermore, although the mechanism of oligomerization appears to be specific for GPI-proteins, it is still unknown whether different raft-associated GPI-APs and transmembrane proteins partition in the same lipid environment or whether they segregate at the level of the TGN or travel together toward the apical surface. Moreover, it will be interesting to clarify whether endosomal compartments play a similar role in the sorting of GPI-APs as they do for some transmembrane proteins.

The second concept that has emerged in recent years is that the GPI anchor is not just a way to attach proteins to membranes, but through its modifications plays an active role in the trafficking of GPI-APs by regulating their transport at the different stations of the secretory pathway. While it has been demonstrated that GPI remodeling is critical for ER exit, its role for Golgi sorting could be postulated on the basis that differences in the lipid anchor could mediate a different affinity of GPI-APs for lipid rafts. Further studies are required to elucidate this aspect as well as to understand whether the remodeling of lipid, the glycan portion, or both could contribute to GPI-AP sorting at the TGN.

Finally, proper trafficking and membrane organization are crucial for the functions that GPI-APs exert at the cell surface. Recent findings highlighted this relationship in polarized epithelial cells where the Golgi-dependent GPI-AP homoclusters are essential for the clustered and functional organization of GPI-APs at the apical plasma membrane. Because the dysregulation of these processes can result in development of different diseases, it will be fundamental to better investigate the molecular mechanism regulating the traffic of GPI-APs as well as the formation and maintenance of the GPI-AP clusters and their dynamics. Thus, these should be the subjects of future studies.

ACKNOWLEDGMENTS

This work was supported by grants from ANR (Agence Nationale de la recherche), European Union FP7, and FRM (Fondation Recherche Médicale) to CZ; AIRC grant (MFAG 2007–2009) to SP.

REFERENCES

Almeida, A., Layton, M., & Karadimitris, A. (2009). Inherited glycosylphosphatidyl inositol deficiency: a treatable CDG. *Biochimica et Biophysica Acta, 1792*(9), 874–880.

Almeida, A. M., Murakami, Y., Layton, D. M., Hillmen, P., Sellick, G. S., Maeda, Y., et al. (2006). Hypomorphic promoter mutation in PIGM causes inherited glycosylphosphatidylinositol deficiency. *Nature Medicine, 12*(7), 846–851.

Ang, A. L., Taguchi, T., Francis, S., Folsch, H., Murrells, L. J., Pypaert, M., et al. (2004). Recycling endosomes can serve as intermediates during transport from the Golgi to the plasma membrane of MDCK cells. *Journal of Cell Biology, 167*(3), 531–543.

Benting, J., Rietveld, A., Ansorge, I., & Simons, K. (1999). Acyl and alkyl chain length of GPI-anchors is critical for raft association in vitro. *FEBS Letters, 462*(1–2), 47–50.

Benting, J. H., Rietveld, A. G., & Simons, K. (1999). N-Glycans mediate the apical sorting of a GPI-anchored, raft-associated protein in Madin-Darby canine kidney cells. *Journal of Cell Biology, 146*(2), 313–320.

Bonnon, C., Wendeler, M. W., Paccaud, J. P., & Hauri, H. P. (2010). Selective export of human GPI-anchored proteins from the endoplasmic reticulum. *Journal of Cell Science, 123*(Pt 10), 1705–1715.

Brameshuber, M., Weghuber, J., Ruprecht, V., Gombos, I., Horvath, I., Vigh, L., et al. (2010). Imaging of mobile long-lived nanoplatforms in the live cell plasma membrane. *Journal of Biological Chemistry, 285*(53), 41765–41771.

Brodsky, R. A. (2014). Paroxysmal nocturnal hemoglobinuria. *Blood, 124*(18), 2804–2811.

Brown, D. A. (2006). Lipid rafts, detergent-resistant membranes, and raft targeting signals. *Physiology (Bethesda), 21*, 430–439.

Brown, D. A., Crise, B., & Rose, J. K. (1989). Mechanism of membrane anchoring affects polarized expression of two proteins in MDCK cells. *Science, 245*(4925), 1499–1501.

Brown, D. A., & Rose, J. K. (1992). Sorting of GPI-anchored proteins to glycolipid-enriched membrane subdomains during transport to the apical cell surface. *Cell, 68*(3), 533–544.

Cao, X., Coskun, U., Rossle, M., Buschhorn, S. B., Grzybek, M., Dafforn, T. R., et al. (2009). Golgi protein FAPP2 tubulates membranes. *Proceedings of the National Academy of Sciences of the United States of America, 106*(50), 21121–21125.

Cao, X., Surma, M. A., & Simons, K. (2012). Polarized sorting and trafficking in epithelial cells. *Cell Research, 22*(5), 793–805.

Castillon, G. A., Watanabe, R., Taylor, M., Schwabe, T. M., & Riezman, H. (2009). Concentration of GPI-anchored proteins upon ER exit in yeast. *Traffic, 10*(2), 186–200.

Catino, M. A., Paladino, S., Tivodar, S., Pocard, T., & Zurzolo, C. (2008). N- and O-glycans are not directly involved in the oligomerization and apical sorting of GPI proteins. *Traffic, 9*(12), 2141–2150.

Chatterjee, S., & Mayor, S. (2001). The GPI-anchor and protein sorting. *Cellular and Molecular Life Sciences, 58*(14), 1969–1987.

Cheong, K. H., Zacchetti, D., Schneeberger, E. E., & Simons, K. (1999). VIP17/MAL, a lipid raft-associated protein, is involved in apical transport in MDCK cells. *Proceedings of the National Academy of Sciences of the United States of America, 96*(11), 6241–6248.

Chubinskiy-Nadezhdin, V. I., Efremova, T. N., Khaitlina, S. Y., & Morachevskaya, E. A. (2013). Functional impact of cholesterol sequestration on actin cytoskeleton in normal and transformed fibroblasts. *Cell Biology International, 37*(6), 617–623.

Cox, J. S., Chapman, R. E., & Walter, P. (1997). The unfolded protein response coordinates the production of endoplasmic reticulum protein and endoplasmic reticulum membrane. *Molecular Biology of the Cell, 8*(9), 1805–1814.

Cramm-Behrens, C. I., Dienst, M., & Jacob, R. (2008). Apical cargo traverses endosomal compartments on the passage to the cell surface. *Traffic, 9*(12), 2206–2220.

Cresawn, K. O., Potter, B. A., Oztan, A., Guerriero, C. J., Ihrke, G., Goldenring, J. R., et al. (2007). Differential involvement of endocytic compartments in the biosynthetic traffic of apical proteins. *EMBO Journal, 26*(16), 3737–3748.

D'Angelo, G., Polishchuk, E., Di Tullio, G., Santoro, M., Di Campli, A., Godi, A., et al. (2007). Glycosphingolipid synthesis requires FAPP2 transfer of glucosylceramide. *Nature, 449*(7158), 62–67.

Delacour, D., Cramm-Behrens, C. I., Drobecq, H., Le Bivic, A., Naim, H. Y., & Jacob, R. (2006). Requirement for galectin-3 in apical protein sorting. *Current Biology, 16*(4), 408–414.

Delacour, D., Gouyer, V., Zanetta, J. P., Drobecq, H., Leteurtre, E., Grard, G., et al. (2005). Galectin-4 and sulfatides in apical membrane trafficking in enterocyte-like cells. *Journal of Cell Biology, 169*(3), 491–501.

Delacour, D., Koch, A., Ackermann, W., Eude-Le Parco, I., Elsasser, H. P., Poirier, F., et al. (2008). Loss of galectin-3 impairs membrane polarisation of mouse enterocytes in vivo. *Journal of Cell Science, 121*(Pt 4), 458–465.

Doering, T. L., & Schekman, R. (1996). GPI anchor attachment is required for Gas1p transport from the endoplasmic reticulum in COP II vesicles. *EMBO Journal, 15*(1), 182–191.

Edidin, M. (2003). The state of lipid rafts: from model membranes to cells. *Annual Review of Biophysics and Biomolecular Structure, 32*, 257–283.

Eggeling, C., Ringemann, C., Medda, R., Schwarzmann, G., Sandhoff, K., Polyakova, S., et al. (2009). Direct observation of the nanoscale dynamics of membrane lipids in a living cell. *Nature, 457*(7233), 1159–1162.

Ehehalt, R., Krautter, M., Zorn, M., Sparla, R., Fullekrug, J., Kulaksiz, H., et al. (2008). Increased basolateral sorting of carcinoembryonic antigen in a polarized colon carcinoma cell line after cholesterol depletion-Implications for treatment of inflammatory bowel disease. *World Journal of Gastroenterology, 14*(10), 1528–1533.

Farr, G. A., Hull, M., Mellman, I., & Caplan, M. J. (2009). Membrane proteins follow multiple pathways to the basolateral cell surface in polarized epithelial cells. *Journal of Cell Biology, 186*(2), 269–282.

Ferguson, M. A. J., Kinoshita, T., & Hart, G. W. (2009). *Glycosylphosphatidylinositol anchors.*

Fiedler, K., Lafont, F., Parton, R. G., & Simons, K. (1995). Annexin XIIIb: a novel epithelial specific annexin is implicated in vesicular traffic to the apical plasma membrane. *Journal of Cell Biology, 128*(6), 1043–1053.

Field, M. C., Moran, P., Li, W., Keller, G. A., & Caras, I. W. (1994). Retention and degradation of proteins containing an uncleaved glycosylphosphatidylinositol signal. *Journal of Biological Chemistry, 269*(14), 10830–10837.

Fivaz, M., Vilbois, F., Thurnheer, S., Pasquali, C., Abrami, L., Bickel, P. E., et al. (2002). Differential sorting and fate of endocytosed GPI-anchored proteins. *EMBO Journal, 21*(15), 3989–4000.

Folsch, H., Mattila, P. E., & Weisz, O. A. (2009). Taking the scenic route: biosynthetic traffic to the plasma membrane in polarized epithelial cells. *Traffic, 10*(8), 972–981.

Friedrichson, T., & Kurzchalia, T. V. (1998). Microdomains of GPI-anchored proteins in living cells revealed by crosslinking. *Nature, 394*(6695), 802–805.

Fujita, M., & Jigami, Y. (2008). Lipid remodeling of GPI-anchored proteins and its function. *Biochimica et Biophysica Acta, 1780*(3), 410–420.

Fujita, M., & Kinoshita, T. (2010). Structural remodeling of GPI anchors during biosynthesis and after attachment to proteins. *FEBS Letters, 584*(9), 1670–1677.

Fujita, M., & Kinoshita, T. (2012). GPI-anchor remodeling: potential functions of GPI-anchors in intracellular trafficking and membrane dynamics. *Biochimica et Biophysica Acta, 1821*(8), 1050–1058.

Fujita, M., Maeda, Y., Ra, M., Yamaguchi, Y., Taguchi, R., & Kinoshita, T. (2009). GPI glycan remodeling by PGAP5 regulates transport of GPI-anchored proteins from the ER to the Golgi. *Cell, 139*(2), 352–365.

Fujita, M., Watanabe, R., Jaensch, N., Romanova-Michaelides, M., Satoh, T., Kato, M., et al. (2011). Sorting of GPI-anchored proteins into ER exit sites by p24 proteins is dependent on remodeled GPI. *Journal of Cell Biology, 194*(1), 61–75.

Fujiwara, T., Ritchie, K., Murakoshi, H., Jacobson, K., & Kusumi, A. (2002). Phospholipids undergo hop diffusion in compartmentalized cell membrane. *Journal of Cell Biology, 157*(6), 1071–1081.

Galbraith, C. G., & Galbraith, J. A. (2011). Super-resolution microscopy at a glance. *Journal of Cell Science, 124*(Pt 10), 1607–1611.

Gamage, D. G., & Hendrickson, T. L. (2013). GPI transamidase and GPI anchored proteins: oncogenes and biomarkers for cancer. *Critical Reviews in Biochemistry and Molecular Biology, 48*(5), 446–464.

Goswami, D., Gowrishankar, K., Bilgrami, S., Ghosh, S., Raghupathy, R., Chadda, R., et al. (2008). Nanoclusters of GPI-anchored proteins are formed by cortical actin-driven activity. *Cell, 135*(6), 1085–1097.

Gowrishankar, K., Ghosh, S., Saha, S. C. R., Mayor, S., & Rao, M. (2009). Active remodeling of cortical actin regulates spatiotemporal organization of cell surface molecules. *Cell, 149*(6), 1353–1367.

Guo, Z., Linn, J. F., Wu, G., Anzick, S. L., Eisenberger, C. F., Halachmi, S., et al. (2004). CDC91L1 (PIG-U) is a newly discovered oncogene in human bladder cancer. *Nature Medicine, 10*(4), 374–381.

Hansen, L., Tawamie, H., Murakami, Y., Mang, Y., ur Rehman, S., Buchert, R., et al. (2013). Hypomorphic mutations in PGAP2, encoding a GPI-anchor-remodeling protein, cause autosomal-recessive intellectual disability. *American Journal of Human Genetics, 92*(4), 575–583.

Harder, T., & Sangani, D. (2009). Plasma membrane rafts engaged in T cell signalling: new developments in an old concept. *Cell Communication and Signaling, 7*, 21.

Harder, T., Scheiffele, P., Verkade, P., & Simons, K. (1998). Lipid domain structure of the plasma membrane revealed by patching of membrane components. *Journal of Cell Biology, 141*(4), 929–942.

Harding, H. P., Calfon, M., Urano, F., Novoa, I., & Ron, D. (2002). Transcriptional and translational control in the mammalian unfolded protein response. *Annual Review of Cell and Developmental Biology, 18*, 575–599.

Howard, M. F., Murakami, Y., Pagnamenta, A. T., Daumer-Haas, C., Fischer, B., Hecht, J., et al. (2014). Mutations in PGAP3 impair GPI-anchor maturation, causing a subtype of hyperphosphatasia with mental retardation. *American Journal of Human Genetics, 94*(2), 278–287.

Howell, K. E., Reuter-Carlson, U., Devaney, E., Luzio, J. P., & Fuller, S. D. (1987). One antigen, one gold? A quantitative analysis of immunogold labeling of plasma membrane 5'-nucleotidase in frozen thin sections. *European Journal of Cell Biology, 44*(2), 318–327.

Howes, M. T., Mayor, S., & Parton, R. G. (2010). Molecules, mechanisms, and cellular roles of clathrin-independent endocytosis. *Current Opinion in Cell Biology, 22*(4), 519–527.

Hua, W., Sheff, D., Toomre, D., & Mellman, I. (2006). Vectorial insertion of apical and basolateral membrane proteins in polarized epithelial cells revealed by quantitative 3D live cell imaging. *Journal of Cell Biology, 172*(7), 1035–1044.

Huttner, W. B., & Zimmerberg, J. (2001). Implications of lipid microdomains for membrane curvature, budding and fission. *Current Opinion in Cell Biology, 13*(4), 478–484.

Ikezawa, H. (2002). Glycosylphosphatidylinositol (GPI)-anchored proteins. *Biological & Pharmaceutical Bulletin, 25*(4), 409–417.

Ilantzis, C., DeMarte, L., Screaton, R. A., & Stanners, C. P. (2002). Deregulated expression of the human tumor marker CEA and CEA family member CEACAM6 disrupts tissue architecture and blocks colonocyte differentiation. *Neoplasia, 4*(2), 151–163.

Imjeti, N. S., Lebreton, S., Paladino, S., de la Fuente, E., Gonzales, A., & Zurzolo, C. (2011). N-glycosylation instead of cholesterol mediates oligomerization and apical sorting of Gpi-Aps in Frt cells. *Molecular Biology of the Cell, 22*(23), 4621–4634.

Jacob, R., Heine, M., Eikemeyer, J., Frerker, N., Zimmer, K. P., Rescher, U., et al. (2004). Annexin II is required for apical transport in polarized epithelial cells. *Journal of Biological Chemistry, 279*(5), 3680–3684.

Jaensch, N., Correa, I. R., Jr., & Watanabe, R. (2014). Stable cell surface expression of GPI-anchored proteins, but not intracellular transport, depends on their fatty acid structure. *Traffic, 15*(12), 1305–1329.

Kalli, K. R., Oberg, A. L., Keeney, G. L., Christianson, T. J., Low, P. S., Knutson, K. L., et al. (2008). Folate receptor alpha as a tumor target in epithelial ovarian cancer. *Gynecologic Oncology, 108*(3), 619–626.

Karadimitris, A., & Luzzatto, L. (2001). The cellular pathogenesis of paroxysmal nocturnal haemoglobinuria. *Leukemia, 15*(8), 1148–1152.

Keller, P., Toomre, D., Diaz, E., White, J., & Simons, K. (2001). Multicolour imaging of post-Golgi sorting and trafficking in live cells. *Nature Cell Biology, 3*(2), 140–149.

Kenworthy, A. K., & Edidin, M. (1998). Distribution of a glycosylphosphatidylinositol-anchored protein at the apical surface of MDCK cells examined at a resolution of <100 A using imaging fluorescence resonance energy transfer. *Journal of Cell Biology, 142*(1), 69–84.

Kenworthy, A. K., Nichols, B. J., Remmert, C. L., Hendrix, G. M., Kumar, M., Zimmerberg, J., et al. (2004). Dynamics of putative raft-associated proteins at the cell surface. *Journal of Cell Biology, 165*(5), 735–746.

Kinoshita, T., Fujita, M., & Maeda, Y. (2008). Biosynthesis, remodelling and functions of mammalian GPI-anchored proteins: recent progress. *Journal of Biochemistry, 144*(3), 287–294.

Kinoshita, T., Maeda, Y., & Fujita, M. (2013). Transport of glycosylphosphatidylinositol-anchored proteins from the endoplasmic reticulum. *Biochimica et Biophysica Acta, 1833*(11), 2473–2478.

Kirkham, M., & Parton, R. G. (2005). Clathrin-independent endocytosis: new insights into caveolae and non-caveolar lipid raft carriers. *Biochimica et Biophysica Acta, 1746*(3), 349–363.

Krawitz, P. M., Murakami, Y., Hecht, J., Kruger, U., Holder, S. E., Mortier, G. R., et al. (2012). Mutations in PIGO, a member of the GPI-anchor-synthesis pathway, cause hyperphosphatasia with mental retardation. *American Journal of Human Genetics, 91*(1), 146–151.

Krawitz, P. M., Murakami, Y., Riess, A., Hietala, M., Kruger, U., Zhu, N., et al. (2013). PGAP2 mutations, affecting the GPI-anchor-synthesis pathway, cause hyperphosphatasia with mental retardation syndrome. *American Journal of Human Genetics, 92*(4), 584–589.

Krawitz, P. M., Schweiger, M. R., Rodelsperger, C., Marcelis, C., Kolsch, U., Meisel, C., et al. (2010). Identity-by-descent filtering of exome sequence data identifies PIGV mutations in hyperphosphatasia mental retardation syndrome. *Nature Genetics, 42*(10), 827–829.

Kusumi, A., Suzuki, K. G., Kasai, R. S., Ritchie, K., & Fujiwara, T. K. (2011). Hierarchical mesoscale domain organization of the plasma membrane. *Trends in Biochemical Sciences, 36*(11), 604–615.

Kvarnung, M., Nilsson, D., Lindstrand, A., Korenke, G. C., Chiang, S. C., Blennow, E., et al. (2013). A novel intellectual disability syndrome caused by GPI anchor deficiency due to homozygous mutations in PIGT. *Journal of Medical Genetics, 50*(8), 521–528.

Kwik, J., Boyle, S., Fooksman, D., Margolis, L., Sheetz, M. P., & Edidin, M. (2003). Membrane cholesterol, lateral mobility, and the phosphatidylinositol 4,5-bisphosphate-dependent organization of cell actin. *Proceedings of the National Academy of Sciences of the United States of America, 100*(24), 13964–13969.

Lafont, F., Lecat, S., Verkade, P., & Simons, K. (1998). Annexin XIIIb associates with lipid microdomains to function in apical delivery. *Journal of Cell Biology, 142*(6), 1413–1427.

Lakhan, S. E., Sabharanjak, S., & De, A. (2009). Endocytosis of glycosylphosphatidylinositol-anchored proteins. *Journal of Biomedical Sciences, 16*, 93.

Lebreton, S., Paladino, S., & Zurzolo, C. (2008). Selective roles for cholesterol and actin in compartmentalization of different proteins in the Golgi and plasma membrane of polarized cells. *Journal of Biological Chemistry, 283*(43), 29545–29553.

Lecat, S., Verkade, P., Thiele, C., Fiedler, K., Simons, K., & Lafont, F. (2000). Different properties of two isoforms of annexin XIII in MDCK cells. *Journal of Cell Science, 113*(Pt 14), 2607–2618.

Legler, D. F., Doucey, M. A., Schneider, P., Chapatte, L., Bender, F. C., & Bron, C. (2005). Differential insertion of GPI-anchored GFPs into lipid rafts of live cells. *FASEB Journal, 19*(1), 73–75.

Lenne, P. F., Wawrezinieck, L., Conchonaud, F., Wurtz, O., Boned, A., Guo, X. J., et al. (2006). Dynamic molecular confinement in the plasma membrane by microdomains and the cytoskeleton meshwork. *EMBO Journal, 25*(14), 3245–3256.

Lipardi, C., Nitsch, L., & Zurzolo, C. (2000). Detergent-insoluble GPI-anchored proteins are apically sorted in fischer rat thyroid cells, but interference with cholesterol or sphingolipids differentially affects detergent insolubility and apical sorting. *Molecular Biology of the Cell, 11*(2), 531–542.

Lisanti, M. P., Caras, I. W., Davitz, M. A., & Rodriguez-Boulan, E. (1989). A glycophospholipid membrane anchor acts as an apical targeting signal in polarized epithelial cells. *Journal of Cell Biology, 109*(5), 2145–2156.

Liu, J., Kaksonen, M., Drubin, D. G., & Oster, G. (2006). Endocytic vesicle scission by lipid phase boundary forces. *Proceedings of the National Academy of Sciences of the United States of America, 103*(27), 10277–10282.

Lock, J. G., & Stow, J. L. (2005). Rab11 in recycling endosomes regulates the sorting and basolateral transport of E-cadherin. *Molecular Biology of the Cell, 16*(4), 1744–1755.

Maeda, Y., & Kinoshita, T. (2011). Structural remodeling, trafficking and functions of glycosylphosphatidylinositol-anchored proteins. *Progress in Lipid Research, 50*(4), 411–424.

Maeda, Y., Tashima, Y., Houjou, T., Fujita, M., Yoko-o, T., Jigami, Y., et al. (2007). Fatty acid remodeling of GPI-anchored proteins is required for their raft association. *Molecular Biology of the Cell, 18*(4), 1497–1506.

Manninen, A., Verkade, P., Le Lay, S., Torkko, J., Kasper, M., Fullekrug, J., et al. (2005). Caveolin-1 is not essential for biosynthetic apical membrane transport. *Molecular and Cellular Biology, 25*(22), 10087–10096.

Martin-Belmonte, F., Puertollano, R., Millan, J., Alonso, M. A., de Marco, M. C., Albar, J. P., et al. (2000). The MAL proteolipid is necessary for the overall apical delivery of membrane proteins in the polarized epithelial Madin-Darby canine kidney and fischer rat thyroid cell lines. *Molecular Biology of the Cell, 11*(6), 2033–2045.

Martin, K. C., & Kandel, E. R. (1996). Cell adhesion molecules, CREB, and the formation of new synaptic connections. *Neuron, 17*(4), 567–570.

Maydan, G., Noyman, I., Har-Zahav, A., Neriah, Z. B., Pasmanik-Chor, M., Yeheskel, A., et al. (2011). Multiple congenital anomalies-hypotonia-seizures syndrome is caused by a mutation in PIGN. *Journal of Medical Genetics, 48*(6), 383–389.

Mayor, S., Rothberg, K. G., & Maxfield, F. R. (1994). Sequestration of GPI-anchored proteins in caveolae triggered by cross-linking. *Science, 264*(5167), 1948–1951.

McConville, M. J., & Menon, A. K. (2000). Recent developments in the cell biology and biochemistry of glycosylphosphatidylinositol lipids (review). *Molecular Membrane Biology, 17*(1), 1–16.

McMahon, H. T., & Gallop, J. L. (2005). Membrane curvature and mechanisms of dynamic cell membrane remodelling. *Nature, 438*(7068), 590–596.

Meder, D., Moreno, M. J., Verkade, P., Vaz, W. L., & Simons, K. (2006). Phase coexistence and connectivity in the apical membrane of polarized epithelial cells. *Proceedings of the National Academy of Sciences of the United States of America, 103*(2), 329–334.

Mellman, I., & Nelson, W. J. (2008). Coordinated protein sorting, targeting and distribution in polarized cells. *Nature Reviews Molecular Cell Biology, 9*(11), 833–845.

Mishra, R., Grzybek, M., Niki, T., Hirashima, M., & Simons, K. (2010). Galectin-9 trafficking regulates apical-basal polarity in Madin-Darby canine kidney epithelial cells. *Proceedings of the National Academy of Sciences of the United States of America, 107*(41), 17633–17638.

Mizuno, H., Abe, M., Dedecker, P., Makino, A., Rocha, S., Ohno-Iwashita, Y., et al. (2011). Fluorescent probes for superresolution imaging of lipid domains on the plasma membrane. *Chemical Science, 2*, 1548–1553.

Morsomme, P., Prescianotto-Baschong, C., & Riezman, H. (2003). The ER v-SNAREs are required for GPI-anchored protein sorting from other secretory proteins upon exit from the ER. *Journal of Cell Biology, 162*(3), 403–412.

Morsomme, P., & Riezman, H. (2002). The Rab GTPase Ypt1p and tethering factors couple protein sorting at the ER to vesicle targeting to the Golgi apparatus. *Developmental Cell, 2*(3), 307–317.

Mostov, K. E., Verges, M., & Altschuler, Y. (2000). Membrane traffic in polarized epithelial cells. *Current Opinion in Cell Biology, 12*(4), 483–490.

Mueller, V., Ringemann, C., Honigmann, A., Schwarzmann, G., Medda, R., Leutenegger, M., et al. (2010). STED nanoscopy reveals molecular details of cholesterol- and cytoskeleton-modulated lipid interactions in living cells. *Biophysical Journal, 101*(7), 1651–1660.

Muniz, M., Morsomme, P., & Riezman, H. (2001). Protein sorting upon exit from the endoplasmic reticulum. *Cell, 104*(2), 313–320.

Muniz, M., & Zurzolo, C. (2014). Sorting of GPI-anchored proteins from yeast to mammals—common pathways at different sites? *Journal of Cell Science, 127*(Pt 13), 2793–2801.

Murakami, Y., Tawamie, H., Maeda, Y., Buttner, C., Buchert, R., Radwan, F., et al. (2014). Null mutation in PGAP1 impairing Gpi-anchor maturation in patients with intellectual disability and encephalopathy. *PLoS Genetics, 10*(5), e1004320.

Nagpal, J. K., Dasgupta, S., Jadallah, S., Chae, Y. K., Ratovitski, E. A., Toubaji, A., et al. (2008). Profiling the expression pattern of GPI transamidase complex subunits in human cancer. *Modern Pathology, 21*(8), 979–991.

Nakamura, K., Osaka, H., Murakami, Y., Anzai, R., Nishiyama, K., Kodera, H., et al. (2014). PIGO mutations in intractable epilepsy and severe developmental delay with mild elevation of alkaline phosphatase levels. *Epilepsia, 55*(2), e13–17.

Ng, B. G., Hackmann, K., Jones, M. A., Eroshkin, A. M., He, P., Wiliams, R., et al. (2012). Mutations in the glycosylphosphatidylinositol gene PIGL cause CHIME syndrome. *American Journal of Human Genetics, 90*(4), 685–688.

Nozaki, M., Ohishi, K., Yamada, N., Kinoshita, T., Nagy, A., & Takeda, J. (1999). Developmental abnormalities of glycosylphosphatidylinositol-anchor-deficient embryos revealed by Cre/loxP system. *Laboratory Investigation, 79*(3), 293–299.

Ohba, C., Okamoto, N., Murakami, Y., Suzuki, Y., Tsurusaki, Y., Nakashima, M., et al. (2013). PIGN mutations cause congenital anomalies, developmental delay, hypotonia, epilepsy, and progressive cerebellar atrophy. *Neurogenetics, 15*(2), 85–92.

Orlean, P., & Menon, A. K. (2007). Thematic review series: lipid posttranslational modifications. GPI anchoring of protein in yeast and mammalian cells, or: how we learned to stop worrying and love glycophospholipids. *Journal of Lipid Research, 48*(5), 993–1011.

Owen, D. M., Magenau, A., Williamson, D., & Gaus, K. (2012a). The lipid raft hypothesis revisited—new insights on raft composition and function from super-resolution fluorescence microscopy. *Bioessays, 34*(9), 739–747.

Owen, D. M., Magenau, A., Williamson, D. J., & Gaus, K. (2013). Super-resolution imaging by localization microscopy. *Methods in Molecular Biology, 950*, 81–93.

Owen, D. M., Rentero, C., Rossy, J., Magenau, A., Williamson, D., Rodriguez, M., et al. (2010). PALM imaging and cluster analysis of protein heterogeneity at the cell surface. *Journal of Biophotonics, 3*(7), 446–454.

Owen, D. M., Williamson, D., Magenau, A., & Gaus, K. (2012b). Optical techniques for imaging membrane domains in live cells (live-cell palm of protein clustering). *Methods in Enzymology, 504*, 221–235.

Paladino, S., Lebreton, S., Tivodar, S., Campana, V., Tempre, R., & Zurzolo, C. (2008). Different GPI-attachment signals affect the oligomerisation of GPI-anchored proteins and their apical sorting. *Journal of Cell Science, 121*(Pt 24), 4001–4007.

Paladino, S., Lebreton, S., Tivodar, S., Formiggini, F., Ossato, G., Gratton, E., et al. (2014). Golgi sorting regulates organization and activity of GPI proteins at apical membranes. *Nature Chemical Biology, 10*(5), 350–357.

Paladino, S., Pocard, T., Catino, M. A., & Zurzolo, C. (2006). GPI-anchored proteins are directly targeted to the apical surface in fully polarized MDCK cells. *Journal of Cell Biology, 172*(7), 1023–1034.

Paladino, S., Sarnataro, D., Pillich, R., Tivodar, S., Nitsch, L., & Zurzolo, C. (2004). Protein oligomerization modulates raft partitioning and apical sorting of GPI-anchored proteins. *Journal of Cell Biology, 167*(4), 699–709.

Paladino, S., Sarnataro, D., Tivodar, S., & Zurzolo, C. (2007). Oligomerization is a specific requirement for apical sorting of glycosyl-phosphatidylinositol-anchored proteins but not for non-raft-associated apical proteins. *Traffic, 8*(3), 251–258.

Paladino, S., & Zurzolo, C. (2009). Mechanism of polarized sorting of GPI-proteins in epithelial cells. *The Enzymes, 26*, 289–319 (Elsevier Inc.).

Pang, S., Urquhart, P., & Hooper, N. M. (2004). N-glycans, not the GPI anchor, mediate the apical targeting of a naturally glycosylated, GPI-anchored protein in polarised epithelial cells. *Journal of Cell Science, 117*(Pt 21), 5079–5086.

Parker, N., Turk, M. J., Westrick, E., Lewis, J. D., Low, P. S., & Leamon, C. P. (2005). Folate receptor expression in carcinomas and normal tissues determined by a quantitative radioligand binding assay. *Analytical Biochemistry, 338*(2), 284–293.

Parton, R. G., & Richards, A. A. (2003). Lipid rafts and caveolae as portals for endocytosis: new insights and common mechanisms. *Traffic, 4*(11), 724–738.

Pike, L. J. (2003). Lipid rafts: bringing order to chaos. *Journal of Lipid Research, 44*(4), 655–667.

Pike, L. J. (2006). Rafts defined: a report on the keystone symposium on lipid rafts and cell Function. *Journal of Lipid Research, 47*(7), 1597–1598.

Potter, B. A., Ihrke, G., Bruns, J. R., Weixel, K. M., & Weisz, O. A. (2004). Specific N-glycans direct apical delivery of transmembrane, but not soluble or glycosylphosphatidylinositol-anchored forms of endolyn in Madin-Darby canine kidney cells. *Molecular Biology of the Cell, 15*(3), 1407–1416.

Ritter, T. E., Fajardo, O., Matsue, H., Anderson, R. G., & Lacey, S. W. (1995). Folate receptors targeted to clathrin-coated pits cannot regulate vitamin uptake. *Proceedings of the National Academy of Sciences of the United States of America, 92*(9), 3824–3828.

Rivier, A. S., Castillon, G. A., Michon, L., Fukasawa, M., Romanova-Michaelides, M., Jaensch, N., et al. (2010). Exit of GPI-anchored proteins from the ER differs in yeast and mammalian cells. *Traffic, 11*(8), 1017–1033.

Rodriguez-Boulan, E., Kreitzer, G., & Musch, A. (2005). Organization of vesicular trafficking in epithelia. *Nature Reviews Molecular Cell Biology, 6*(3), 233–247.

Rodriguez-Boulan, E., & Macara, I. G. (2014). Organization and execution of the epithelial polarity programme. *Nature Reviews Molecular Cell Biology, 15*(4), 225–242.

Rodriguez-Boulan, E., & Musch, A. (2005). Protein sorting in the Golgi complex: shifting paradigms. *Biochimica et Biophysica Acta, 1744*(3), 455–464.

Rodriguez-Boulan, E., Musch, A., & Le Bivic, A. (2004). Epithelial trafficking: new routes to familiar places. *Current Opinion in Cell Biology, 16*(4), 436–442.

Rustom, A., Bajohrs, M., Kaether, C., Keller, P., Toomre, D., Corbeil, D., et al. (2002). Selective delivery of secretory cargo in Golgi-derived carriers of nonepithelial cells. *Traffic, 3*(4), 279–288.

Sabharanjak, S., & Mayor, S. (2004). Folate receptor endocytosis and trafficking. *Advanced Drug Delivery Reviews, 56*(8), 1099–1109.

Sarnataro, D., Paladino, S., Campana, V., Grassi, J., Nitsch, L., & Zurzolo, C. (2002). PrPC is sorted to the basolateral membrane of epithelial cells independently of its association with rafts. *Traffic, 3*(11), 810–821.

Schuck, S., & Simons, K. (2004). Polarized sorting in epithelial cells: raft clustering and the biogenesis of the apical membrane. *Journal of Cell Science, 117*(Pt 25), 5955–5964.

Schutz, G. J., Kada, G., Pastushenko, V. P., & Schindler, H. (2000). Properties of lipid microdomains in a muscle cell membrane visualized by single molecule microscopy. *EMBO Journal, 19*(5), 892–901.

Screaton, R. A., DeMarte, L., Draber, P., & Stanners, C. P. (2000). The specificity for the differentiation blocking activity of carcinoembryonic antigen resides in its glycophosphatidylinositol anchor. *Journal of Cell Biology, 150*(3), 613–626.

Sengupta, P., Jovanovic-Talisman, T., & Lippincott-Schwartz, J. (2013). Quantifying spatial organization in point-localization superresolution images using pair correlation analysis. *Nature Protocols, 8*(2), 345–354.

Sengupta, P., Jovanovic-Talisman, T., Skoko, D., Renz, M., Veatch, S. L., & Lippincott-Schwartz, J. (2011). Probing protein heterogeneity in the plasma membrane using PALM and pair correlation analysis. *Nature Methods, 8*(11), 969–975.

Sengupta, P., & Lippincott-Schwartz, J. (2012). Quantitative analysis of photoactivated localization microscopy (PALM) datasets using pair-correlation analysis. *Bioessays, 34*(5), 396–405.

Sengupta, P., Van Engelenburg, S., & Lippincott-Schwartz, J. (2012). Visualizing cell structure and function with point-localization superresolution imaging. *Developmental Cell, 23*(6), 1092–1102.

Seong, J., Wang, Y., Kinoshita, T., & Maeda, Y. (2013). Implications of lipid moiety in oligomerization and immunoreactivities of GPI-anchored proteins. *Journal of Lipid Research, 54*(4), 1077–1091.

Sharma, P., Varma, R., Sarasij, R. C., Ira, Gousset, K., Krishnamoorthy, G., et al. (2004). Nanoscale organization of multiple GPI-anchored proteins in living cell membranes. *Cell, 116*(4), 577–589.

Sheets, E. D., Lee, G. M., Simson, R., & Jacobson, K. (1997). Transient confinement of a glycosylphosphatidylinositol-anchored protein in the plasma membrane. *Biochemistry, 36*(41), 12449–12458.

Shia, J., Klimstra, D. S., Nitzkorski, J. R., Low, P. S., Gonen, M., Landmann, R., et al. (2008). Immunohistochemical expression of folate receptor alpha in colorectal carcinoma: patterns and biological significance. *Human Pathology, 39*(4), 498–505.

Simons, K., & Gerl, M. J. (2010). Revitalizing membrane rafts: new tools and insights. *Nature Reviews Molecular Cell Biology, 11*(10), 688–699.

Simons, K., & Ikonen, E. (1997). Functional rafts in cell membranes. *Nature, 387*(6633), 569–572.

Simons, K., & Sampaio, J. L. (2011). Membrane organization and lipid rafts. *Cold Spring Harbor Perspectives in Biology, 3*(10), a004697.

Simons, K., & Toomre, D. (2000). Lipid rafts and signal transduction. *Nature Reviews Molecular Cell Biology, 1*(1), 31–39.

Simons, K., & Vaz, W. L. (2004). Model systems, lipid rafts, and cell membranes. *Annual Review of Biophysics and Biomolecular Structure, 33*, 269–295.

Singer, S. J., & Nicolson, G. L. (1972). The fluid mosaic model of the structure of cell membranes. *Science, 175*(23), 720–731.

Smart, E. J., Mineo, C., & Anderson, R. G. (1996). Clustered folate receptors deliver 5-methyltetrahydrofolate to cytoplasm of MA104 cells. *Journal of Cell Biology, 134*(5), 1169–1177.

Stephens, D. J., & Pepperkok, R. (2004). Differential effects of a GTP-restricted mutant of Sar1p on segregation of cargo during export from the endoplasmic reticulum. *Journal of Cell Science, 117*(Pt 16), 3635–3644.

Strating, J. R., van Bakel, N. H., Leunissen, J. A., & Martens, G. J. (2009). A comprehensive overview of the vertebrate p24 family: identification of a novel tissue-specifically expressed member. *Molecular Biology and Evolution, 26*(8), 1707–1714.

Strating, J. R., Hafmans, T. G., & Martens, G. J. (2009). Functional diversity among p24 subfamily members. *Biology of the Cell, 101*(4), 207–219.

Strating, J. R., & Martens, G. J. (2009). The p24 family and selective transport processes at the ER-Golgi interface. *Biology of the Cell, 101*(9), 495–509.

Sun, M., Northup, N., Marga, F., Huber, T., Byfield, F. J., Levitan, I., et al. (2007). The effect of cellular cholesterol on membrane-cytoskeleton adhesion. *Journal of Cell Science, 120*(Pt 13), 2223–2231.

Surma, M. A., Klose, C., & Simons, K. (2012). Lipid-dependent protein sorting at the trans-Golgi network. *Biochimica et Biophysica Acta, 1821*(8), 1059–1067.

Sutterlin, C., Doering, T. L., Schimmoller, F., Schroder, S., & Riezman, H. (1997). Specific requirements for the ER to Golgi transport of GPI-anchored proteins in yeast. *Journal of Cell Science, 110*(Pt 21), 2703–2714.

Suzuki, K. G., Fujiwara, T. K., Edidin, M., & Kusumi, A. (2007). Dynamic recruitment of phospholipase C gamma at transiently immobilized GPI-anchored receptor clusters induces IP3-Ca^{2+} signaling: single-molecule tracking study 2. *Journal of Cell Biology, 177*(4), 731–742.

Suzuki, K. G., Fujiwara, T. K., Sanematsu, F., Iino, R., Edidin, M., & Kusumi, A. (2007). GPI-anchored receptor clusters transiently recruit Lyn and G alpha for temporary cluster immobilization and Lyn activation: single-molecule tracking study 1. *Journal of Cell Biology, 177*(4), 717–730.

Suzuki, K. G., Kasai, R. S., Hirosawa, K. M., Nemoto, Y. L., Ishibashi, M., Miwa, Y., et al. (2012). Transient GPI-anchored protein homodimers are units for raft organization and function. *Nature Chemical Biology, 8*(9), 774–783.

Takida, S., Maeda, Y., & Kinoshita, T. (2008). Mammalian GPI-anchored proteins require p24 proteins for their efficient transport from the ER to the plasma membrane. *Biochemical Journal, 409*(2), 555–562.

Tanaka, S., Maeda, Y., Tashima, Y., & Kinoshita, T. (2004). Inositol deacylation of glycosylphosphatidylinositol-anchored proteins is mediated by mammalian PGAP1 and yeast Bst1p. *Journal of Biological Chemistry, 279*(14), 14256–14263.

Tashima, Y., Taguchi, R., Murata, C., Ashida, H., Kinoshita, T., & Maeda, Y. (2006). PGAP2 is essential for correct processing and stable expression of GPI-anchored proteins. *Molecular Biology of the Cell, 17*(3), 1410–1420.

Theiler, R., Fujita, M., Nagae, M., Yamaguchi, Y., Maeda, Y., & Kinoshita, T. (2014). The alpha-helical region in p24gamma2 subunit of p24 protein cargo receptor is pivotal for the recognition and transport of glycosylphosphatidylinositol-anchored proteins. *Journal of Biological Chemistry, 289*(24), 16835–16843.

Tian, A., & Baumgart, T. (2009). Sorting of lipids and proteins in membrane curvature gradients. *Biophysical Journal, 96*(7), 2676–2688.

Varma, R., & Mayor, S. (1998). GPI-anchored proteins are organized in submicron domains at the cell surface. *Nature, 394*(6695), 798–801.

Vieira, O. V., Verkade, P., Manninen, A., & Simons, K. (2005). FAPP2 is involved in the transport of apical cargo in polarized MDCK cells. *Journal of Cell Biology, 170*(4), 521–526.

Watanabe, R., & Riezman, H. (2004). Differential ER exit in yeast and mammalian cells. *Current Opinion in Cell Biology, 16*(4), 350–355.
Weisz, O. A., & Rodriguez-Boulan, E. (2009). Apical trafficking in epithelial cells: signals, clusters and motors. *Journal of Cell Science, 122*(Pt 23), 4253–4266.
Yuan, X., Braunstein, E. M., Ye, Z., Liu, C. F., Chen, G., Zou, J., et al. (2013). Generation of glycosylphosphatidylinositol anchor protein-deficient blood cells from human induced pluripotent stem cells. *Stem Cells Translational Medicine, 2*(11), 819–829.
van Zanten, T. S., Cambi, A., Koopman, M., Joosten, B., Figdor, C. G., & Garcia-Parajo, M. F. (2009). Hotspots of GPI-anchored proteins and integrin nanoclusters function as nucleation sites for cell adhesion. *Proceedings of the National Academy of Sciences of the United States of America, 106*(44), 18557–18562.
Zurzolo, C., Lisanti, M. P., Caras, I. W., Nitsch, L., & Rodriguez-Boulan, E. (1993). Glycosylphosphatidylinositol-anchored proteins are preferentially targeted to the basolateral surface in Fischer rat thyroid epithelial cells. *Journal of Cell Biology, 121*(5), 1031–1039.

CHAPTER NINE

Recent Progress in the Topology, Structure, and Oligomerization of Caveolin: A Building Block of Caveolae

Kyle T. Root[a], Sarah M. Plucinsky[a] and Kerney Jebrell Glover*
Department of Chemistry, Lehigh University, Bethlehem, PA, USA
*Corresponding author: E-mail: kjg206@lehigh.edu

Contents

Abstract

Caveolae are cholesterol-rich plasma membrane invaginations that are found in a plethora of cell types. They play many roles including signal transduction, endocytosis, and mechanoprotection. The most critical protein in caveolae is the integral membrane

[a] These two authors contributed equally to this work.

Current Topics in Membranes, Volume 75
ISSN 1063-5823
http://dx.doi.org/10.1016/bs.ctm.2015.03.007

protein, caveolin, which has been shown to be necessary for caveolae formation, and governs the major functions attributed to caveolae. Caveolin is postulated to act as a scaffold in the high molecular weight striated coat that surrounds the caveolar bulb, stabilizing it. Caveolin interacts, both directly and indirectly, with a large number of signaling molecules, and presides over the endocytosis of molecular cargo by caveolae. However, many of the key biophysical aspects of the caveolin protein, its structure, topology, and oligomeric behavior, are just beginning to come to light. Herein is an up-to-date summary and critique of the progress that has been made in understanding caveolin on a molecular and atomic level.

1. INTRODUCTION

Caveolae are 50–100 nm invaginations in the plasma membrane that are found in many cell types such as adipocytes, endothelial cells, fibroblasts, and smooth muscle cells (Mobley & Eisenberg, 1975; Palade, 1953; Thorn et al., 2003; Wheaton et al., 2001). Using electron microscopy, caveolae were first observed in the 1950s as flask-shaped domains within the heart and also the gall bladder (Palade, 1953; Yamada, 1955). Caveolae are proposed to have three major functions in the cell: mechanoprotection; recruiting, concentrating, and effecting signaling molecules; and endocytosis (Chang et al., 1994; Lisanti et al., 1994; Liu, Oh, Horner, Rogers, & Schnitzer, 1997; Parton, Joggerst, & Simons, 1994). Caveolae act as membrane reservoirs that allow cells to sustain mechanical stresses (Sinha et al., 2011). Facilitated by their highly curved nature, caveolae can bud off from the membrane in an endocytic manner and form vesicle-like structures that can carry cargo from the cell membrane to other parts of the cell (Kiss, 2012).

At the center of all of these functions is the integral membrane protein caveolin. Initially, there were two key proteins that were associated with caveolae, VIP-21 and caveolin. Caveolin was first isolated as a marker protein for caveolae, while VIP-21 was identified as part of the *trans*-golgi network (Kurzchalia et al., 1992; Rothberg et al., 1992). Shortly afterward, VIP-21 was sequenced and shown to be the same protein as caveolin (Glenney, 1992). This led to the consensus of calling the protein VIP-21/caveolin and now the protein is referred to simply as caveolin.

Caveolin interacts and effects, both directly and indirectly, a large number of signaling molecules including Src-like kinases, Ha-Ras, endothelial nitric oxide synthase (eNOS), and G-proteins (Carman, Lisanti, & Benovic, 1999; Couet, Sargiacomo, & Lisanti, 1997; Ju, Zou, Venema, & Venema, 1997; Lisanti et al., 1994; Okamoto, Schlegel, Scherer, & Lisanti, 1998;

Roy et al., 1999; Sargiacomo, Sudol, Tang, & Lisanti, 1993). One well-studied interaction is between caveolin (specifically caveolin-1) and eNOS (García-Cardeña et al., 1997; Ghosh et al., 1998; Ju et al., 1997). The disruption of the binding between caveolin-1 and eNOS has negative effects on cellular function by causing a buildup of nitric oxide within the cell, as caveolin acts as a negative regulator of eNOS (Feron, Saldana, Michel, & Michel, 1998). Initially, many of the caveolin-signaling protein interactions were proposed to occur directly through a consensus sequence dubbed the caveolin-binding motif (CBM). However, recent structural data and bioinformatic analyses suggest that this motif (CBM) is not indicative of direct caveolin-partner interactions (Byrne, Dart, & Rigden, 2012; Collins, Davis, Hancock, & Parton, 2012). Although these studies call into question the direct interaction of caveolin with signaling molecules through a well-defined consensus motif, they do not preclude a specific nonconsensus interface for caveolin-partner interactions. Clearly, more work will be needed to establish which caveolin-partner interactions are direct versus indirect.

Caveolin-1 has been shown to play an integral role in regulating and mediating the endocytosis of caveolae. For example, the overexpression of caveolin-1 appears to inhibit the endocytosis of caveolae (Le, Guay, Altschuler, & Nabi, 2002). Caveolin-1 has also been implicated in the endocytosis of virus particles through caveolae (Pelkmans, Kartenbeck, & Helenius, 2001). Clearly, caveolin is involved in how molecules are transported from the plasma membrane to other parts of the cell through non-clathrin-mediated pathways.

Caveolae allow for a cell to experience abrupt mechanical stress without being deformed or lysed (Sinha et al., 2011). For example, when a cell is subjected to mechanical force, caveolae rapidly dissipate to allow the cell to accommodate the additional force. When the force is removed, caveolae are reformed.

Caveolin is an approximately 22 kDa protein that can be described by four domains: the N-terminal domain, the scaffolding domain, the intramembrane domain (IMD), and the C-terminal domain (Figure 1). Caveolin has three isoforms, caveolin-1, -2, and -3 (Figure 2). Caveolin-1 is the most ubiquitous, being found in most terminally differentiated cell types, and the protein has been shown to be necessary for caveolae formation (Drab et al., 2001; Park et al., 2002; Razani et al., 2002, 2001). Caveolin-1 also has two specific isoforms referred to as caveolin-1α and caveolin-1β. The caveolin-1β isoform can be thought of as a truncated construct of the α-isoform lacking the first 31 amino acid residues (Scherer et al., 1995). Caveolin-2 is coexpressed

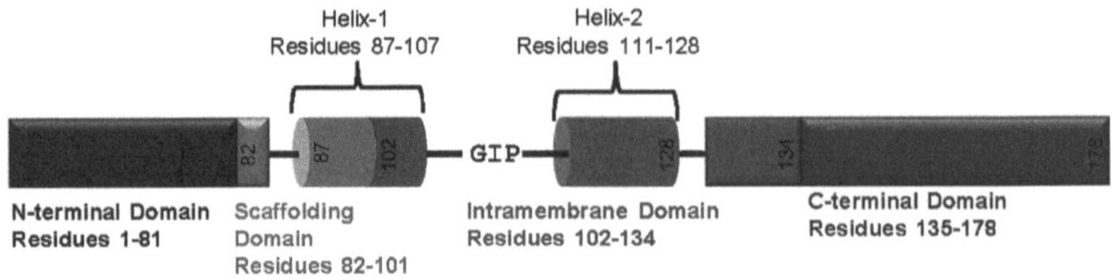

Figure 1 Diagram of the four domains of caveolin-1. Residues 87–128 have been shown as α-helical (Rui et al., 2014). (See color plate)

with caveolin-1 and the two are thought to form heterooligomeric complexes that are part of the striated coat (Scherer et al., 1996). However, caveolin-2 appears to lack the ability to form caveolae in the absence of caveolin-1 (Mora et al., 1999; Parolini et al., 1999). Caveolin-3 is expressed mostly in muscle tissue and also has the ability to form caveolae (Song et al., 1996; Tang et al., 1996; Way & Parton, 1995). This is supported by the high degree of sequence identity between caveolin-1 and -3. In contrast, the sequence of caveolin-2 deviates from that of the other isoforms, particularly in its scaffolding domain (Figure 2). In addition, the lengths of the N-terminal domains vary significantly between the isoforms, with caveolin-1 having the longest and caveolin-3 having the shortest (Figure 2).

Caveolin-1 has been shown to be palmitoylated at all three of its cysteine residues located in or near the C-terminal domain (C133, C143, and C156).

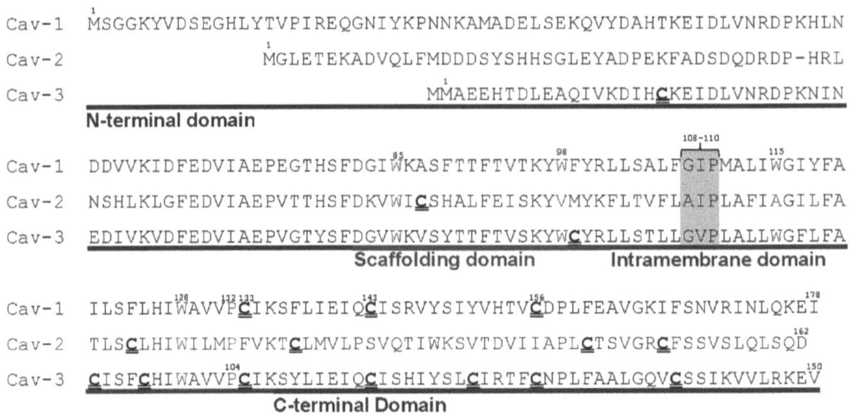

Figure 2 Sequence alignment of the three caveolin isoforms (human): caveolin-1, -2, and -3. The colored bar beneath the alignment denotes the domain. Highlighted in red are important residues that are referred to in the manuscript. Bolded and underlined are the cysteine residues in caveolin-1, caveolin-2, and caveolin-3. Residues shaded in gray are part of the break region (residues 108–110) implicated as the intramembrane turn. (See color plate)

The removal of this posttranslational modification by mutation of the cysteines to serines has been shown to have no effect on the trafficking of caveolin-1 to the plasma membrane (Dietzen, Hastings, & Lublin, 1995). Caveolin-2 has five cysteines and it has recently been shown that three of its cysteines are palmitoylated (Kwon, Lee, Jeong, Jang, & Pak, 2015). Caveolin-3 contains nine cysteines and has been shown to be heavily palmitoylated, but the exact sites have not been experimentally verified (Galbiati, Volonte, Minetti, Chu, & Lisanti, 1999). However, because of the sequence similarity to caveolin-1, it has been assumed that caveolin-3 is palmitoylated at the analogous three sites (Figure 2) (Dietzen et al., 1995; Galbiati et al., 1999).

The composition of caveolae is distinct compared to the bulk membrane because it is enriched in cholesterol and sphingomyelin (Ortegren et al., 2004). This has led to the postulation that caveolae have significantly less fluidity than the bulk membrane, a composition not unlike that of lipid rafts. The relationship between cholesterol and caveolin appears to be very strong and has been implicated in the proper folding and behavior of caveolin (Murata et al., 1995). In particular, the scaffolding domain of caveolin contains a CRAC (cholesterol recognition amino acid consensus) motif that is purported to directly bind cholesterol (Epand, Sayer, & Epand, 2005).

Mutations in the caveolin proteins as well as cellular expression levels have been linked to many diseases such as cancer, diabetes, cardiovascular disease, artherosclerosis, pulmonary fibrosis, and many muscular degenerative diseases (Capozza et al., 2005; Cohen et al., 2003; Fernández-Hernando, Yu, Dávalos, Prendergast, & Sessa, 2010; Komers et al., 2006; Sotgia et al., 2005; Weiss et al., 2008; Williams et al., 2004; Xia et al., 2010). The literature on the links between caveolin and disease is extensive and is beyond the scope and purpose of this review; therefore, only a few examples will be highlighted (For extensive reviews of caveolin and disease, see Engelman et al., 1998; Razani & Lisanti, 2001; Williams & Lisanti, 2004, and others). The knockout of caveolin-1 in mice can cause an increase in tumor proliferation indicating that caveolin-1 is a tumor suppressor (Sloan, Stanley, & Anderson, 2004, Williams et al., 2003). However, in prostate cancer it is found that caveolin-1 is upregulated and that it assists in the metastasis of the cancer (Williams et al., 2005). This dual role of caveolin-1 in cancer highlights that normal levels of caveolin-1 expressed within specific cell types is critical for maintaining proper cell function. Other studies have also shown that the overexpression of caveolin-1 can mirror the phenotype of a putative disease mutant (P132L) implicated in breast cancer by causing a disruption of caveolin migration from the Golgi to the plasma membrane (Bonuccelli et al., 2009; Hanson et al., 2013). However the true role and extent that this mutation appears in affected patients is unclear (Koike, Kodera, Nakao, Iwata, & Yatabe, 2010; Lacroix-Triki, Geyer, & Reis-Filho, 2010). Caveolin-3 has been linked to many muscular diseases such as heart disease and muscular dystrophy (Horikawa et al., 2008; Weiss et al., 2008). P104 in caveolin-3 which is analogous to P132 in caveolin-1 is also implicated in increased disease states and misregulation of the protein (Ohsawa et al., 2004; Stoppani et al., 2011).

Despite caveolin's preeminent role in cell homeostasis, many of the biophysical details surrounding caveolin function remain elusive. Much like other proteins whose improper function has been linked to disease states, a complete understanding of the determinants of cellular disruption will only come to fruition once a clear picture of the caveolin protein is obtained in molecular and atomic detail. This review will strive to report the progress that has been made in the field of caveolin structure, topology, and oligomerization and to highlight the work that still needs to be done.

2. TOPOLOGY OF CAVEOLIN

2.1 Caveolins have an unusual topology

Caveolin-1 has been shown to be an integral membrane protein using carbonate extraction, a method which distinguishes between peripherally bound membrane proteins and integral membrane proteins (Dupree, Parton, Raposo, Kurzchalia, & Simons, 1993; Schlegel, Schwab, Scherer, & Lisanti, 1999). Primary sequence analysis of caveolin-1 indicates that the protein has an unusually long 33-amino acid hydrophobic segment (residues 102–134, IMD, Figure 1). The length of the caveolin hydrophobic segment is much longer than the typical 15–20 amino acid residues required to span the hydrophobic core of the bilayer, and there are no soluble (polar amino acid residues) interruptions that would suggest a multipass topology. Therefore it is clear that caveolin may have a topology that is unlike that of a typical membrane protein.

Early work on caveolin-1 topology focused on determining the location of the N- and C-termini. Using cell surface biotinylation probes it was shown that caveolin-1 completely spanned the membrane with the N-terminus being cytoplasmic and the C-terminus being extracellular (Sargiacomo et al., 1993). This datum was incongruent with another study performed around the same time in which a pair of antibodies, one directed against the

N-terminus and one directed against the C-terminus of caveolin-1, did not bind unless the cells were permeabilized, suggesting that the location of *both* termini were intracellular (Dupree et al., 1993). When this discrepancy was probed further, it was revealed that *both* the N-and C-termini were indeed cytoplasmic and that the presence of a protein impurity was giving a false positive in the biotinylation experiments (Sargiacomo et al., 1995). In agreement with these findings, when a modified caveolin-1 construct containing two potential glycosylation sites at both the N- and C-termini was expressed using an in vitro transcription/translation system in the presence of dog pancrease microsomes, neither site was glycosylated, thereby implying that the N- and C-termini were cytoplasmic (Monier et al., 1995). Taken together, the unusually long hydrophobic stretch, the cytoplasmic location of the N- and C-termini, and the fact that the protein is deeply membrane embedded, the most realistic model is one in which the protein makes a turn within the hydrophobic core of the bilayer allowing both termini to reside on the same side of the membrane (Figure 3).

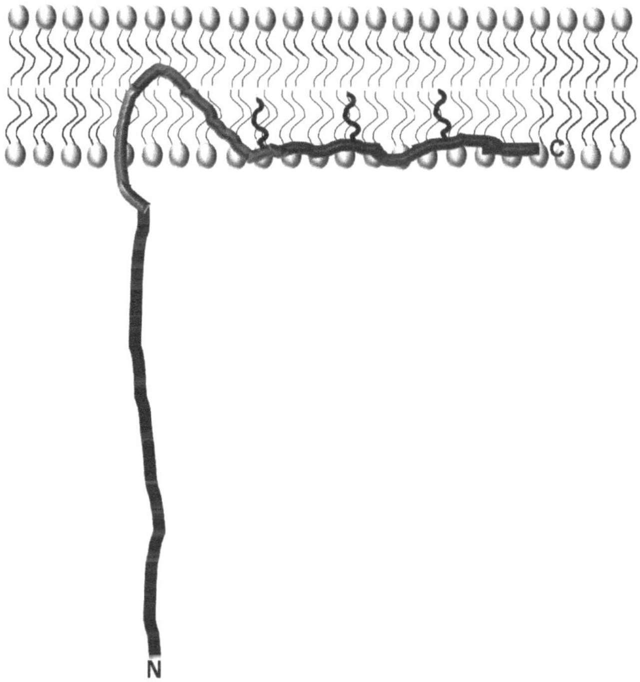

Figure 3 Topological model of the four caveolin-1 domains: N-terminal (purple), scaffolding domain (green), intramembrane domain (red), and C-terminal domain (blue). Palmitoylation is shown in black. (See color plate)

2.2 Proline 110 may be critical to the native caveolin-1 topology

In support of this model, secondary structural data from NMR experiments showed that the IMD is composed of a helix-break-helix motif where the break region, composed of glycine 108, isoleucine 109, and proline 110, was postulated to be the location of the intramembrane turn (Lee & Glover, 2012; Rui, Root, Lee, Glover, & Im, 2014). An in vivo experiment using a caveolin-1 construct with a FLAG epitope at its N-terminus, revealed that wild-type caveolin-1 was not accessible to antibody binding (consistent with prior studies showing the intracellular location of the N-terminus), but became accessible when proline 110 was mutated to alanine (Aoki, Thomas, Decaffmeyer, Brasseur, & Epand, 2010). This has lead to the postulation that this mutation causes caveolin to adopt a bitopic topology. In support of this postulation, molecular dynamics simulations of the IMD lent credence to this dramatic conformational change (Aoki et al., 2010). However, the location of the C-terminus upon this mutation was not investigated; therefore, it is unclear if the mutation of P110A causes *both* the N- and C-termini to be exposed extracellularly or the N-terminus to be exposed whereas the C-terminus remains on the cytoplasmic side of the membrane (one long single spanning α-helix).

In a study which compared the solvent accessibility of wild-type and P110A IMD constructs (residues 102–122) using fluorescence quenching and λ_{max} determinations, it was found that the overall tryptophan environment appeared to differ between constructs (with the P110A mutant becoming less solvent accessible), consistent with a topological change for the mutant (Aoki & Epand, 2012). Additionally, circular dichroism spectroscopy was used to compare the secondary structure of the wild-type and mutant and showed an increase in the α-helical character on the order of 1.5- to 2-fold in the case of the mutant, consistent with the reported linearization of the protein upon P110A mutation. However, the construct used in these studies had four and five lysine residues added to its N- and C-termini respectively to increase the solubility, therefore it is unclear how these lysines would affect the conclusions drawn. In contrast, another study revisited the glycosylation studies performed by Monier et al., this time comparing the wild-type and P110A mutant. It was found that *both* the wild-type *and* P110A mutant could not be glycosylated under these conditions, suggesting that the orientation was not disrupted by the P110A mutation (Norholm, Shulga, Aoki, Epand, & von Heijne, 2011). However, the removal of residues 5–95 resulted in both wild-type and P110A caveolin constructs that were dually and singly glycosylated. The results of this study are in direct contrast to the earlier work performed on the P110A mutant, which showed some type of topological change. This discrepancy could be due to the experimental setup of the glycosylation studies, which is open to multiple interpretations. For example, it is unclear as whether the lack of glycosylation is due to inaccessibility to the glycosylation machinery or inability of the sequence itself to be glycosylated. Additionally, the result observed from the truncated construct could be due to a large number of possible changes to the protein including misfolding, a missing signal sequence, etc. Nonetheless, NMR experiments have suggested a significant change in the structure of the IMD when this mutation was made (Lee & Glover, 2012). Taken together, these studies provide evidence that P110 is likely important to the structure of caveolin, but more work will have to be done to characterize the topological changes that arise upon the P110A mutation.

2.3 Caveolin-1 penetrates deeply into bilayers

Another key question pertaining to caveolin topology is how deeply does the protein penetrate the bilayer. The positioning of the four tryptophans in caveolin-1 was recently evaluated using fluorescence spectroscopy (Rui et al., 2014). In this study, single tryptophan mutants of a caveolin construct containing the entire scaffolding domain and IMD (residues 82–136) were reconstituted into phospholipid bicelles (native-like discoidal lipid-detergent aggregates). The study showed that W85 was solvent exposed, W98 and W115 resided in the hydrophobic core, and W128 was interfacial (consistent with a hydrophobic core/headgroup border location). Of particular interest is W115, which is only one helical turn away from the turn region and therefore is likely the most accurate reporter of the environment in which the turn is experiencing. This tryptophan is deeply buried and suggests that the turn would be membrane buried also. This finding confirms the previous results from carbonate extraction experiments that showed that caveolin is deeply buried in the bilayer, and is unlikely to be a membrane surface-associated protein (Dupree et al., 1993; Schlegel et al., 1999). Strong agreement is observed when these results are compared to the model put forth by Parton et al., with the exception of W85, which was predicted to reside within the membrane (Parton, Hanzal-Bayer, & Hancock, 2006).

2.4 Caveolin-1 contains a membrane-embedded turn

To date, little has been done to characterize exactly where the putative turn region of caveolin-1 resides. The motif of an intramembrane turn is especially controversial due to the unfavorable breaking of strong i to $i + 4$ hydrogen bonds of the helix within the membrane. In early models put forth by Parton et al. that were based solely on sequence analysis, the postulated glycine—isoleucine—proline—methionine turn was placed at the interface between the extracellular headgroup region and the hydrophobic core of the bilayer (Parton et al., 2006). The most complete picture of location of the intramembrane turn arises from molecular dynamics simulations that were performed on the entire intact scaffolding domain and IMD of the protein (residues 82—136) in a planar 1,2-dimyristoyl-*sn*-glycero-3-phosphocholine bilayer (40 Å) to investigate the protein orientation and energetic favorability of burying the turn region within the core of the bilayer (Rui et al., 2014). These studies revealed that the turn region was stable within the hydrophobic core of the membrane over a 1 μs time course, placing the turn residues at depths of −5.9 Å, −6.9 Å, and −9.3 Å from the bilayer center for glycine, isoleucine, and proline, respectively. This placement puts the turn close to, but not in the headgroup region of the second leaflet of the bilayer. At this location, the lipid headgroups are close enough to stabilize the turn via local backbone hydrogen bonding interactions. Helix-1 (Figures 1 and 3) was found to have a nearly vertical disposition in the membrane, whereas Helix-2 displayed a high degree of tilt. The presence of two conserved positively charged residues in the center of Helix-1 (lysine 96 and arginine 101) could play a role in determining the depth of this helix. Interestingly, Helix-2 of the protein had not exited the hydrophobic core of the membrane by the terminal residue (W128), and it is therefore still an open question as to at which residue the polypeptide chain returns to the membrane surface.

2.5 Cholesterol plays a role in modulating caveolin-1 topology

The topology of caveolin may be somewhat responsive to the inclusion of cholesterol within membranes. Coarse-grain MD simulations performed on a truncated construct of caveolin (residues 94—122) in 1,2-dipalmitoyl-*sn*-glycero-3-phosphocholine and 1,2-dipalmitoyl-*sn*-glycero-3-phosphocholine-cholesterol (20%) membranes revealed that caveolin adopted a hairpin configuration, was more deeply buried in the absence of cholesterol, and was near parallel to the surface when cholesterol was present (Sengupta, 2012). This change in topology occurred without a specific interaction between the helices and cholesterol, indicating that the conformational change was likely related to changes in bilayer fluidity/order. Interestingly, tryptophan fluorescence quenching studies performed on a slightly longer construct (residues 102–134) in the presence of vesicles composed of equimolar amounts of 1,2-dimyristoyl-*sn*-glycero-3-phosphocholine, 1,2-dimyristoyl-*sn*-glycero-3-phosphoethanolamine, sphingomyelin, and cholesterol showed a slight decrease in tryptophan accessibility compared to the construct in pure 1,2-dimyristoyl-sn-glycero-3-phosphocholine vesicles, in contrast to the MD simulations (Lokappa & Nagaraj, 2012). Therefore, there is still a great need to flesh out how cholesterol is involved in modulating caveolin-1 topology.

To date there is no data pertaining to the topology of the other caveolin isoforms. It is assumed that based on sequence similarity these proteins would have a near identical topology. However, in vivo, caveolin-2 appears to be unable to form caveolae in the absence of caveolin-1, and this may indicate a different topology for this isoform (Mora et al., 1999; Parolini et al., 1999).

Although much has been gained in generating a topological map of the caveolin protein, there are still some open questions remaining. With a better understanding of the caveolin-1 topology, aspects of its structural role in generating membrane curvature will undoubtedly be made clearer.

3. CAVEOLIN STRUCTURE

The structure of caveolin clearly has implications for how it governs its many proposed functions. Within this section the progress of understanding the structure of caveolin-1, -2, and -3 will be discussed.

3.1 Caveolin-1

The initial inspection into the structure of caveolin-1 was presented by Parton et al. (2006). Using extensive primary sequence analysis, a model of caveolin-1 structure was built. In this model, the N-terminal domain (residues 1–81) contained a single amphipathic helix (residues 30–50) lying on the membrane surface with the remaining residues being unstructured. The scaffolding domain and the IMD were comprised predominantly of three helices: an amphipathic helix containing residues 80–92 resting on the membrane surface, a short helical membrane-embedded segment containing residues 97–110, and a longer membrane-embedded helix containing residues 110–130. The C-terminal domain was also described as an amphipathic helix resting on the membrane surface encompassing residues 160–178. The region from residues 131–159 was shown to be a solvent exposed, unstructured loop. Importantly, this study identified a break between the two intramembrane helices that contained four residues: G108, I109, P110, and M111 (Parton et al., 2006). While this study gave the first insights into the structure of caveolin-1, the basis of the study was on mathematical predictions and lacked experimental verification.

The first experimental study exploring the secondary structure of caveolin-1 was conducted by Le Lan et al. This study analyzed two fragments of the caveolin-1 protein by circular dichroism spectroscopy, the first containing the scaffolding domain (residues 82–101) and the second, an extended construct, containing the scaffolding domain plus half of the IMD (residues 82–109) (Le Lan, Neumann, & Jamin, 2006). The study revealed that the shorter peptide contained a mixture of helical and extended conformations in the presence of dodecylphosphocholine micelles, while the longer peptide (containing part of the IMD) showed predominantly helical character. The helical character of the extended fragment (residues 82–109) was also confirmed using solution-state NMR (nuclear magnetic resonance) spectroscopy (Le Lan et al., 2006). A later study performed by Le Lan et al., using a shorter fragment containing only the C-terminal region of the scaffolding domain (residues 94–101), was also shown to adopt a helical confirmation (Le Lan et al., 2010). These studies showed that the structure of the scaffolding domain appears to be dependent on the presence of the IMD and suggests that the scaffolding domain should not be viewed as an independently folded domain.

The secondary structure of the full IMD of caveolin-1 (residues 96–136) was analyzed using NMR spectroscopy in lysomyristoylphosphatidylglycerol micelles (Lee & Glover, 2012). The study revealed that the IMD contained a helix-break-helix motif. The first helix (Helix-1) began at residue 97 and extended to residue 107. The second helix (Helix-2) encompassed residues 111–128. The two helices were separated by a three-residue break containing residues 108,109, and 110. This finding supported the model presented by Parton et al., which showed these residues being part of the intramembrane turn. However, in Parton's model M111 was included in the turn and these studies show that this residue is unlikely to be part of this region (Parton et al., 2006). The three break residues were further analyzed for their structural importance using site-directed mutagenesis. The mutation of glycine at position 108 revealed that only small side chain amino acids are tolerated. Position 109 was shown to require a β-branched amino acid and the proline residue at position 110 appears to be critical (Figure 1) (Lee & Glover, 2012).

The same NMR analysis was performed on an elongated construct (residues 82−136), which contains the intact scaffolding and IMDs. The analysis of the elongated construct revealed that Helix-1 extends to include almost the entire scaffolding domain (residues 87−107), while Helix-2 was unaffected (residues 111−128) (Figure 1) (Rui et al., 2014). This experimental evidence is in contrast with the model put forth by Parton et al., which showed three helices comprising the scaffolding and IMDs: in actuality there are only two. This also supports Le Lan's studies that show that the scaffolding domain is helical when the IMD is present (Le Lan et al., 2006). Therefore this region can be postulated to be two equal length helices separated by a three-residue break, which is the putative membrane turn.

In a similar study to that conducted by Le Lan et al., Hoop et al. also probed the secondary structure of the caveolin-1 scaffolding domain using two fragments, one containing just the scaffolding domain (residues 83−102) and the other containing the scaffolding and the first helix of the IMD (residues 82−109) (Hoop, Sivanandam, Kodali, Srnec, & van der Wel, 2012). The important difference between the Hoop et al. and the Le Lan et al. studies was that the Hoop et al. studies were conducted in 1-Palmitoyl-2-oleoyl-sn-glycero-3-phosphocholine bilayers containing cholesterol. NMR, circular dichroism, and Fourier transform infrared spectroscopy were used to show that the scaffolding domain shows a β-strand conformation (Hoop et al., 2012). The extension of the construct to include part of the first helix of the IMD did show an increase in the helical character. However, further analysis using solid-state NMR techniques showed that the helical structure was isolated to the IMD and that the scaffolding domain did indeed still form a β-strand (Hoop et al., 2012). This is in contrast to the aforementioned studies, and could be attributed to the presence of cholesterol. It is also important to note that the existence of a β-strand structure can be indicative of aggregation which may or may not be biologically relevant.

A construct containing the N-terminal domain (residues 1−81) has been shown to be unstructured in aqueous media using circular dichroism spectroscopy (Fernandez, Ying, Albanesi, & Anderson, 2002). In contrast, a construct containing the N-terminal and scaffolding domains (residues 1–101) was shown to have increased helical content. This increase in helical content was attributed to the presence of the scaffolding domain. This suggests that the N-terminal domain may also have an effect on the structure of the scaffolding domain, and that differences seen in these various studies may be due to truncation artifacts. There has been no structural characterization of the C-terminal domain, although as stated before, it may be an amphipathic helix resting on the membrane surface.

The most recent study of caveolin-1 structure examined two peptides, the first corresponding to a portion of the IMD (residues 101–126) and the second including a portion of the intramembrane and scaffolding domains (residues 93–126) (Yang et al., 2015). When analyzed by circular dichroism spectroscopy, both peptides showed characteristic α-helical signatures in 60% 1,1,1,3,3,3-hexafluoro-2-propanol (HFIP), sodium dodecyl sulfate micelles, and dodecylphosphocholine micelles (Yang et al., 2015). NMR analysis of the shorter peptide in sodium dodecyl sulfate micelles showed a helix-break-helix motif. This was also seen in the structural analysis of the longer peptide in 60% HFIP. This finding is in agreement with NMR analysis of the full IMD (Lee & Glover, 2012). However, the authors report that in contrast to the study by Lee et al., the break region is significantly longer (residues 108–116), and the two helices are shorter (Helix-1, residues 103–107 and Helix-2, residues 117–126) (Yang et al., 2015). While this is the first report of tertiary structure for the IMD of caveolin-1, it is important to note that the full IMD was not present. Therefore, the discrepancy between this study and Lee et al. is likely due to the choice of construct length, and it is unclear as to why the authors did not use a construct which encompassed the entire IMD.

Taken together, the overall secondary structure of the caveolin-1 scaffolding and IMDs has been described. However, the tertiary structure of the full-length protein still remains elusive. One of the major questions that still persists and is necessary for understanding the structure of this protein, is the angle that exists between the two intramembrane helices. This angle is critical in understanding the tertiary structure of the protein because there could be stabilizing interactions between the domains that allow for the proposed U-shaped topology of the caveolin-1. The only study to investigate the angle between the two helices was a molecular dynamics study by Rui et al. (Rui et al., 2014). This study concluded that the angle between Helix-1 and Helix-2 was dynamic and sampled a distance distribution between 35° and 69°. These studies also revealed that residues in Helix-1 and Helix-2 close to the intramembrane turn interacted intimately with each other (Rui et al., 2014). In particular, the simulations supported the importance of a β-branched amino acid at position 109 that was shown by Lee et al. (Lee & Glover, 2012). These interactions are likely key in limiting the angle between Helix-1 and Helix-2. Another molecular dynamics simulation also probed the angle between two helices this time using a construct that stretched from residues 102−122 and found that the angle between the helices was roughly 35° and did not vary with the inclusion of cholesterol (Sengupta, 2012). While these in silico studies represent a major step forward, the exact angle will have to be validated.

3.2 Caveolin-2

There have only been a few structural studies of caveolin-2. The first study looked at the C-terminal region of the scaffolding domain, residues 79−87. NMR analysis showed that this portion of the protein is helical, which is in agreement with the data on the caveolin-1 protein (Le Lan et al., 2010). The second study was a construct of caveolin-2 that contained residues 1−73 (the N-terminal domain) the authors found that the construct contained 25% α-helicity (Fernandez et al., 2002). More work will be needed to structurally characterize caveolin-2, as it may have a structure that is different than that of caveolin-1 since it appears that caveolin-2 cannot form caveolae on its own (Mora et al., 1999; Parolini et al., 1999).

3.3 Caveolin-3

The secondary structure of small peptides derived from the scaffolding domain of caveolin-3 was examined in water, sodium dodecyl sulfate micelles, and/or trifluoroethanol using circular dichroism spectroscopy (Sowmya, Jagannadham, & Nagaraj, 2006). Although structural information was not the focus of this report, it does represent an attempt to understand caveolin-3 scaffolding domain structure. The major conclusion that followed from these studies was that in sodium dodecyl sulfate micelles, the scaffolding peptides adopted a β-strand conformation. However, the extremely short length of the peptides examined precludes definitive conclusions from being drawn from these studies.

More recently, structural studies were performed on a full-length construct of caveolin-3 with a synthetic palmitoylation strategy (Kim, Peng, Schlebach, Hadziselimovic, & Sanders, 2014). The addition of a thioalkyl chain through a disulfide bond was utilized to determine the effects of palmitoylation on the protein structure. Circular dichroism and NMR spectroscopy were utilized to elucidate the secondary structure of the protein. The overall secondary structure of caveolin-3 was seen as helical based on circular dichroism spectroscopy (Fernandez et al., 2002; Kim et al., 2014). Comparison of the palmitoylated and the non-palmitoylated $^{1}H-^{15}N$ heteronuclear single quantum coherence spectrum of caveolin-3 showed a modest effect of the palmitoylation with the only major chemical shift perturbations being seen in close proximity to the palmitoylated sites (Kim et al., 2014). This may indicate that the structural changes caused by palmitoylation are local changes rather than overall global changes. It is important to note that only three of the nine cysteine residues that are contained within caveolin-3 were alkylated (the three corresponding to caveolin-1 palmitoylation, Figure 2). Although it has been established that caveolin-3 is highly palmitoylated, the exact location of the sites is unknown but is thought to be analogous to the sites in caveolin-1 (Machleidt, Li, Liu, & Anderson, 2000). Therefore, it is unclear as to whether this study captured the full essence of caveolin palmitoylation because it is possible that caveolin-3 is more extensively palmitoylated than caveolin-1. However this work represents a milestone as this is the first study of caveolin looking at the structural effects of palmitoylation.

From the structural data so far, it is clear that all the caveolin isoforms have significant α-helical elements. The IMD is likely to be predominately helical while at this point it is unclear what the secondary structure of the scaffolding domain is, and whether it is dependent on cholesterol. The most compelling area of future research will be in determining the angle between the helices that form the intramembrane turn, which is most likely the predominate tertiary structural feature of this protein, which has been ascribed to its membrane-curving predilection.

4. OLIGOMERIZATION OF CAVEOLIN

4.1 Caveolin-1 homo-oligomerization

The first hint that caveolin may oligomerize arose from the observation that caveolin-1 (VIP21) was part of a complex derived from *trans*-Golgi network vesicles isolated from MDCK (Madin–Darby canine kidney) cells (Kurzchalia et al., 1992). These complexes were isolated based on their insolubility in the mild detergents, Triton X-114 and CHAPS (3-[(3-Cholamidopropyl)dimethylammonio]-1-propanesulfonate). This result was followed shortly afterward with a study that analyzed these complexes further, this time using canine lung tissue and MDCK cell membranes (Monier et al., 1995). After an extensive isolation procedure involving sucrose gradient centrifugation, the caveolin complex was isolated and analyzed by SDS-PAGE (sodium dodecyl sulfate polyacrylamide gel electrophoresis) under reducing conditions (since caveolin has three cysteine residues) with and without sample boiling. When the sample was not boiled, three discrete bands were observed, with rough estimations of their molecular weights at 200, 400, and 600 kDa. On the other hand, when the samples were boiled, a single band at a molecular weight of 22 kDa was observed, indicating that the caveolin-1 oligomer was not held together by covalent interactions. Furthermore, it was found that caveolin was clearly the major component in the complex, as other proteins were not observed by SDS-PAGE analysis. From these findings, it was concluded that caveolin homo-oligomerizes. Around the same time, Sargiacomo et al. also confirmed the presence of high mass caveolin-1 complexes using a similar methodology (Sargiacomo et al., 1995). In these studies, the homo-oligomeric complexes were found to contain 14–15 caveolin-1 monomers. From this data, Song and coworkers conjectured that the oligomerization process occurs in two steps where the oligomers of 14–15 caveolin-1 monomers assemble into larger caveolae-sized clusters (Song, Tang, Li, & Lisanti, 1997). This has been confirmed more recently by Hayer and coworkers who showed that high-order caveolin-1 oligomers assemble in a stepwise fashion beginning with 8S complexes (14–16 caveolin monomers) in the endoplasmic reticulum which then assemble into much larger 70S complexes in the Golgi complex (Hayer, Stoeber, Bissig, & Helenius, 2010). However, the 70S complex appeared to be sensitive to the detergent used in the extraction procedure (Hayer et al., 2010). Studies using chemical cross-linking, which should render the complexes less sensitive to disruption by various detergents, found that 80S oligomers containing caveolin-1 were indeed formed (Ludwig et al., 2013). While immensely insightful, these studies must be taken with pause, as the insolubility of a protein in a specific detergent system could itself trigger nonbiologically relevant aggregation events. Therefore, further investigations of caveolin-1 in vivo oligomerization that do not rely on the principle of detergent insolubility may be warranted.

After observing this fascinating oligomeric behavior, many questioned which portion of caveolin is responsible for oligomerization. In one study, various portions of caveolin-1, expressed as GST (glutathione-*S*-transferase) fusions, were analyzed for oligomeric behavior (Sargiacomo et al., 1995). It was found that residues 61–101 were critical for oligomerization and surprisingly, neither the IMD (102–134) nor the C-terminal domain (135–178) resulted in the appearance of high molecular weight complexes.

Another study probed the role that the C-terminus plays in caveolin–caveolin association (Schlegel & Lisanti, 2000). Various incarnations of the C-terminus (ranging from the full-length C-terminus (135–178) with or without palmitoylation, to a construct which had the last 28 amino acids removed) were expressed as N-terminal fusions to GFP (green fluorescent protein) and were examined for their ability to coimmunoprecipitate with full-length caveolin-1. It was found that association decreased mildly with removal of palmitoylation and substantially with the removal of the distal C-terminal residues. This behavior was specifically linked to a highly conserved serine residue, 10 amino acids prior to the end of the protein, where mutations of this residue to either a negatively charged glutamic acid or alanine resulted in poor association with the full-length protein. Therefore, it seems that residues 61–101 are required for the formation of high-order oligomers and residues 150–178 enhance the strength of this association. Of course, these conclusions are drawn from two separate fusion constructs (GST and GFP), which could be prone to unique experimental artifacts. In particular, GST, which is known to be a dimer, could influence the aggregation state of the protein (Maru, Afar, Witte, & Shibuya, 1996). Additionally, a recent paper evaluated detergent resistance, subcelluar distribution, and the oligomeric state of overexpressed C-terminally tagged caveolin-1 constructs and found that the type of tag (myc, mCherry, or GFP) had dramatic effects on the self-association behavior of the caveolin protein (Han, Tiwari, & Kenworthy, in press). Therefore, conclusions drawn about the oligomeric behavior of caveolin in vivo using fusion tags may warrant revisitation in light of these findings.

Another study characterized the nature of the caveolin-1 oligomerization using a truncated construct containing only residues 1–101 (Fernandez et al., 2002). Gel filtration chromatography and analytical ultracentrifugation indicated that a 450 kDa aggregate was formed, which was shown to be composed purely of caveolin-1 by SDS-PAGE analysis. Evaluation of the aggregates by electron microscopy showed that short filaments and ring structures were formed. It appeared that the filaments were composed of the ring structures, and each ring contained seven copies of the N-terminal caveolin construct. It was found that the corresponding caveolin-2 and caveolin-3 constructs behaved similar to caveolin-1. The authors suggested that these oligomeric bundles represent the basic building block of caveolae. However, these studies were performed on a truncated construct, which precludes definitive conclusions from being drawn. In addition, the studies were performed in the absence of any membrane mimic (detergent, vesicles, etc.), so it is unclear if the structures formed would be similar to caveolin-1 oligomers that are purported to form in the plasma membrane.

4.2 Caveolin-2 hetero-oligomerization

Studies have also investigated the hetero-interactions between caveolins-1, -2, and -3 (Scherer et al., 1997). A series of in vivo coimmunoprecipitation and immunofluorescence experiments were used to investigate the association of isoforms within an oligomer. Coimmunoprecipitation experiments showed that caveolin-1 immunoprecipitates can react with anti-caveolin-2 IgG using Western analysis (Scherer et al., 1997). Caveolin-2 was found to be purely associated with the caveolin-1 immunoprecipitate indicating that caveolin-2 does not oligomerize on its own. Immunofluorescence experiments showed that both the caveolin-1 and caveolin-2 proteins were localized within the same regions of the cell, giving some additional evidence to the existence of hetero-oligomeric complexes (although colocalization does not necessitate a direct interaction). With the region of homo-oligomerization being mapped to the 61–101 region, it was important to know if caveolin-2 binds through the same domain. To isolate the region responsible, 32-ten amino acid in length caveolin-2 peptides were generated to give full sequence coverage, and were incubated on blot paper (Das, Lewis, Scherer, & Lisanti, 1999). To this blot, radio-iodinated caveolin-1 was added and was found to bind tightly to peptides that encompassed the IMD, indicating that caveolin hetero-oligomerization is mediated through a distinct region of the protein and not the same site of homo-oligomerization. In immunoprecipitation experiments performed using caveolin-3 instead of caveolin-1, caveolin-2 did not appear in pull-down assays, which indicated that caveolin-2 likely is not a member of caveolin-3 high molecular weight complexes (Das et al., 1999).

4.3 Caveolin-3 homo-oligomerization

Caveolin-3 is not as well characterized as the other caveolin isoforms in terms of understanding its oligomeric behavior. A study employing electron microscopy to obtain reconstructed images of full-length caveolin-3 complexes purified using a sucrose gradient indicated that a 220 kDa oligomer was formed that consisted of nine caveolin-3 monomers (Whiteley, Collins, & Kitmitto, 2012). These monomers were observed to be arranged into a disk-like aggregate. The configuration of the three-dimensional monomer bundle can be described by two regions. The central region of the disc is shown to be made up of the C-terminal region of the protein with the N-terminus forming an outer ring that protrudes into the cytosolic space. The close proximity of the monomers could support the conjecture that caveolin can interact with many signaling molecules in a small amount of space (Whiteley et al., 2012). While this was the first reported caveolin-3 oligomeric structure, the resolution was only 17 Å. Therefore, more in-depth biophysical techniques will be needed to definitively confirm this arrangement for caveolin-3 oligomers.

4.4 Caveolin-1 does not homo-oligomerize in the absence of cell components

Although it is generally accepted that caveolin homo- and hetero-oligomerizes in vivo and in vitro (using transcription/translation systems), there still is the open question remaining as to if caveolins could oligomerize in vitro in the absence of other proteins and lipid components. In particular, the initial characterization results were based on caveolin preparations that are generated by detergent insolubility, which may or may not exclude important accessory proteins involved in oligomerization. Recent work has addressed the biophysical question of caveolin's ability to homo-oligomerize in the absence of the cellular milieu (Rieth, Lee, & Glover, 2012). A highly purified caveolin construct encompassing residues 62–178 reconstituted into n-dodecylphosphocholine micelles was found to behave as a monomer using the combined techniques of analytical ultracentrifugation and gel filtration chromatography. This behavior was not likely an artifact of the truncated construct because it contained the critical 62–101 and 150–178 regions that were purported to be ultrasensitive to oligomerization. It also is not likely a result of n-dodecylphosphocholine action on the protein as the in vivo caveolin oligomers are SDS (sodium dodecyl sulfate) resistant, and n-dodecylphosphocholine is a much more mild detergent than SDS (Monier et al., 1995). This monomeric behavior indicates that various factors could be required for proper caveolin aggregation. Interestingly, the monomeric behavior was found to rely on the presence of a highly conserved proline residue at position 132. It was found that no amino acid substitutions could be made to this position, or the caveolin protein would behave as a dimer. This is consistent with in vivo observations that a P132L mutant, that is putatively associated with breast cancer, can result in retention of the protein at the level of the Golgi complex, possibly indicating that the dimerization disrupts cellular trafficking of the protein to the plasma membrane (Lee et al., 2002). The dimerization interface was mapped by successive truncations of the P132L mutant and it was found that the only requirement was the IMD. Secondary structure analysis of the P132L mutant by NMR revealed that Helix-2, which normally stretches from 111 to 128, was extended four additional amino acids, providing a plausible mechanistic rational for the appearance of a novel dimerization interface. This work suggests that other players may be involved in facilitating the oligomeric behavior that caveolin displays in vivo.

4.5 Caveolin-1 oligomerization is controlled by acylation and lipid environments

Research has shown that lipids may be critical to determining caveolin's oligomeric activities. Palmitoylation of caveolin after translocation in the membrane has been reported to strengthen oligomeric interactions (Monier, Dietzen, Hastings, Lublin, & Kurzchalia, 1996). A study in which the treatment of in vitro transcribed caveolin with potassium hydroxide and methanol, which would deacylate any cysteine residues, resulted in a complete disruption of oligomeric behavior (Monier et al., 1996). Similarly, oligomer formation could be induced upon palmitoylation of caveolin-1. In vivo experiments employing a caveolin-1 that lacked cysteines (all were mutated to serine) resulted in the disappearance of high-order oligomers, again showing that these acylated sites may improve oligomerization efficiency. This held true when a C143S and C156S double mutant was generated (Figure 2). Interestingly, when only C133S mutation was made, the oligomerization was unaffected. When C133S and C143S mutations were made, the 400 kDa complex was the only one observed. Therefore, it appeared that a fine balance of oligomeric states was being maintained by the presence of C143 and C156. In the same study it was also shown that the inclusion of 25-hydroxy-cholesterol could induce the formation of high-order oligomers in the absence of cytosol using an in vitro transcription/translation system. Therefore, it seems that a number of important factors contribute to the oligomeric behavior of caveolin, such as acylation and association with cholesterol.

4.6 Cavins are an integral part of the caveolin-1 high molecular weight complex

A typical caveola is thought to contain about 140—180 caveolin molecules (Pelkmans & Zerial, 2005). Although it was originally thought that oligomeric complexes formed were composed exclusively of caveolin, recent work has shown that other proteins may be involved in oligomerization as well. It has now been shown that caveolin likely associates with the cavin family of proteins to form hetero-oligomeric complexes, as immunopurification of caveolae from human adipocytes revealed the presence of cavins (Hill et al., 2008). There are four isoforms of the cavin family of proteins that are thought to interact with caveolin (cavin-1 (polymerase I and transcript release factor), -2 (serum deprivation protein response), -3 (sdr-related gene product that binds to C-kinase), and cavin-4 (muscle-restricted coiled-coil protein)) (Bastiani et al., 2009; Hansen, Bright, Howard, & Nichols, 2009; Hill et al., 2008). It has been demonstrated that the cavin isoforms coassociate with each other in the absence of caveolin (Hansen et al., 2009). Cavin—cavin interactions are also hypothesized to be important for the formation of the cavin membrane coat. X-ray crystallographic studies have shown that a homo-trimeric coiled-coil structure is formed by parallel α-helices of a conserved oligomerization domain that is present in all cavin isoforms (Kovtun et al., 2014).

Under conditions where caveolin was expressed, cavin was found to be associated with the cell membrane (Hill et al., 2008). Importantly, cavin-1 expression was shown to be required for caveolae formation (Hansen, Shvets, Howard, Riento, & Nichols, 2013; Hill et al., 2008; Liu et al., 2008; Liu & Pilch, 2008). Overexpression of cavin-1 resulted in no changes to caveolae morphology whereas cavin-2 overexpression resulted in the formation of larger caveolae and also generated caveolae tubules. Cavin-3 has been shown to be important in regulating caveolae-mediated endocytosis (McMahon et al., 2009). Taken together, cavins are thought to modulate caveolae behavior, and cross-linking studies have suggested a direct interaction between caveolin and cavins (Ludwig et al., 2013). Knockouts of specific cavins have also been shown to lead to tissue-specific disruptions in cell function (Hansen et al., 2013). Interestingly, tomographic reconstructions of the caveolar coat have revealed that it is uniform in its distribution of coat proteins, disputing previous claims that the organization of protein species may be different around the neck of the caveolae than in the bulb (Ludwig et al., 2013). The exact details of the formation, organization, and stoichiometry of the caveolin—cavin complex have only recently begun to be uncovered.

The cellular location where caveolin and cavins begin to interact has also been under investigation. It appears that the association of caveolin and cavin occurs at the plasma membrane as caveolin and cavins do not cofractionate in detergent-solubilized cell lysates where Golgi transport activity has been arrested (Hayer et al., 2010). This observation agrees with a previous study using immunofluorescence where caveolin and cavin were colocalized in the plasma membrane but not in the Golgi apparatus in NIH 3T3 fibroblasts (Hill et al., 2008).

The caveolin–cavin complex has been shown to form large aggregates (70S) using sucrose velocity gradients (Hayer et al., 2010). The stoichiometry of these complexes has been investigated by a number of groups (Gambin et al., 2013; Ludwig et al., 2013). Immunoprecipitation results suggested a fixed ratio between cavins and caveolin where the caveolin-1 to cavin-1 ratio is about 4 to 1. Immunofluorescence studies have shown that cavin-2 and cavin-3 may compete for the same binding site in the cavin-1/caveolin complex. The binding site of the cavins-2 and -3 to the cavin-1/caveolin complex is thought to be on cavin-1, as cells expressing cavin-1 with either cavin-2 or cavin-3 resulted in two different-sized species as evaluated by sucrose velocity centrifugation. This result may provide a mechanism for tissue-specific caveolar functions where modulation of caveolae properties occurs through different complements of cavins-2 and -3 in association with the cavin-1/caveolin-1 complex (Ludwig et al., 2013). This idea is supported by the fact that the variations in cavin-2 and cavin-3 expression appears to be tissue specific (Bastiani et al., 2009).

Much less is known about cavin-4. It was shown to be highly expressed in skeletal and cardiac muscle cells but not in other tissue types (Bastiani et al., 2009). High-resolution light microscopy studies have indicated that cavin-4 localizes with caveolin-3 in skeletal muscle cells. Furthermore, the specific immunolabeling of cavin-4 showed that it localizes in pits and vesicles close to the plasma membrane, pointing to a location within caveolae.

Another recent study has employed GFP- and mCherry-tagged cavins and caveolin-1 fusion proteins to study the stoichiometry of membrane-dissociated (solubilized) caveolin–cavin complexes by fluorescence correlation spectroscopy, which can give information about particle size, as well as the amount of different molecules within the complex if they have different fluorescent tags (Gambin et al., 2013). In this study, cavin-1-mCherry and caveolin-1-GFP were overexpressed in MCF-7 (Michigan Cancer Foundation-7) cells. The cells were then mechanically lysed and it was found that soluble caveolar aggregates were released into the cytoplasm. Analysis of the aggregates showed that the typical cavin-1/caveolin-1 aggregate contained about 50 cavin-1 molecules and 150–180 caveolin-1 molecules, yielding a stoichiometry similar to the 1:4 ratio that has been suggested using immunoprecipitation. The size of these particles was estimated to be about

60 nm, similar to the size of a caveolae. A similar experimental setup was used to investigate the relationship between the cavins in these complexes. Transfection of MCF-7 cells with cavin-1-GFP and cavin-2-mCherry was used to characterize the stoichiometry of the two cavin family members and it was found that the cavin-1/cavin-2 ratio was about 2:1. The same ratio was observed for in experiments employing cavin-1-GFP and cavin-3-mCherry. Expression of only cavin-2-GFP and cavin-3-mcherry in MCF-7 cells did not show any association between the two cavins. This result is in agreement with the idea that cavin-2 and cavin-3 compete for the same binding site on cavin-1 and they do not interact with each other (although they are both observed to be present on the same caveolae), suggesting that there are distinct cavin subcomplexes. To examine the size of oligomers made up entirely of cavins, hypoosmotic shock treatment was used to dissociate cavin-1-GFP/cavin-1-mcherry complexes from the membrane, and a particle of 20 nm in size was observed. Interestingly, dissociation of the cavin-1/cavin-2 complex from the caveolae using hypoosmotic treatment resulted in a 30 nm particle being released from the membrane. When this same experiment was performed using cavin-1/cavin-3 expressing cells, hypoosmotic treatment resulted in a 20 nm particle being released from the membrane. Therefore it seems that cavins-2 and -3 are critical in determining the geometric properties of the cavin subcomplexes, in agreement with the sucrose gradient experiments performed by Ludwig et al. (Ludwig et al., 2013).

However, even with this knowledge that caveolin is likely part of a large hetero-oligomeric complex, the nature of the interaction between caveolin and the cavins is still completely uncharacterized from a biophysical perspective as no binding sites have been proposed. It is also important to examine the caveolin/cavin interaction without the use of fluorescent protein fusions, which could alter the oligomeric behavior and oligomeric structures of caveolin and/or cavin (Han et al., in press). Also, one must be skeptical when examining nonmembrane-bound forms of oligomeric membrane protein complexes, as the behavior of soluble aggregates may not reflect accurately the behavior of the natively membrane bound form. Furthermore, studies will be needed to elucidate the role that the homo-trimeric coiled-coil structures formed by cavin homo-oligomers play in caveolae structure.

We are just beginning to understand how all of the molecular components in caveolae play into the formation and stability of these highly important microdomains. It is still unclear as to how other molecules associated with caveolae, such as cholesterol, would associate with these complexes, and must be part of the ongoing effort to characterize caveolin oligomers. Also, a methodology that can be used to isolate caveolin oligomers from cells in a way that does not depend on detergent insolubility still needs to be developed to add confidence to results presented in this review that have relied heavily on this approach.

5. CONCLUSIONS

Caveolae are a critical component of the eukaryotic plasma membrane. In the time since their discovery, the number of the functions and events postulated to be caveolae localized is both diverse and immense. The central player involved in the formation and function of these microdomains is the integral protein caveolin, whose importance is belied by the fact that mutation and misregulation of the protein results in disease states. Much is known about the involvement of this protein in disease states. However, in this review, we focused in on the less well-understood biophysical aspects of the caveolin proteins.

Structural work done in the past 5 to 10 years has lead to a renaissance in terms of understanding caveolin-1 secondary structure. Specifically, studies reveal that the scaffolding domain and IMD are composed of a helix-break-helix motif, which is consistent with the postulated intramembrane turn topology of caveolin. Interestingly, the residues of the break region are extremely sensitive to mutation, indicating that these residues are likely critical for structuring the protein. Even with these gains, determining the tertiary structure of the caveolin protein remains to be the zenith of structural endeavors.

Strides have also been made in uncovering the exact details of the caveolin topology. It is now clear that both N- and C-termini of caveolin are co-cytoplasmic and that the protein is deeply membrane embedded. However, the *exact* location of the intramembrane turn, and the *exact* orientation of caveolin with respect to the membrane, remains to be fully addressed. Much of the recent work addressing these topological issues is from in silico simulations and therefore experimental evidence is needed to validate these conclusions.

The oligomeric properties of the caveolin proteins is still a very complex issue that will require much more work in order to truly parse out the exact details. This issue is particularly difficult to address because lipid composition, posttranslational modification, and other accessory proteins are all proposed to be involved in oligomeric behavior. Recent evidence suggests that although caveolin itself is critical to the observed oligomeric behavior, the protein cavin also plays a preeminent role.

One of the issues that became apparent from the studies of caveolin is the employment of truncated constructs. While necessary for experimental feasibility, artifacts due to the truncations themselves can hinder the clear interpretation of the data, especially when comparing studies using different constructs. In addition, the use of a variety of detergents, lipids, fusion constructs, etc., befuddles the generation of definitive conclusions. Therefore, a clear goal moving forward will be in developing and utilizing methodologies that will allow the full-length protein to be characterized under native conditions.

Clearly, we are just beginning to understand the structural, topological, and oligomeric aspects of caveolin in full detail. Unveiling these aspects will likely be crucial to future developments in understanding the pathogenesis and mechanisms of caveolin/caveolae-related diseases and will be pertinent to the development of therapeutic endeavors, which will target these microdomains.

REFERENCES

Aoki, S., & Epand, R. M. (2012). Caveolin-1 hydrophobic segment peptides insertion into membrane mimetic systems: role of proline residue. *Biochimica et Biophysica Acta, 1818*(1), 12–18.

Aoki, S., Thomas, A., Decaffmeyer, M., Brasseur, R., & Epand, R. (2010). The role of proline in the membrane re-entrant helix of caveolin-1. *The Journal of Biological Chemistry, 285*, 33371–33380.

Bastiani, M., Liu, L., Hill, M. M., Jedrychowski, M. P., Nixon, S. J., Lo, H. P., et al. (2009). MURC/Cavin-4 and cavin family members form tissue-specific caveolar complexes. *The Journal of Cell Biology, 185*(7), 1259–1273.

Bonuccelli, G., Casimiro, M. C., Sotgia, F., Wang, C., Liu, M., Katiyar, S., et al. (2009). Caveolin-1 (P132L), a common breast cancer mutation, confers mammary cell invasiveness and defines a novel stem cell/metastasis-associated gene signature. *The American Journal of Pathology, 174*(5), 1650–1662.

Byrne, D. P., Dart, C., & Rigden, D. J. (2012). Evaluating caveolin interactions: do proteins interact with the caveolin scaffolding domain through a widespread aromatic residue-rich motif? *PloS One,* 7(9), e44879.

Capozza, F., Combs, T. P., Cohen, A. W., Cho, Y. R., Park, S. Y., Schubert, W., et al. (2005). Caveolin-3 knockout mice show increased adiposity and whole body insulin resistance, with ligand-induced insulin receptor instability in skeletal muscle. *American Journal of Physiology–Cell Physiology, 288*(6), C1317–C1331.

Carman, C. V., Lisanti, M. P., & Benovic, J. L. (1999). Regulation of G protein-coupled receptor kinases by caveolin. *Journal of Biological Chemistry, 274*, 8858–8864.

Chang, W. J., Ying, Y. S., Rothberg, K. G., Hooper, N. M., Turner, A. J., Gambliel, H. A., et al. (1994). Purification and characterization of smooth muscle cell caveolae. *Journal of Cell Biology, 126*, 127–138.

Cohen, A. W., Park, D. S., Woodman, S. E., Williams, T. M., Chandra, M., Shirani, J., et al. (2003). Caveolin-1 null mice develop cardiac hypertrophy with hyperactivation of p42/44 MAP kinase in cardiac fibroblasts. *American Journal of Physiology—Cell Physiology, 284*(2), C457—C474.

Collins, B. M., Davis, M. J., Hancock, J. F., & Parton, R. G. (2012). Structure-based reassessment of the caveolin signaling model: do caveolae regulate signaling through caveolin-protein interactions? *Developmental Cell, 23*(1), 11—20.

Couet, J., Sargiacomo, M., & Lisanti, M. P. (1997). Interaction of a receptor tyrosine kinase, EGF-R, with caveolins — caveolin binding negatively regulates tyrosine and serine/threonine kinase activities. *The Journal of Biological Chemistry, 272*(48), 30429—30438.

Das, K., Lewis, R. Y., Scherer, P. E., & Lisanti, M. P. (1999). The membrane-spanning domains of caveolins-1 and -2 mediate the formation of caveolin hetero-oligomers. Implications for the assembly of caveolae membranes in vivo. *The Journal of Biological Chemistry, 274*(26), 18721—18728.

Dietzen, D. J., Hastings, W. R., & Lublin, D. M. (1995). Caveolin is palmitoylated on multiple cysteine residues. Palmitoylation is not necessary for localization of caveolin to caveolae. *The Journal of Biological Chemistry, 270*(12), 6838—6842.

Drab, M., Verkade, P., Elger, M., Kasper, M., Lohn, M., Lauterbach, B., et al. (2001). Loss of caveolae, vascular dysfunction, and pulmonary defects in caveolin-1 gene-disrupted mice. *Science, 293*(5539), 2449—2452.

Dupree, P., Parton, R. G., Raposo, G., Kurzchalia, T. V., & Simons, K. (1993). Caveolae and sorting in the trans-golgi network of epithelial cells. *The EMBO Journal, 12*(4), 1597—1605.

Engelman, J. A., Zhang, X. L., Galbiati, F., Volonte, D., Sotgia, F., Pestell, R. G., et al. (1998). Molecular genetics of the caveolin gene family: Implications for human cancers, diabetes, Alzheimer disease, and muscular dystrophy. *American Journal of Human Genetics, 63*(6), 1578—1587.

Epand, R. M., Sayer, B. G., & Epand, R. F. (2005). Caveolin scaffolding region and cholesterol-rich domains in membranes. *Journal of Molecular Biology, 345*(2), 339—350.

Fernandez, I., Ying, Y., Albanesi, J., & Anderson, R. G. (2002). Mechanism of caveolin filament assembly. *Proceedings of the National Academy of Sciences of the United States of America, 99*(17), 11193—11198.

Fernández-Hernando, C., Yu, J., Dávalos, A., Prendergast, J., & Sessa, W. C. (2010). Endothelial-specific overexpression of caveolin-1 accelerates atherosclerosis in apolipoprotein E-deficient mice. *The American Journal of Pathology, 177*(2), 998—1003.

Feron, O., Saldana, F., Michel, J. B., & Michel, T. (1998). The endothelial nitric-oxide synthase-caveolin regulatory cycle. *The Journal of Biological Chemistry, 273*(6), 3125—3128.

Galbiati, F., Volonte, D., Minetti, C., Chu, J. B., & Lisanti, M. P. (1999). Phenotypic behavior of caveolin-3 mutations that cause autosomal dominant limb girdle muscular dystrophy (LGMD-1C). Retention of LGMD-1C caveolin-3 mutants within the golgi complex. *Journal of Biological Chemistry, 274*(36), 25632—25641.

Gambin, Y., Ariotti, N., McMahon, K. A., Bastiani, M., Sierecki, E., Kovtun, O., et al. (2013). Single-molecule analysis reveals self assembly and nanoscale segregation of two distinct cavin subcomplexes on caveolae. *Elife, 3*, e01434.

García-Cardeña, G., Martasek, P., Masters, B. S., Skidd, P. M., Couet, J., Li, S., et al. (1997). Dissecting the interaction between nitric oxide synthase (NOS) and caveolin: functional significance of the NOS caveolin binding domain in vivo. *The Journal of Biological Chemistry, 272*(41), 25437—25440.

Ghosh, S., Gachhui, R., Crooks, C., Wu, C. Q., Lisanti, M. P., & Stuehr, D. J. (1998). Interaction between caveolin-1 and the reductase domain of endothelial nitric-oxide synthase. Consequences for catalysis. *The Journal of Biological Chemistry, 273*(35), 22267—22271.

Glenney, J. R. (1992). The sequence of human caveolin reveals identity with VIP21, a component of transport vesicles. *FEBS Letters, 314*, 45–48.

Han, B., Tiwari, A., & Kenworthy, A. K. Tagging strategies strongly impact the fate of overexpressed caveolin-1. *Traffic*, in press.

Hansen, C. G., Bright, N. A., Howard, G., & Nichols, B. J. (2009). SDPR induces membrane curvature and functions in the formation of caveolae. *Nature Cell Biology, 11*(7), 807–814.

Hansen, C. G., Shvets, E., Howard, G., Riento, K., & Nichols, B. J. (2013). Deletion of cavin genes reveals tissue-specific mechanisms for morphogenesis of endothelial caveolae. *Nature Communications, 4*, 1831.

Hanson, C. A., Drake, K. R., Baird, M. A., Han, B., Kraft, L. J., Davidson, M. W., et al. (2013). Overexpression of caveolin-1 is sufficient to phenocopy the behavior of a disease-associated mutant. *Traffic, 14*(6), 663–667.

Hayer, A., Stoeber, M., Bissig, C., & Helenius, A. (2010). Biogenesis of caveolae: stepwise assembly of large caveolin and cavin complexes. *Traffic, 11*(3), 361–382.

Hill, M., Bastiani, M., Luetterforst, R., Kirkham, M., Kirkham, A., Nixon, S., et al. (2008). PTRF-cavin, a conserved cytoplasmic protein required for caveola formation and function. *Cell, 132*, 113–124.

Hoop, C. L., Sivanandam, V. N., Kodali, R., Srnec, M. N., & van der Wel, P. C. (2012). Structural characterization of the caveolin scaffolding domain in association with cholesterol-rich membranes. *Biochemistry, 51*(1), 90–99.

Horikawa, Y. T., Patel, H. H., Tsutsumi, Y. M., Jennings, M. M., Kidd, M. W., Hagiwara, Y., et al. (2008). Caveolin-3 expression and caveolae are required for isoflurane-induced cardiac protection from hypoxia and ischemia/reperfusion injury. *Journal of Molecular and Cell Cardiology, 44*(1), 123–130.

Ju, H., Zou, R., Venema, V. J., & Venema, R. C. (1997). Direct interaction of endothelial nitric-oxide synthase and caveolin-1 inhibits synthase activity. *The Journal of Biological Chemistry, 272*(30), 18522–18525.

Kim, J., Peng, D., Schlebach, J. P., Hadziselimovic, A., & Sanders, C. R. (2014). Modest effects of lipid modifications on the structure of caveolin-3. *Biochemistry, 53*(27), 4320–4322.

Kiss, A. (2012). Caveolae and the regulation of endocytosis. *Advances in Experimental Medicine and Biology, 2012*(729), 14–28.

Koike, S., Kodera, Y., Nakao, A., Iwata, H., & Yatabe, Y. (2010). Absence of the caveolin-1 P132L mutation in cancers of the breast and other organs. *The Journal of Molecular Diagnostics, 12*(5), 712–717.

Komers, R., Schutzer, W. E., Reed, J. F., Lindsley, J. N., Oyama, T. T., Buck, D. C., et al. (2006). Altered endothelial nitric oxide synthase targeting and conformation and caveolin-1 expression in the diabetic kidney. *Diabetes, 55*, 1651–1658.

Kovtun, O., Tillu, V. A., Jung, W., Leneva, N., Ariotti, N., Chaudhary, N., et al. (2014). Structural insights into the organization of the cavin membrane coat complex. *Developmental Cell, 31*(4), 405–419.

Kurzchalia, T. V., Dupree, P., Parton, R. G., Kellner, R., Virta, H., Lehnert, M., et al. (1992). VIP21, a 21-kD membrane protein is an integral component of trans-golgi-network-derived transport vesicles. *The Journal of Cell Biology, 118*(5), 1003–1014.

Kwon, H., Lee, J., Jeong, K., Jang, D., & Pak, Y. (2015). Fatty acylated caveolin-2 is a substrate of insulin receptor tyrosine kinase for insulin receptor substrate-1-directed signaling activation. *Biochimica et Biophysica Acta, 1853*(5), 1022–1034.

Lacroix-Triki, M., Geyer, F. C., & Reis-Filho, J. S. (2010). Caveolin-1 P132L mutation in human cancers: 1 CAVeat to be voiced. *The Journal of Molecular Diagnostics, 12*(5), 562–565.

Le, P. U., Guay, G., Altschuler, Y., & Nabi, I. R. (2002). Caveolin-1 is a negative regulator of caveolae-mediated endocytosis to the endoplasmic reticulum. *The Journal of Biological Chemistry, 277*(5), 3371–3379.

Le Lan, C., Gallay, J., Vincent, M., Neumann, J. M., de Foresta, B., & Jamin, N. (2010). Structural and dynamic properties of juxta-membrane segments of caveolin-1 and caveolin-2 at the membrane interface. *European Biophysics Journal, 39*(2), 307–325.
Le Lan, C., Neumann, J. M., & Jamin, N. (2006). Role of the membrane interface on the conformation of the caveolin scaffolding domain: a CD and NMR study. *FEBS Letters, 580*(22), 5301–5305.
Lee, J., & Glover, K. J. (2012). The transmembrane domain of caveolin-1 exibits a helix-break-helix structure. *Biochimica et Biophysica Acta, 1818*(5), 1158–1164.
Lee, H., Park, D. S., Razani, B., Russell, R. G., Pestell, R. G., & Lisanti, M. P. (2002). Caveolin-1 mutations (P132L and null) and the pathogenesis of breast cancer — caveolin-1 (P132L) behaves in a dominant-negative manner and caveolin-1 (−/−) null mice show mammary epithelial cell hyperplasia. *American Journal of Pathology, 161*(4), 1357–1369.
Lisanti, M. P., Scherer, P. E., Vidugiriene, J., Tang, Z. L., Hermanowskivosatka, A., Tu, Y. H., et al. (1994). Characterization of caveolin-rich membrane domains isolated from an endothelial-rich source: implications for human-disease. *The Journal of Cell Biology, 126*(1), 111–126.
Liu, L., Brown, D., McKee, M., Lebrasseur, N. K., Yang, D., Albrecht, K. H., et al. (2008). Deletion of Cavin/PTRF causes global loss of caveolae, dyslipidemia, and glucose intolerance. *Cell Metabolism, 8*(4), 310–317.
Liu, J., Oh, P., Horner, T., Rogers, R. A., & Schnitzer, J. E. (1997). Organized endothelial cell surface signal transduction in caveolae distinct from glycosylphosphatidylinositol-anchored protein microdomains. *The Journal of Biological Chemistry, 272*(11), 7211–7222.
Liu, L., & Pilch, P. F. (2008). A critical role of cavin (polymerase I and transcript release factor) in caveolae formation and organization. *The Journal of Biological Chemistry, 283*(7), 4314–4322.
Lokappa, S. B., & Nagaraj, R. (2012). Interaction of peptides spanning the transmembrane domain of caveolin-1 with model membranes. *Journal of Peptide Science, 18*(11), 696–703.
Ludwig, A., Howard, G., Mendoza-Topaz, C., Deerinck, T., Mackey, M., Sandin, S., et al. (2013). Molecular composition and ultrastructure of the caveolar coat complex. *PLoS Biology, 11*(8), e1001640.
Machleidt, T., Li, W., Liu, P., & Anderson, R. G. (2000). Multiple domains in caveolin-1 control its intracellular traffic. *The Journal of Cell Biology, 148*(1), 17–28.
Maru, Y., Afar, D. E., Witte, O. N., & Shibuya, M. (1996). The dimerization property of glutathione S-transferase partially reactivates bcr-abl lacking the oligomerization domain. *The Journal of Biological Chemistry, 271*(26), 15353–15357.
McMahon, K. A., Zajicek, H., Li, W. P., Peyton, M. J., Minna, J. D., Hernandez, V. J., et al. (2009). SRBC/cavin-3 is a caveolin adapter protein that regulates caveolae function. *The EMBO Journal, 28*(8), 1001–1015.
Mobley, B. A., & Eisenberg, B. R. (1975). Sizes of components in frog skeletal muscle measured by methods of stereology. *The Journal of General Physiology, 66*, 31–45.
Monier, S., Dietzen, D. J., Hastings, W. R., Lublin, D. M., & Kurzchalia, T. V. (1996). Oligomerization of VIP21-caveolin in vitro is stabilized by long chain fatty acylation or cholesterol. *FEBS Letters, 388*(2–3), 143–149.
Monier, S., Parton, R. G., Vogel, F., Behlke, J., Henske, A., & Kurzchalia, T. V. (1995). VIP21-caveolin, a membrane protein constituent of the caveolar coat, oligomerizes in vivo and in vitro. *Molecular Biology of the Cell, 6*(7), 911–927.
Mora, R., Bonilha, V. L., Marmorstein, A., Scherer, P. E., Brown, D., Lisanti, M. P., et al. (1999). Caveolin-2 localizes to the golgi complex but redistributes to plasma membrane, caveolae, and rafts when co-expressed with caveolin-1. *The Journal of Biological Chemistry, 274*(36), 25708–25717.
Murata, M., Peranen, J., Schreiner, R., Wieland, F., Kurzchalia, T. V., & Simons, K. (1995). Vip21/Caveolin is a cholesterol-binding protein. *Proceedings of the National Academy of Sciences of the United States of America, 92*(22), 10339–10343.

Norholm, M. H., Shulga, Y. V., Aoki, S., Epand, R. M., & von Heijne, G. (2011). Flanking residues help determine whether a hydrophobic segment adopts a monotopic or bitopic topology in the endoplasmic reticulum membrane. *The Journal of Biological Chemistry, 286*(28), 25284–25290.

Ohsawa, Y., Toko, H., Katsura, M., Morimoto, K., Yamada, H., Ichikawa, Y., et al. (2004). Overexpression of P104L mutant caveolin-3 in mice develops hypertrophic cardiomyopathy with enhanced contractility in association with increased endothelial nitric oxide synthase activity. *Human Molecular Genetics, 13*(2), 151–157.

Okamoto, T., Schlegel, A., Scherer, P. E., & Lisanti, M. P. (1998). Caveolins, a family of scaffolding proteins for organizing preassembled signaling complexes at the plasma membrane. *The Journal of Biological Chemistry, 273*(10), 5419–5422.

Ortegren, U., Karlsson, M., Blazic, N., Blomqvist, M., Nystrom, F. H., Gustavsson, J., et al. (2004). Lipids and glycosphingolipids in caveolae and surrounding plasma membrane of primary rat adipocytes. *European Journal of Biochemistry, 271*(10), 2028–2036.

Palade, G. E. (1953). Fine structure of blood capillaries. *Journal of Applied Physiology, 24,* 1424.

Park, D. S., Woodman, S. E., Schubert, W., Cohen, A. W., Frank, P. G., Chandra, M., et al. (2002). Caveolin-1/3 double-knockout mice are viable, but lack both muscle and non-muscle caveolae, and develop a severe cardiomyopathic phenotype. *American Journal of Pathology, 160*(6), 2207–2217.

Parolini, I., Sargiacomo, M., Galbiati, F., Rizzo, G., Grignani, F., Engelman, J. A., et al. (1999). Expression of caveolin-1 is required for the transport of caveolin-2 to the plasma membrane. Retention of caveolin-2 at the level of the golgi complex. *The Journal of Biological Chemistry, 274*(36), 25718–25725.

Parton, R. G., Hanzal-Bayer, M., & Hancock, J. F. (2006). Biogenesis of caveolae: a structural model for caveolin-induced domain formation. *Journal of Cell Science, 119*(Pt 5), 787–796.

Parton, R. G., Joggerst, B., & Simons, K. (1994). Regulated internalization of caveolae. *The Journal of Cell Biology, 127,* 1199–1215.

Pelkmans, L., Kartenbeck, J., & Helenius, A. (2001). Caveolar endocytosis of simian virus 40 reveals a new two-step vesicular-transport pathway to the ER. *Nature Cell Biology, 3,* 473–483.

Pelkmans, L., & Zerial, M. (2005). Kinase-regulated quantal assemblies and kiss-and-run recycling of caveolae. *Nature, 436*(7047), 128–133.

Razani, B., Combs, T. P., Wang, X. B., Frank, P. G., Park, D. S., Russell, R. G., et al. (2002). Caveolin-1-deficient mice are lean, resistant to diet-induced obesity, and show hypertriglyceridemia with adipocyte abnormalities. *The Journal of Biological Chemistry, 277*(10), 8635–8647.

Razani, B., Engelman, J. A., Wang, X. B., Schubert, W., Zhang, X. L., Marks, C. B., et al. (2001). Caveolin-1 null mice are viable but show evidence of hyperproliferative and vascular abnormalities. *The Journal of Biological Chemistry, 276*(41), 38121–38138.

Razani, B., & Lisanti, M. P. (2001). Caveolin-deficient mice: Insights into caveolar function and human disease. *Journal of Clinical Investigation, 108*(11), 1553–1561.

Rieth, M. D., Lee, J., & Glover, K. J. (2012). Probing the caveolin-1 P132L mutant: critical insights into its oligomeric behavior and structure. *Biochemistry, 51*(18), 3911–3918.

Rothberg, K. G., Heuser, J. E., Donzell, W. C., Ying, Y. S., Glenney, J. R., & Anderson, R. G. (1992). Caveolin, a protein component of caveolae membrane coats. *Cell, 68,* 673–682.

Roy, S., Luetterforst, R., Harding, A., Apolloni, A., Etheridge, M., Stang, E., et al. (1999). Dominant-negative caveolin inhibits H-Ras function by disrupting cholesterol-rich plasma membrane domains. *Nature Cell Biology, 1*(2), 98–105.

Rui, H., Root, K. T., Lee, J., Glover, K. J., & Im, W. (2014). Probing the U-shaped conformation of caveolin-1 in a bilayer. *Biophysical Journal, 106*(6), 1371–1380.

Sargiacomo, M., Scherer, P. E., Tang, Z., Kübler, E., Song, K. S., Sanders, M. C., et al. (1995). Oligomeric structure of caveolin: Implications for caveolae membrane organization. *Proceedings of the National Academy of Sciences of the United States of America, 92*(20), 9407–9411.

Sargiacomo, M., Sudol, M., Tang, Z., & Lisanti, M. P. (1993). Signal transducing molecules and glycosyl-phosphatidylinositol-linked proteins form a caveolin-rich insoluble complex in MDCK cells. *The Journal of Cell Biology, 122*(4), 789–807.

Scherer, P. E., Lewis, R. Y., Volonte, D., Engelman, J. A., Galbiati, F., Couet, J., et al. (1997). Cell-type and tissue-specific expression of caveolin-2. Caveolins 1 and 2 co-localize and form a stable hetero-oligomeric complex in vivo. *The Journal of Biological Chemistry, 272*(46), 29337–29346.

Scherer, P. E., Okamoto, T., Chun, M. Y., Nishimoto, I., Lodish, H. F., & Lisanti, M. P. (1996). Identification, sequence, and expression of caveolin-2 defines a caveolin gene family. *Proceedings of the National Academy of Sciences of the United States of America, 93*(1), 131–135.

Scherer, P. E., Tang, Z. L., Chun, M. Y., Sargiacomo, M., Lodish, H. F., & Lisanti, M. P. (1995). Caveolin isoforms differ in their N-terminal protein-sequence and subcellular-distribution. Identification and epitope mapping of an isoform-specific monoclonal-antibody probe. *The Journal of Biological Chemistry, 270*(27), 16395–16401.

Schlegel, A., & Lisanti, M. P. (2000). A molecular dissection of caveolin-1 membrane attachment and oligomerization. Two separate regions of the caveolin-1 C-terminal domain mediate membrane binding and oligomer/oligomer interactions in vivo. *The Journal of Biological Chemistry, 275*(28), 21605–21617.

Schlegel, A., Schwab, R. B., Scherer, P. E., & Lisanti, M. P. (1999). A role for the caveolin scaffolding domain in mediating the membrane attachment of caveolin-1. The caveolin scaffolding domain is both necessary and sufficient for membrane binding in vitro. *The Journal of Biological Chemistry, 274*(32), 22660–22667.

Sengupta, D. (2012). Cholesterol modulates the structure, binding modes, and energetics of caveolin-membrane interactions. *The Journal of Physical Chemistry B, 116*(50), 14556–14564.

Sinha, B., Köster, D., Ruez, R., Gonnord, P., Bastiani, M., Abankwa, D., et al. (2011). Cells respond to mechanical stress by rapid disassembly of caveolae. *Cell, 144*(3), 402–413.

Sloan, E. K., Stanley, K. L., & Anderson, R. L. (2004). Caveolin-1 inhibits breast cancer growth and metastasis. *Oncogene, 23*(47), 7893–7897.

Song, K. S., Scherer, P. E., Tang, Z. L., Okamoto, T., Li, S. W., Chafel, M., et al. (1996). Expression of caveolin-3 in skeletal, cardiac, and smooth muscle cells. Caveolin-3 is a component of the sarcolemma and co-fractionates with dystrophin and dystrophin-associated glycoproteins. *The Journal of Biological Chemistry, 271*(25), 15160–15165.

Song, K. S., Tang, Z. L., Li, S. W., & Lisanti, M. P. (1997). Mutational analysis of the properties of caveolin-1. A novel role for the C-terminal domain in mediating homo-typic caveolin-caveolin interactions. *The Journal of Biological Chemistry, 272*(7), 4398–4403.

Sotgia, F., Williams, T. M., Cohen, A. W., Minetti, C., Pestell, R. G., & Lisanti, M. P. (2005). Caveolin-1-deficient mice have an increased mammary stem cell population with upregulation of Wnt/beta-catenin signaling. *Cell Cycle, 4*(12), 1808–1816.

Sowmya, B. L., Jagannadham, M. V., & Nagaraj, R. (2006). Interaction of synthetic peptides corresponding to the scaffolding domain of caveolin-3 with model membranes. *Biopolymers, 84*(6), 615–624.

Stoppani, E., Rossi, S., Meacci, E., Penna, F., Costelli, P., Bellucci, A., et al. (2011). Point mutated caveolin-3 form (P104L) impairs myoblast differentiation via akt and p38 signalling reduction, leading to an immature cell signature. *Biochimica et Biophysica Acta (BBA) – Molecular Basis of Disease, 1812*(4), 468–479.

Tang, Z. L., Scherer, P. E., Okamoto, T., Song, K., Chu, C., Kohtz, D. S., et al. (1996). Molecular cloning of caveolin-3, a novel member of the caveolin gene family expressed predominantly in muscle. *The Journal of Biological Chemistry, 271*(4), 2255–2261.

Thorn, H., Stenkula, K. G., Karlsson, M., Ortegren, U., Nystorm, F., Gustavsson, J., et al. (2003). Cell surface orifices of caveolae and localization of caveolin to the necks of caveolae in adipocytes. *Molecular Biology of the Cell, 14*, 3967–3976.

Way, M., & Parton, R. G. (1995). M-caveolin, a muscle-specific caveolin-related protein. *FEBS Letters, 376*, 108–112.

Weiss, N., Couchoux, H., Legrand, C., Berthier, C., Allard, B., & Jacquemond, V. (2008). Expression of the muscular dystrophy-associated caveolin-3(P104L) mutant in adult mouse skeletal muscle specifically alters the Ca(2+) channel function of the dihydropyridine receptor. *Pflügers Archiv: European Journal of Physiology, 457*(2), 361–375.

Wheaton, K., Sampsel, K., Boisvert, F., Davy, A., Robbins, S., & Riabowol, K. (2001). Loss of functional caveolae during senescence of human fibroblasts. *Journal of Cellular Physiology, 187*(2), 226–235.

Whiteley, G., Collins, R. F., & Kitmitto, A. (2012). Characterization of the molecular architecture of human caveolin-3 and interaction with the skeletal muscle ryanodine receptor. *The Journal of Biological Chemistry, 287*(48), 40302–40316.

Williams, T. M., Cheung, M. W. C., Park, D. S., Razani, B., Cohen, A. W., Muller, W. J., et al. (2003). Loss of caveolin-1 gene expression accelerates the development of dysplastic mammary lesions in tumor-prone transgenic mice. *Molecular Biology of the Cell, 14*(3), 1027–1042.

Williams, T. M., Hassan, G. S., Li, J. W., Cohen, A. W., Medina, F., Frank, P. G., et al. (2005). Caveolin-1 promotes tumor progression in an autochthonous mouse model of prostate cancer – genetic ablation of cav-1 delays advanced prostate tumor development in tramp mice. *The Journal of Biological Chemistry, 280*(26), 25134–25145.

Williams, T. M., & Lisanti, M. (2004). The caveolin genes: from cell biology to medicine. *Annals of Medicine, 36*, 584–595.

Williams, T. M., Medina, F., Badano, I., Hazan, R. B., Hutchinson, J., Muller, W. J., et al. (2004). Caveolin-1 gene disruption promotes mammary tumorigenesis and dramatically enhances lung metastasis in vivo – role of cav-1 in cell invasiveness and matrix metalloproteinase (MMP-2/9) secretion. *The Journal of Biological Chemistry, 279*(49), 51630–51646.

Xia, H., Khalil, W., Kahm, J., Jessurun, J., Kleidon, J., & Henke, C. A. (2010). Pathologic caveolin-1 regulation of PTEN in idiopathic pulmonary fibrosis. *The American Journal of Pathology, 176*(6), 2626–2637.

Yamada, E. (1955). The fine structure of the gall bladder epithelium of the mouse. *The Journal of Biophysical and Biochemical Cytology, 1*, 445–447.

Yang, G., Dong, Z., Xu, H., Wang, C., Li, H., Li, Z., et al. (2015). Structural study of caveolin-1 intramembrane domain by circular dichroism and nuclear magnetic resonance. *Peptide Science, 104*(1), 11–20.

Index

Note: Page numbers followed by "f" and "t" indicate figures and tables, respectively.

G

N

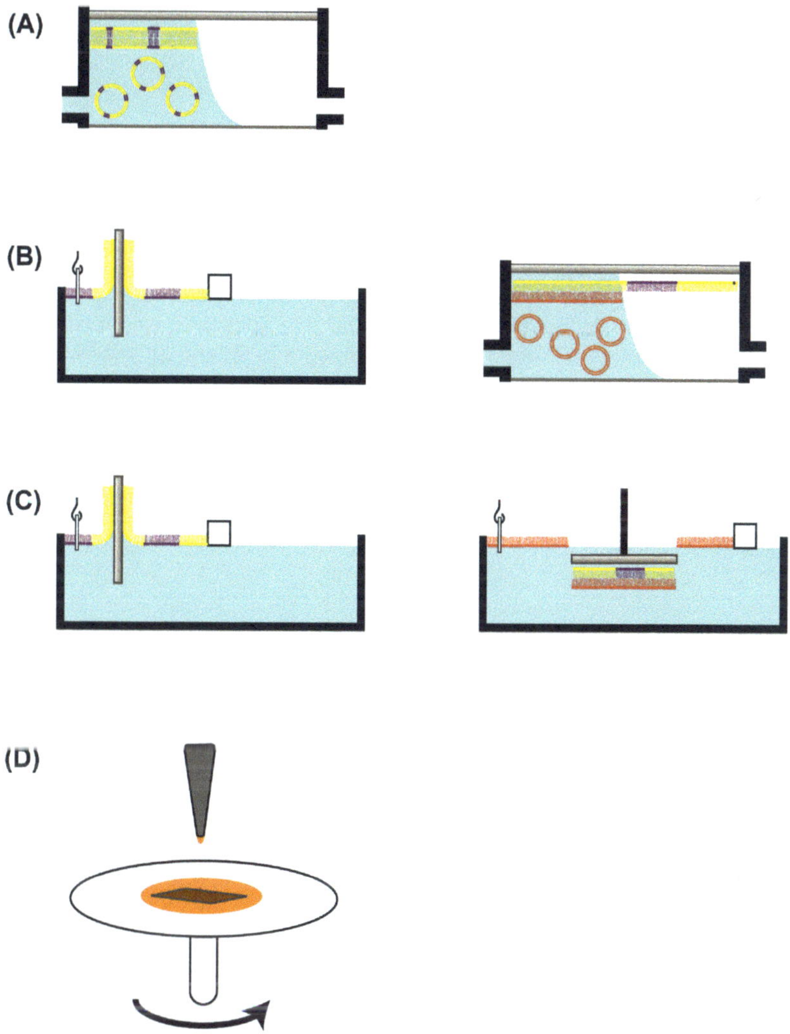

Volker Kiessling *et al.*, Figure 1 ***Methods for supported lipid bilayer preparation.*** (A) Direct vesicle fusion (VF). (B) Langmuir–Blodgett/vesicle fusion (LB/VF). (C) Langmuir–Blodgett/Langmuir–Schäfer (LB/LS). (D) Spin coating.

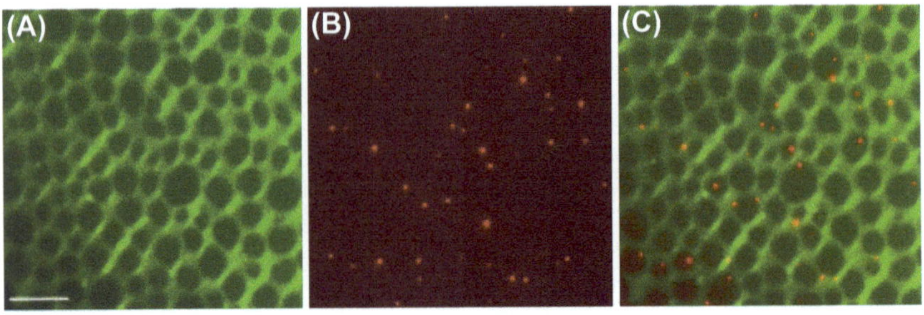

Volker Kiessling *et al.*, Figure 2 ***HIV-fusion peptide-mediated vesicle binding to Lo/Ld phase boundary regions.*** (A) Fluorescence micrograph of a supported lipid bilayer composed of brain SM/brain PS/cholesterol (50:25:25) and labeled with rhodamine-PE, which partitions into the Ld phase. (B) Total internal reflection fluorescence micrograph of membrane-bound liposomes composed of brain PC/brain PS (75:25) and labeled with the fluorescent lipid analog DiD. The liposomes were preincubated with HIV-fusion peptide. (C) The overlay of the images in (A) and (B) shows that most liposomes target the domain boundaries. Scale bar is 10 μm.

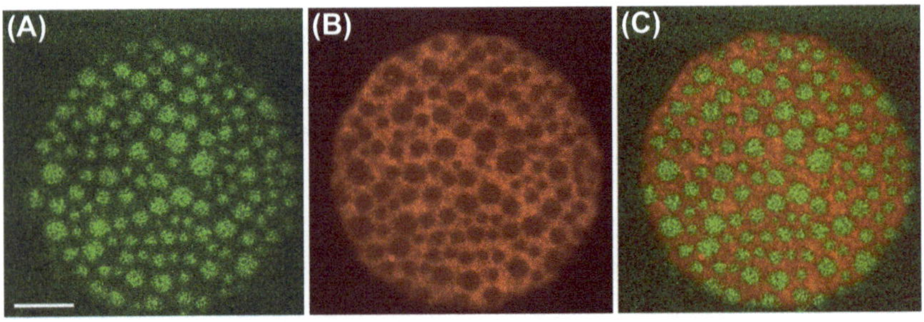

Volker Kiessling *et al.*, Figure 3 ***Lateral organization of syntaxin 1a in asymmetric supported lipid bilayers with Lo/Ld phase separation.*** (A) Fluorescence micrograph of nitro-benzoxadiazol-aminoethyl-(NBD)-labeled DPPE in an asymmetric supported lipid bilayer formed by the LB/VF technique. The proximal leaflet was composed of brain SM/brain PS/cholesterol (40:40:20) and labeled with NBD-DPPE which partitions into the Lo phase. (B) Fluorescence micrograph of Alexa-546-labeled syntaxin 1a of the supported lipid bilayer in (A). Syntaxin 1a labeled with Alexa-546 was reconstituted into vesicles composed of brain PC/brain PE/brain PS/brain PIP_2/cholesterol (32:30:15:3:20). The vesicles were added to the LB monolayer to form the distal leaflet of the supported bilayer and integrate syntaxin 1a into the membrane. (C) The overlay image shows that syntaxin 1a partitions preferentially into the Ld phase. Scale bar is 10 μm.

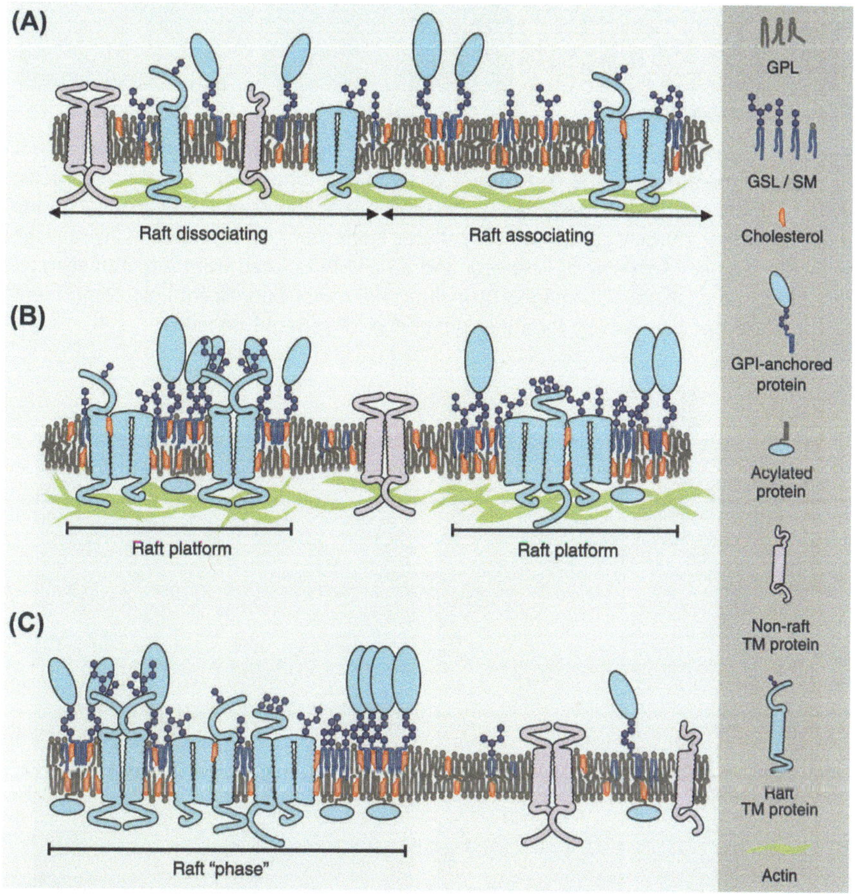

Kandice R. Levental and Ilya Levental, Figure 1 ***Lipid rafts as functional organizers of cellular membranes.*** This figure *(adapted from Lingwood & Simons, 2010)* shows the various possible organization states of cholesterol-rich mammalian membranes. (A) Preferential interactions between raft lipids and proteins (including cholesterol, sphingolipids, GPI-anchored proteins, acylated proteins) manifest themselves as domains of various sizes and stabilities. (B) Specific interactions between proteins can stabilize smaller, more dynamic domains into larger, functional raft platforms. (C) Under certain conditions, e.g., release from constraint by the actin cytoskeleton, domains coalesce into microscopic phases. *From Lingwood, D., & Simons, K. (2010). Lipid rafts as a membrane-organizing principle.* Science, *327(5961), 46–50. Reprinted with permission from AAAS.*

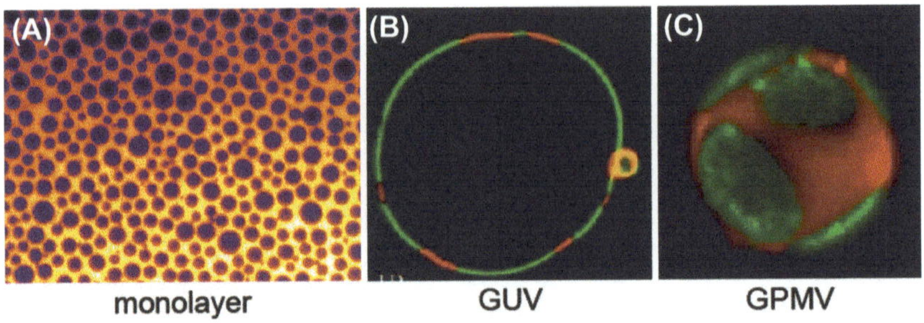

Kandice R. Levental and Ilya Levental, Figure 2 ***Liquid–liquid phase separation in biomimetic membranes.*** The coexistence of two liquid phases can be observed across various model membrane modalities in the presence of cholesterol. Shown are (A) Langmuir monolayer, (B) GUV composed of synthetic lipids *(from Korlach et al., 1999; copyright © by the National Academy of Sciences)*, and (C) GPMV isolated from a mammalian cell *(from Levental et al., 2010b)*, imaged by inclusion of minor fluorescent lipid components with preferential partitioning for either ordered or disordered domains.

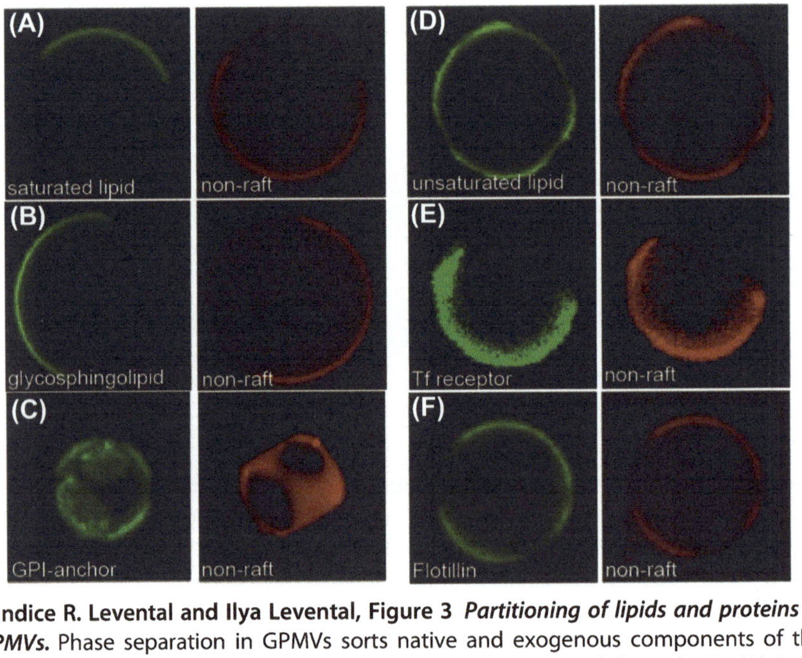

Kandice R. Levental and Ilya Levental, Figure 3 ***Partitioning of lipids and proteins in GPMVs.*** Phase separation in GPMVs sorts native and exogenous components of the membranes, often in accordance with predictions from the raft hypothesis. (A) Saturated lipids, (B) glycosphingolipids, and (C) GPI-anchored proteins partition strongly to the raft phase (in this figure, the nonraft phase is marked by the fluorescent unsaturated phospholipid rhodamine-DOPE). (D) Unsaturated lipids and (E) the transferrin receptor partition to the nonraft phase, consistent with their predicted exclusion from lipid rafts. However, (F) flotillin and some other predicted raft proteins are sometimes excluded from the raft phase. *Figures adapted from Sengupta et al. (2008) and Levental et al. (2010b). Reprinted from* Biochimica et Biophysica Acta, 1778*(1), Sengupta P, Hammond A, Holowka D, Baird B, Structural determinants for partitioning of lipids and proteins between coexisting fluid phases in giant plasma membrane vesicles, 20–32, Copyright (2008), with permission from Elsevier.*

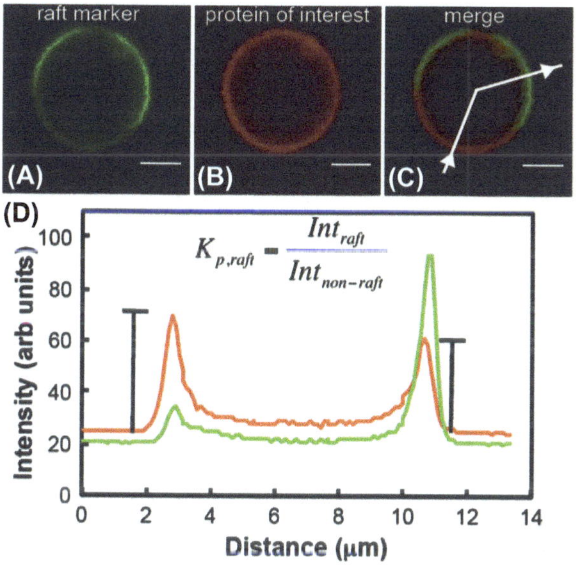

Kandice R. Levental and Ilya Levental, Figure 4 ***Quantification of protein partitioning in GPMVs.*** Component partitioning in GPMVs can be quantified by standard fluorescent microscopy approaches. GPMVs stained with (A) a marker for the raft domain can be isolated from cells expressing (B) a fluorescently tagged protein of interest. (C) The merged image can be used to (D) calculate the ratio of fluorescent intensity of the protein in the raft and nonraft phases, which is a direct quantification of the raft partition coefficient ($K_{p,raft}$) of the protein. *Figure adapted from Levental et al. (2010b).*

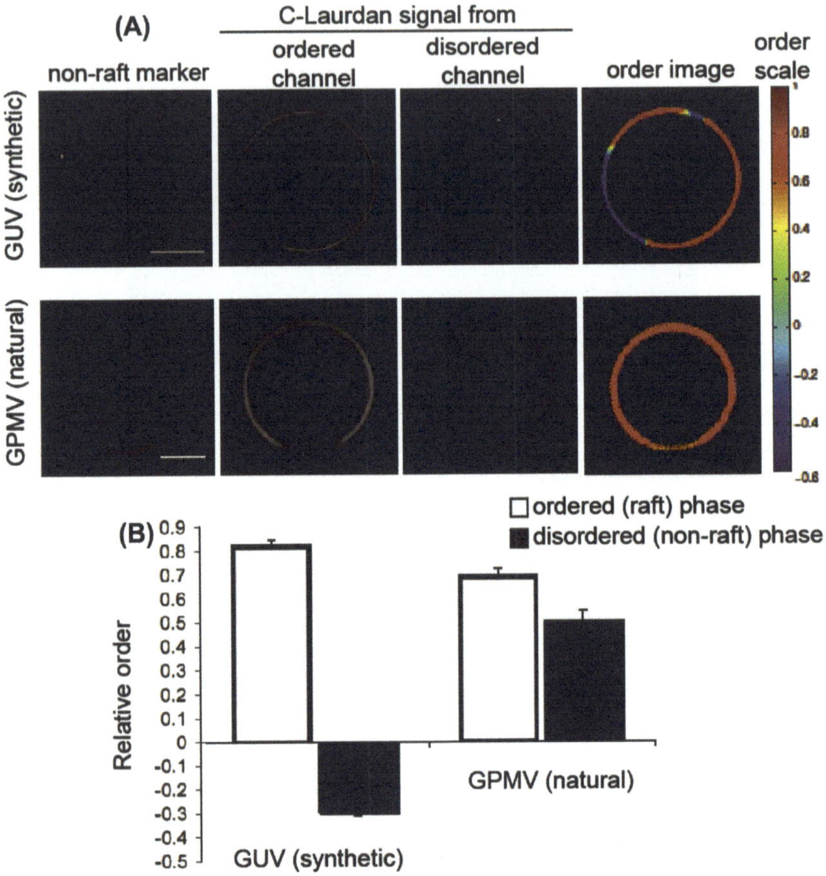

Kandice R. Levental and Ilya Levental, Figure 5 ***Quantification of domain order in model membranes.*** Microscopy of the environment-sensitive dye C-Laurdan can be used to measure the relative order differences between phases. (A) Images of the ordered (~440 nm) and disordered (~490 nm) emission channels are processed to yield pixel-by-pixel order maps. (B) The disparity between ordered and disordered domains is much greater in synthetic GUVs than in cell-derived GPMVs, with the relatively disordered, nonraft phase responsible for most of the difference. *Figure adapted from Kaiser et al. (2009).*

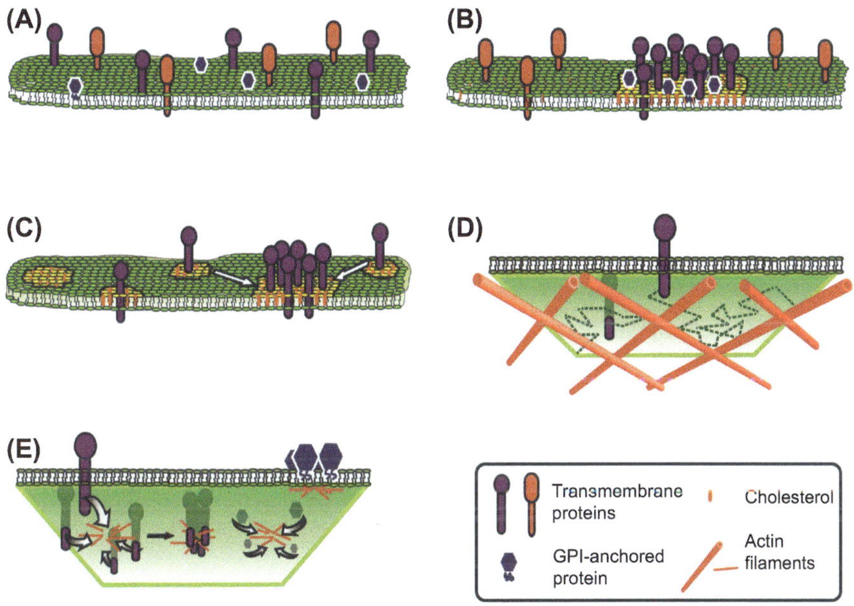

Nikki M. Curthoys *et al.*, Figure 1 Models of cell membrane organization discussed in Section 1.1. (A) "Fluid mosaic" model. Proteins are distributed randomly through a homogenous phospholipid bilayer. (B) "Lipid raft" model. Sphingolipid and cholesterol patches are populated with proteins which have an affinity for these patches. Protein species can be raft associated or nonraft associated. (C) "Lipid shell" model. Some proteins will be targeted to self-assembled cholesterol and sphingolipid complexes which form a "lipid shell" around the protein. These "lipid shells" have an affinity for, and can coalesce with, larger lipid domains. (D) "Picket Fence" model. Transmembrane proteins are restricted in their diffusion by actin filaments (the "fence") which appose and run parallel to the cytoplasmic leaflet of the membrane, and by other transmembrane proteins bound to these filaments ("pickets", not shown). (E) "Active composite" model. Short actin filaments adjacent to the cytoplasmic membrane leaflet are arranged in "asters". Transmembrane proteins and GPI-anchored proteins are advected by actin and myosin to the centers of these asters, resulting in protein nanoclustering. See Section 1.1 for more detail. Readers please note that depictions of cell membranes here do not show as much protein (relative to lipid) as would be found in actual cell membranes.

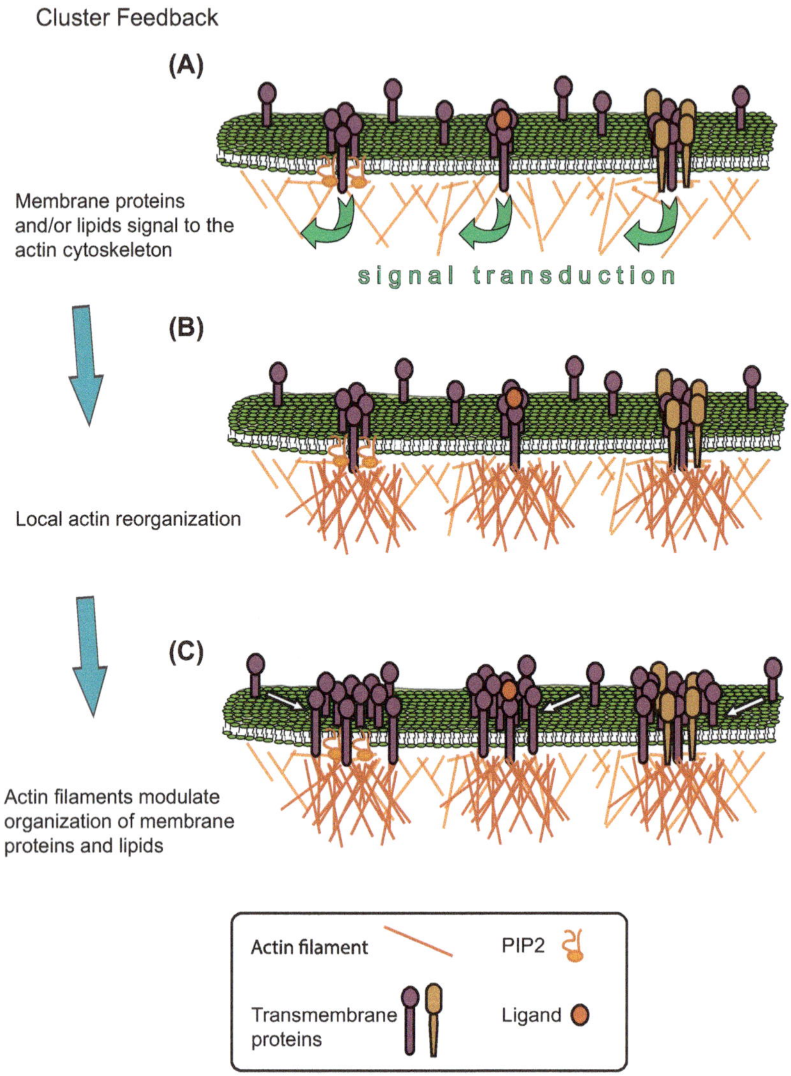

Cluster Feedback
(A)
Membrane proteins and/or lipids signal to the actin cytoskeleton
signal transduction
(B)
Local actin reorganization
(C)
Actin filaments modulate organization of membrane proteins and lipids
Actin filament
PIP2
Transmembrane proteins
Ligand

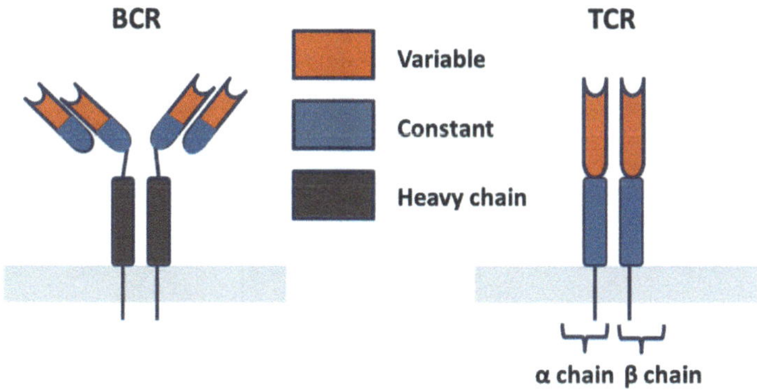

Juliette Griffié *et al.*, Figure 1 B cell receptor (BCR) and T cell receptor (TCR) main subunits, highlighting their variable components that enable the creation of a vast repertoire.

Nikki M. Curthoys *et al.*, Figure 2 "Cluster Feedback" model of membrane organization. (A) Proteins cluster at the nanoscale in the plasma membrane. These clusters are protein–lipid (left, transmembrane protein and PIP2), mediated by ligand binding (middle), or protein–protein (right; depicted as a heterocluster; a homocluster oligomer is not shown). Of course, in clusters of any protein there exists the potential for local lipid clustering also. These protein–protein or protein–lipid clusters are collectively referred to hereafter as "membrane clusters." In each case, signaling to the actin cytoskeleton initiated by proteins and/or lipids in membrane clusters elicits the local remodeling of actin (B), either through the recruitment and binding of actin filaments to the membrane lipids or proteins, the de novo nucleation of new filaments or branches thereof, or both. ABPs which mediate these interactions are not shown, but see Section 3.1 for more detail. The increase in actin density immediately adjacent to the membrane cluster then acts as a recruitment platform for other proteins (or lipids, not shown) diffusing in the membrane to join existing membrane clusters (C). This results in changes in the size, density, perimeter, and/or number of molecules within membrane clusters. Readers should note, the numbers of proteins depicted above are only for ease of communication—we do not hypothesize about the specific sizes (or numbers of molecules within) any of the membrane clusters here. Clusters may exist on the nano-, meso-, or micro scale; the Cluster Feedback model only predicts changes (here shown as increases) in cluster sizes from (A) to (C). See Sections 1.1 and 3.1 for more detail.

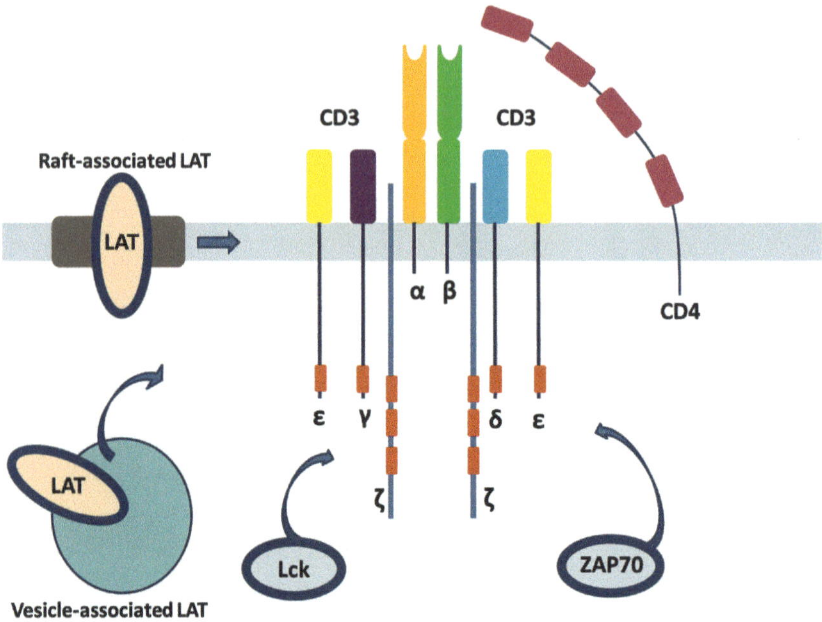

Juliette Griffié *et al.*, Figure 2 The T cell receptor early signaling pathway. Illustration of the recruitment of key signaling proteins in the form of nanoclusters either sustained by protein–protein interactions or protein-lipid interactions in the case of linker for activation of T cells (LAT). The 10 intracellular tyrosine-based activation motif domains have also been highlighted in red on the CD3 and zeta subunits.

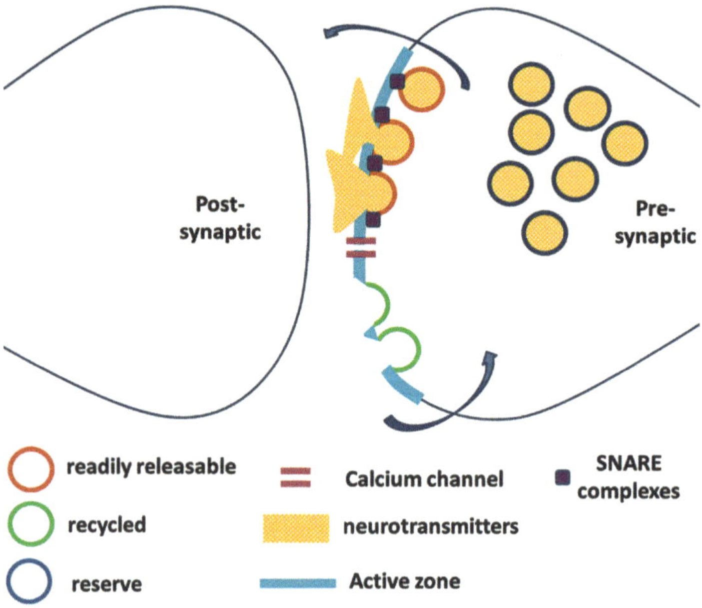

Juliette Griffié *et al.*, Figure 4 Synaptic cleft after opening of calcium channels showing the main three pools of synaptic vesicles.

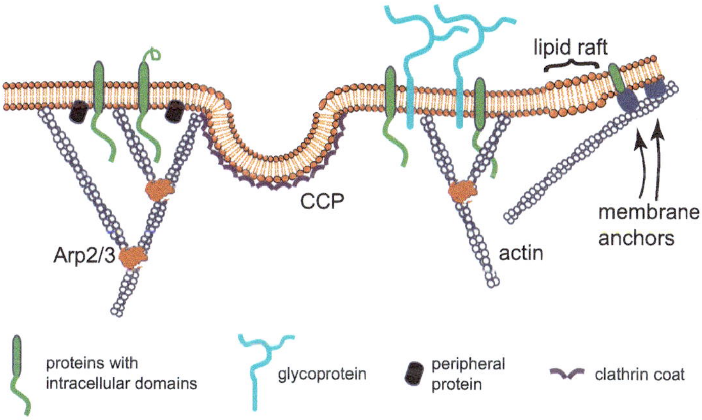

Diego Krapf, Figure 3 Largely simplified sketch of the plasma membrane. In this illustration, actin filaments can link to the plasma membrane and obstruct the motion of proteins. Two types of membrane anchors are shown, proteins that bind actin directly to the membrane (e.g., annexins) and to proteins, such as talin and vinculin, that connect actin to cell adhesion receptors, such as integrins (Winder & Ayscough, 2005). Membrane proteins may also have intracellular domains that contribute to interactions with cytoplasmic filaments such as the cortical actin. Extracellular carbohydrate chains (e.g., glycoproteins) can also contribute to interactions among themselves or with the extracellular matrix. A clathrin-coated pit (CCP) is shown where molecules with specific recognition sites can be captured. In addition, the high density of different membrane proteins contributes to its complexity. Transient functional domains such as cholesterol-rich lipid rafts also partition the membrane into regions with different diffusive behavior.

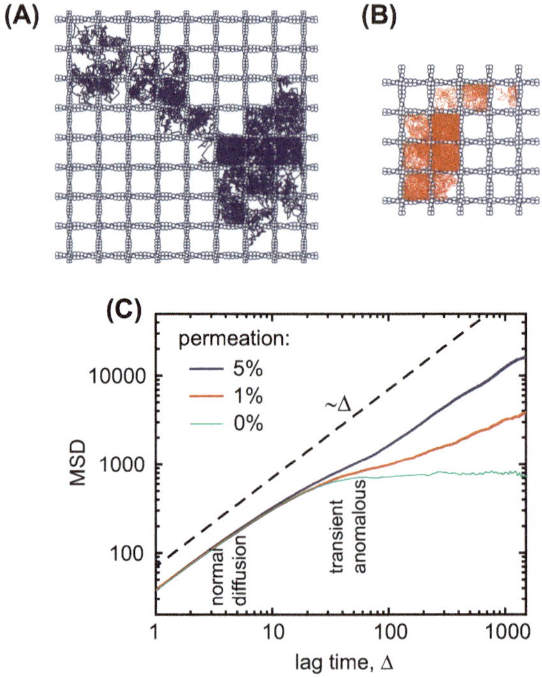

Diego Krapf, Figure 4 Fence model for transient anomalous diffusion. (A,B) Numerical simulations for random walks in the presence of permeable fences. The fences are sketched as actin filaments. These simulations are performed off-lattice with jumps drawn from a normal distribution with standard deviation $\sigma = 5$ in both x and y directions. The fences are placed a distance of 50 units from each other. Both trajectories are 10,000 jumps. The fences are permeable with a given probability of permeation: (A) 5% and (B) 1%. (C) Time-averaged mean square displacement (MSD) $\left(\overline{\delta^2}\right)$ for the trajectories shown in the upper panels as a function of lag time. At lag times much shorter than the hopping time, the MSD is linear. At intermediate times, a sublinear regime is observed and diffusion becomes normal again at longer times. The recovery of linear MSD is much slower for 1% permeation probability than for 5% permeation probability. A trajectory with no permeation through the fences (0% permeation probability), so that the trajectory is confined to a compartment, is also shown as a thin line. The dashed line indicates linear behavior (free diffusion), i.e., MSD $\sim \Delta$.

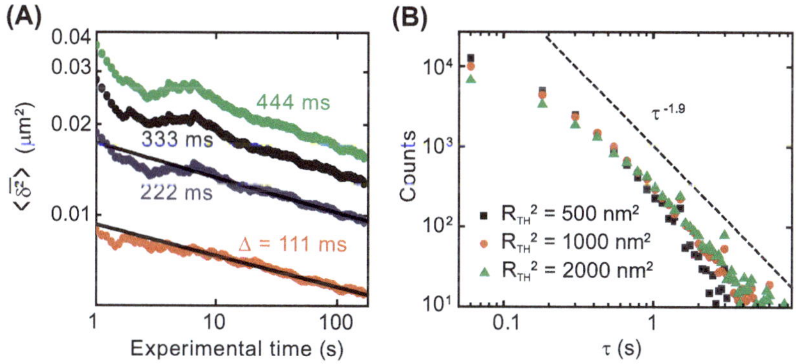

Diego Krapf, Figure 5 The ion channel Kv2.1 exhibits features of a heavy-tailed CTRW (Weigel, Simon, et al., 2011). (A) Aging. The ensemble average of the temporal mean square displacement (MSD) ($\langle\overline{\delta^2}\rangle$) is shown as a function of experimental time for four different lag times, namely $\Delta = 111$, 222, 333, and 444 ms. The averaged MSD depends on the experimental time with $\langle\overline{\delta^2}\rangle \sim 1/T^{0.1}$. Following Eqn (15), this observation is consistent with a CTRW with $\alpha = 0.9$. (B) Distribution of immobilization times of the Kv2.1 channel. Immobilizations are identified as a particle being temporarily confined within a radius R_{TH} and analyses employing different thresholds are shown. The distribution exhibits a power-law dependence $\psi(\tau) \sim \tau^{-(1+\alpha)}$, also with $\alpha = 0.9$.

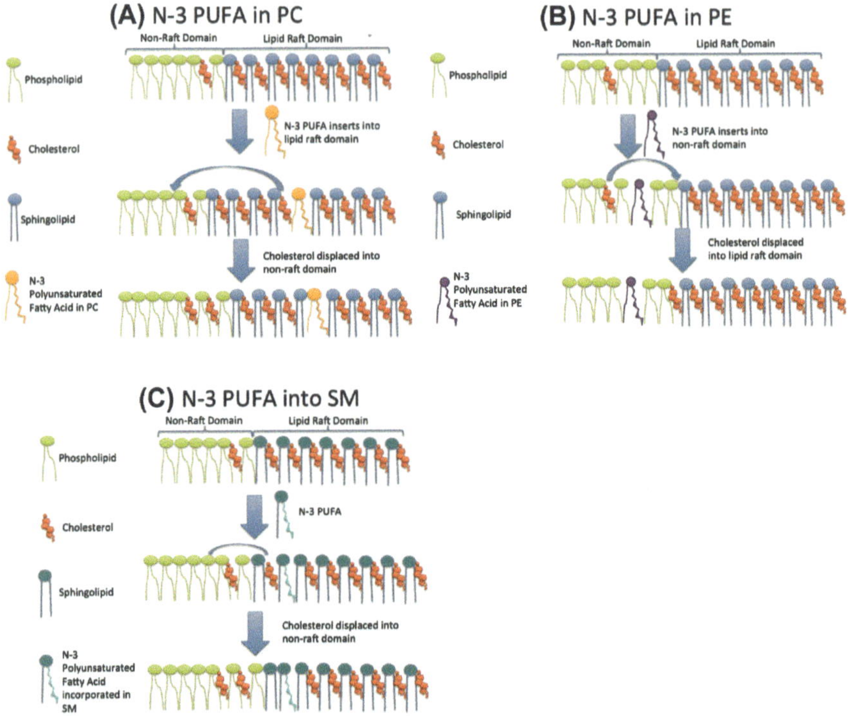

Saame Raza Shaikh *et al.*, Figure 2 ***Potential mechanisms by which n-3 PUFAs could disrupt lipid microdomain distribution.*** Based on the existing literature, we propose that n-3 PUFA acyl chains esterified to (A) phosphatidylcholines (PCs) will incorporate largely into raft domains, which will displace cholesterol into nonrafts. In contrast, if n-3 PUFA acyl chains are esterified into (B) phosphatidylethanolamines (PEs), then these molecules will infiltrate nonrafts to promote cholesterol displacement into rafts. (C) There is some suggestion that n-3 PUFAs could incorporate as acyl chains of sphingomyelin (SM), which would prevent the appropriate packing of sphingolipid–cholesterol-enriched and sphingolipid–cholesterol-free domains. Finally, it is entirely plausible that n-3 PUFAs can esterify to cholesterol (not depicted), which could influence lipid raft packing but would have more pronounced effects on intracellular lipid pools (not shown). For simplicity, we only present a monolayer and proteins are not shown. The overall notion is that changes in lipid microcluster organization with n-3 PUFAs will fundamentally change the environment of lipid raft associated proteins and thereby downstream cellular function, potentially explaining some of the effects of the fatty acids on human health.

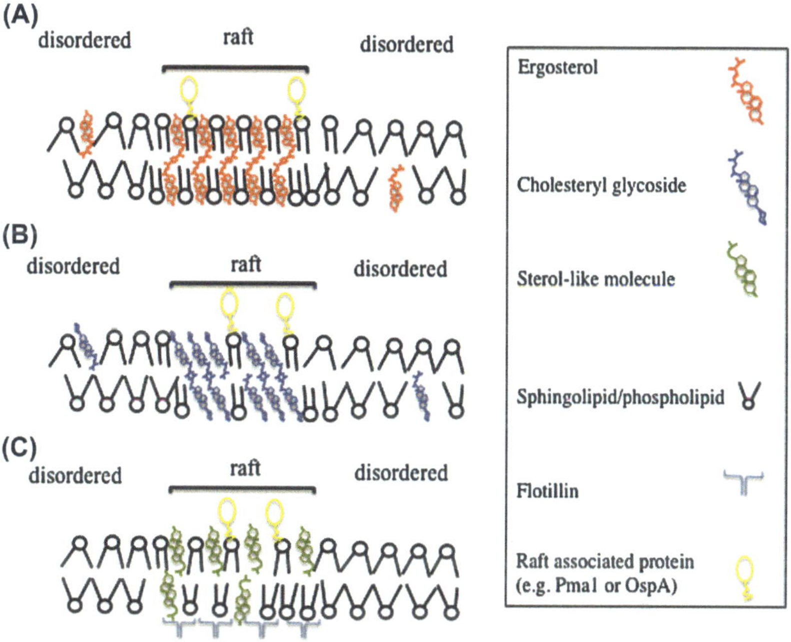

Amir M. Farnoud *et al.*, Figure 2 A hypothetical schematic of lipid rafts in (A) fungal, (B) sterol-containing and (C) sterol-lacking bacteria: (A) in fungal cells, ergosterol and glycosphingolipids pack to form membrane microdomains, (B) in sterol-containing bacteria, cholesteryl glycosides play the role of both sterols and sphingolipids (much of the cholesteryl glycosides are acylated, which is not shown in the schematic), and (C) in sterol-lacking bacteria, flotillin and sterol-like molecules mediate raft formation. It should be emphasized that the lipid composition of lipid rafts in bacteria is only poorly characterized. Certain membrane lipids and proteins are not included for simplicity.

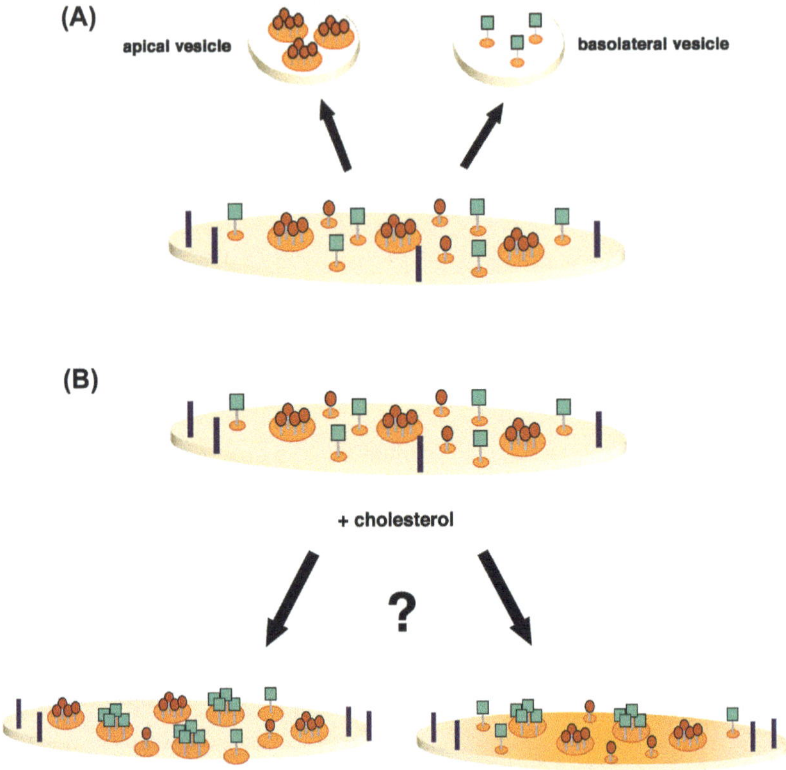

Simona Paladino *et al.*, Figure 1 ***TGN sorting of GPI-APs in polarized epithelial cells.*** (A) At the TGN, the GPI anchor, both apical (red) and basolateral (cyan) GPI-APs, partition into lipid rafts and segregate from nonraft-transmembrane proteins (blue). Only apical GPI-APs are stabilized into rafts by protein oligomerization/clustering and are able to be included in a raft-enriched apical vesicle. Thus, two requirements are necessary for GPI-AP apical sorting. GPI-AP homoclustering might promote raft coalescence that reinforces the segregation between different lipid phases and consequently favors the formation of apical vesicles. Alternatively, a putative apical receptor (not shown here) could drive the raft coalescence by interacting directly with GPI-APs (by recognition of some modifications on the protein or more likely on the GPI anchor) or binding specific raft lipids. (B) Cholesterol has a crucial role in GPI-AP sorting in the TGN. In addition to cholesterol being necessary for oligomer formation, the addition of exogenous cholesterol is sufficient to lead to the oligomerization of basolateral GPI-APs and to redirect them to apical surface. Increased levels of cholesterol could stabilize the interaction of basolateral GPI-AP with rafts by favoring their oligomerization (left). Alternatively, cholesterol could change the properties of the Golgi lipid environment, thus enabling GPI-AP oligomerization and apical sorting (right).

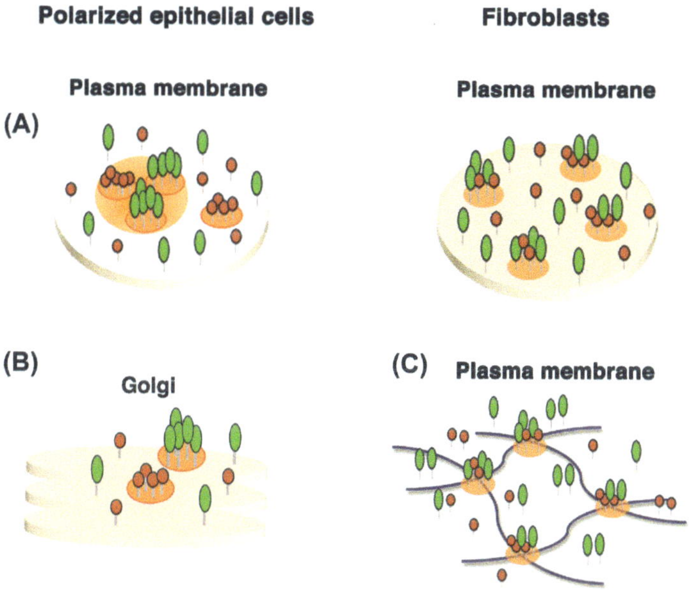

Simona Paladino *et al.*, Figure 2 ***Plasma membrane organization of GPI-anchored proteins in polarized epithelial cells and in fibroblasts.*** The plasma membrane organization of GPI-APs is different in epithelial cells and fibroblasts. (A) While in fibroblasts, GPI-AP cholesterol-dependent nanoclusters accommodate three to five different proteins (red and green), in polarized epithelial cells apical GPI-APs are organized into cholesterol-independent homoclusters of the same GPI-AP species (red and green aggregates), and cholesterol-dependent heteroclusters are derived from coalescence of homoclusters. (B) Homoclusters form in the Golgi and are a prerequisite for heterocluster formation. (C) Cortical actin regulates the formation/maintenance of GPI-AP clusters at the plasma membrane of fibroblasts, while it is unknown whether it plays a role in epithelial cells.

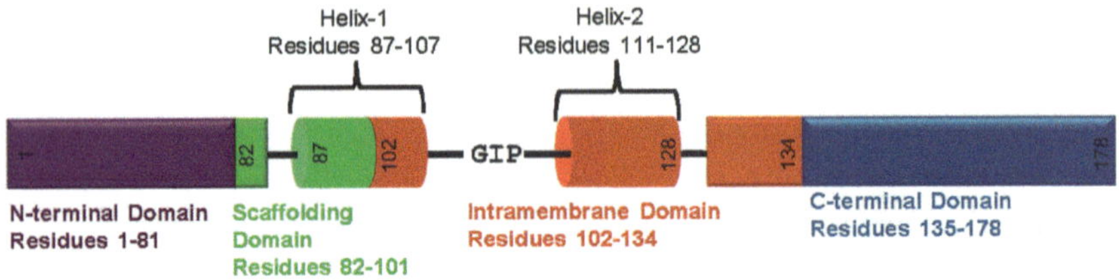

Kyle T. Root *et al.*, Figure 1 Diagram of the four domains of caveolin-1. Residues 87–128 have been shown as α-helical (Rui et al., 2014).

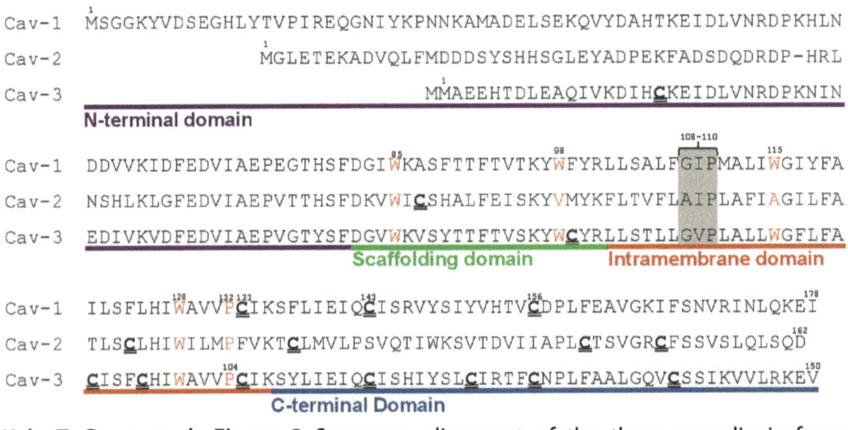

Kyle T. Root *et al.*, Figure 2 Sequence alignment of the three caveolin isoforms (human): caveolin-1, -2, and -3. The colored bar beneath the alignment denotes the domain. Highlighted in red are important residues that are referred to in the manuscript. Bolded and underlined are the cysteine residues in caveolin-1, caveolin-2, and caveolin-3. Residues shaded in gray are part of the break region (residues 108–110) implicated as the intramembrane turn.

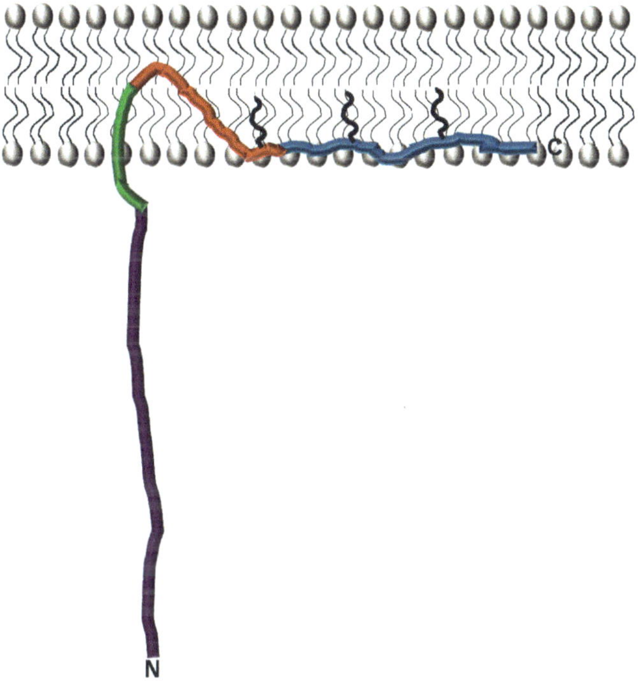

Kyle T. Root *et al.*, Figure 3 Topological model of the four caveolin-1 domains: N-terminal (purple), scaffolding domain (green), intramembrane domain (red), and C-terminal domain (blue). Palmitoylation is shown in black.

Kolibri 360 GmbH
Ferdinand-Jühlke-Straße 7
99095 Erfurt, Deutschland
produktsicherheit@kolibri360.de